The Effect of Nutrients on Neurological Disorders

The Effect of Nutrients on Neurological Disorders

Guest Editors

Lorena Perrone
William B. Grant

MDPI

Basel • Beijing • Wuhan • Barcelona • Belgrade • Novi Sad • Cluj • Manchester

Guest Editors
Lorena Perrone
Department of Life Science,
Health and Health Professions
Link Campus University
Rome
Italy

William B. Grant
Sunlight, Nutrition, and
Health Research Center
San Francisco
USA

Editorial Office
MDPI AG
Grosspeteranlage 5
4052 Basel, Switzerland

This is a reprint of the Special Issue, published open access by the journal *Nutrients* (ISSN 2072-6643), freely accessible at: https://www.mdpi.com/journal/nutrients/special_issues/70M3BF23E9.

For citation purposes, cite each article independently as indicated on the article page online and as indicated below:

Lastname, A.A.; Lastname, B.B. Article Title. *Journal Name* **Year**, *Volume Number*, Page Range.

ISBN 978-3-7258-6480-5 (Hbk)
ISBN 978-3-7258-6481-2 (PDF)
https://doi.org/10.3390/books978-3-7258-6481-2

Contents

About the Editors

Lorena Perrone

Lorena Perrone is an associate professor in clinical biochemistry and clinical molecular biology at Link Campus University in Rome, Italy. Her research focuses on the molecular mechanisms involved in neurodegenerative disease progression. She is also interested in the role of nutrition as a preventive strategy against Alzheimer's disease.

William B. Grant

William B. Grant is the director of the Sunlight, Nutrition and Health Research Center, San Francisco. He conducts research on the health benefits of solar UVB radiation, vitamin D, and nutrition. He has published landmark papers regarding diet and risk of Alzheimer's disease, solar UVB and prevention of cancer, and vitamin D in the prevention of COVID-19.

nutrients

Editorial

The Effect of Nutrients on Neurological Disorders

Lorena Perrone [1,*] and William B. Grant [2,*]

1 Department of Medicine and Surgery, University KORE of Enna, 94100 Enna, Italy
2 Sunlight, Nutrition, and Health Research Center, 1745 Pacific Ave., Suite 504, San Francisco, CA 94109, USA
* Correspondence: lorena.perrone@unikore.it (L.P.); wbgrant@infionline.net (W.B.G.)

The prevalence of neurological disorders (NDs) is increasing, with great cost to public health [1]. In addition, the molecular pathways that cause several NDs are not fully understood, leading to delayed diagnosis and therapeutic treatments that treat symptoms instead of targeting the causative molecular alterations [2]. Both epidemiological studies and prospective cohort studies indicate that good nutrition exerts a beneficial effect on health, including its positive role in brain function [3]. Increasing evidence also shows that diet and nutrients play an important role in preventing and managing several NDs. In particular, data suggest that defined dietary patterns are beneficial against NDs, mostly through their anti-inflammatory function [4]. Indeed, data indicate that chronic low-grade inflammation contributes to the progression of several NDs, including psychiatric disorders and neurodegenerative diseases [5,6]. However, foods rich in high saturated fats, trans fats, processed meats, and refined grains and sugar promote an inflammatory response, leading to chronic inflammation, and their consumption correlates with higher risk of NDs [7]. Thus, dietary intervention is a promising therapeutic strategy to prevent and manage NDs. Studies showing how dietary patterns and nutrients affect ND progression, therefore, play an important role in managing diseases with such a high impact on well-being and public health.

This Special Issue, titled "The Effect of Nutrients on Neurological Disorders", collects original articles illuminating how diet and nutrients affect NDs.

Cognitive impairment is a burdensome disease that strongly affects people's well-being, as well as having an economic impact on public health. Cognitive impairment affects mostly older people. Owing to higher life expectancy, the number of people suffering from cognitive impairment is rapidly increasing [8]. Several studies showed that environmental factors and lifestyle contribute to the onset of cognitive decline [9]. This Special Issue includes four articles analyzing four aspects of environmental factors that modulate the progression of cognitive decline, including heavy metal exposure, nutritional supplements, and dietary habits.

Several studies indicate that heavy metal exposure affects human health, including cognitive function [10]. Enhanced heavy metal exposure comes from cigarette smoke, air pollution, and the contamination of water and food. Such exposure then contributes to the pathophysiology of cognitive impairment [10]. In this Special Issue, Song and colleagues offer a study analyzing the combined effects of five mixed metals on cognitive function. The researchers also investigate the correlation between a person's sex and the effects of exposure to five metals on cognitive function [11]. That large cross-sectional cohort study includes 1833 older Americans (883 males and 950 females) and analyzes blood levels of mercury, cadmium, lead, manganese, and selenium in correlation with cognitive performance assessed via four cognitive tests. The study reports that blood levels of lead and cadmium correlate with diminished cognitive performance, whereas selenium blood levels correlate with better cognitive performance. High levels of selenium counteract the other metals' negative effects on cognitive function [11]. The study also shows the gender specificity of the correlation between the ratio of levels of metals in the blood and cognitive performance [11], suggesting that environmental heavy metals affect cognitive function through sex-specific mechanisms.

Several studies analyze how dietary supplements and nutraceuticals affect cognitive function, suggesting a beneficial effect, proposing them as ways to alleviate the progression of cognitive impairment [12]. In that regard, Leonard and colleagues, in this Special Issue, investigate how ashwagandha supplementation affects levels of biomarkers for cognitive and psychological function [13]. For more than 3000 years, ashwagandha (*Withania somnifera*) has been used as an anti-inflammatory, immunomodulatory, and antioxidant agent to manage mood disorders, traumatic brain diseases, and neurodegenerative diseases [14–19]. The researchers carry out a prospective clinical trial on 59 males and 59 females, along with age- and body mass index-matched control subjects, treated with ashwagandha supplementation, with liposomes used as the carrier. Cognitive performance and mood states are compared in treated and control subjects. Ashwagandha supplementation enhances memory and other cognitive functions and improves some mood aspects such as tension and fatigue [13]. Those data indicate that ashwagandha could be a beneficial nutraceutical capable of improving cognitive performance and mood.

The active metabolic form of vitamin D_3, 1,25-dihydroxyvitamin D_3 [1,25$(OH)_2D_3$; also called calcitriol], exhibits beneficial effects on various NDs, including cognitive disfunction and Alzheimer's disease (AD). Calcitriol modulates inflammation, oxidative stress, and energy balance, as well as having a neuroprotective function [20]. Supplementation with vitamin D is proposed as a complementary treatment for various NDs, including AD [21]. To correlate the effect of vitamin D supplementation, researchers can measure two metabolites of vitamin D_3: blood levels of 1,25$(OH)_2D_3$, the active form, and 25$(OH)D_3$, which has a higher blood concentration and is more stable than the active form with its short half-life. Indeed, 25$(OH)D_3$ levels remain stable for almost 2 weeks [22]. However, clinical studies analyzing blood levels of 25$(OH)D_3$ are hard to compare because the studies are carried out in different laboratories using different assays. Measuring 25$(OH)D_3$ requires further standardization because the assays employed in different laboratories yield significant differences [22]. Since 1999, it has been known that prospective cohort studies are subject to the underestimation of the effect of health outcomes due to changes in the variables after baseline. Such underestimation is called "regression dilution" [23]. Grant's review, in this Special Issue, looks at regression dilution after long follow-up times and shows its effects on results from various prospective cohort clinical studies (nine for all-cause dementia, six for AD, and nine for cognitive impairment), analyzing the relationship between vitamin D deficiency and cognitive disorders [24]. His study shows that in prospective cohort studies, the vitamin D-dependent risk of developing cognitive diseases exhibits an inverse correlation with the mean follow-up period [24]. When the follow-up period is taken into account, the regression fit to the shortest period finds that the association of high vs. low 2(OH)D is about twice as high as that calculated by averaging the findings from all of the studies. Those data are relevant for setting up cohort studies aimed at analyzing how vitamin D affects the risk of developing dementia and cognitive impairment, as well as many other health outcomes.

An increasing number of studies show that the Mediterranean diet (MeDi) and the Mediterranean lifestyle reduce the risk of several chronic diseases, including cognitive impairment and dementia [25–31]. The MeDi consists of high consumption of vegetables, fruits, legumes, nuts and seeds, whole grains, and olive oil; moderate consumption of fish; and very low consumption of red meat [32]. The MeDi is enriched in natural molecules that benefit health and brain function [33], and the diet shows reduced levels of proinflammatory and pro-oxidant molecules, such as the advanced glycation end-products involved in the progression of several chronic diseases [34]. Dominguez and colleagues present, in this Special Issue, an original study investigating adherence to the MeDi and the level of physical activity in 73 patients affected by mild-to-moderate AD and 73 age-matched control subjects [35]. Their study shows that the dietary pattern of AD patients is one of low adherence to the MeDi; AD patients also had less physical activity than control subjects. By using a multivariate analysis, the researchers show that only AD significantly correlates with adherence to the MeDi, whereas sex, physical activity, polypharmacy, and comorbidities exhibit no correlation with MeDi adherence [35]. The study further shows the value of high adherence to the MeDi to prevent AD.

The review article of Sbai and colleagues, in this Special Issue, summarizes recent data investigating the MeDi's role in the progression of diabetic retinopathy (DR), age-related macular degeneration (AMD), and glaucoma [36]. Visual impairment adversely affects health, quality of life, and cognitive and psychological development, and the therapies are expensive [37]. This is why researchers have shown increased interest in dietary habits that can prevent retinal diseases. Sbai and colleagues effectively describe the characteristics of the MeDi and the beneficial natural molecules enriched therein. The article underlines the differences among the MeDi scoring systems used to analyze MeDi adherence in cohort studies—variations that cause difficulties in comparing results from different scoring systems [36]. Moreover, the review summarizes the molecular pathways induced by the MeDi, in particular the activity of Nrf2, which counteracts the reactive oxygen species-induced cell damage that characterizes retinal diseases [38]. Finally, their review gives an accurate overview of recent data analyzing the role that MeDi adherence plays in preventing DR, AMD, and glaucoma. The researchers do so by summarizing findings from cohort studies, animal models, and in vitro analysis showing that high MeDi adherence plays a crucial role in preventing those diseases by lowering risk and delaying onset and progression. Moreover, supplements can help as adjuvant therapies in preventing those diseases but cannot be a substitute for the beneficial effects of the MeDi dietary habits and lifestyle [36].

Recent studies show the relevance of the gut microbiome's modulation of immune system activity and the nervous system. Dysbiosis affects the equilibrium between the gut microbiota and the host, promoting chronic inflammation and contributing to several diseases, including NDs [39–41]. Dietary habits and nutrients play an important role in determining the composition of the gut microbiota, ultimately modulating the gut–brain axis [42]. Hrnciarova and colleagues, in this Special Issue, investigate nutritional supplementation's role in the gut microbiota and its effect on autism spectrum disorder (ASD) in children [43]. One of the most common neurodevelopmental disorders in children, ASD is characterized by psychiatric and behavioral dysfunction. At present, no known biomarkers are diagnostic for the disorder [44]. The molecular mechanisms responsible for ASD remain unknown. However, recent data show that ASD patients exhibit a peculiar composition of the gut microbiota, indicating that altering the microbiota–gut axis may contribute to ASD pathophysiology [45,46]. Because nutritional supplements modulate the gut microbiota, researchers have investigated using nutritional supplements as a way to improve microbiota composition in people with ASD [47]. Hrnciarova and colleagues analyze an interventional clinical study by supplementing children for 3 months (eight patients with ASD and eight placebo-treated control subjects). This pilot study shows that juvenile supplementation modifies the gut microbiota in children with ASD, ameliorating their symptoms [43]. Those results are promising, opening the way for therapeutic interventions aimed at better treating ASD.

Taste influences dietary habits. The gut microbiota modulates the neurosensorial response and taste, thereby influencing dietary habits, and in turn can contribute to obesity [48]. Thus, dietary patterns can substantially affect public health [49]. Obesity is associated with the onset of multiple sclerosis (MS) in children and young people [49]. MS is an autoimmune disease characterized by chronic inflammation and affects the central nervous system [50]. Papetti and colleagues examine, in this Special Issue, how obesity affects the onset of MS in children [51]. This prospective clinical study shows that obesity presages the onset of MS in pediatric patients, suggesting that interventions aimed at blocking the development of childhood obesity, such as nutritional intervention, could delay or prevent MS [51].

With their very low prevalence, rare NDs (RNDs) are poorly investigated. Indeed, only a few RNDs have a clear diagnosis that allows for the use of therapeutic strategies that can address the symptoms [52]. Briglia and colleagues present, in this Special Issue, an innovative review summarizing the results of dietary and nutritional intervention in RNDs [53]. Only a few clinical studies and in vitro studies have investigated the role of diet and nutrients in RNDs. Therefore, their narrative review can help with carrying out further investigations in the field. Their study focuses on Angelman syndrome, Rett syndrome, rare leukodystrophies (Krabbe disease and Pelizaeus–Merzbacher disease), rare epilepsy, rare forms of ataxia, and rare brain tumors. Although the literature includes only a few clinical studies of those diseases, Briglia and colleagues highlight the positive results obtained in preclinical studies, which can serve as the basis for further clinical studies.

Author Contributions: Conceptualization, L.P. and W.B.G.; methodology, L.P.; formal analysis, L.P.; writing—original draft preparation, L.P.; writing—review and editing, L.P. and W.B.G. All authors have read and agreed to the published version of the manuscript.

Funding: This research received no external funding.

Acknowledgments: We acknowledge Gabe Waggoner for undertaking the critical reading and revision of the manuscript.

Conflicts of Interest: The authors declare no conflicts of interest.

References

1. Brown, G.C. The endotoxin hypothesis of neurodegeneration. *J. Neuroinflamm.* **2019**, *16*, 180. [CrossRef] [PubMed]
2. Kearns, R. Gut-Brain Axis and Neuroinflammation: The Role of Gut Permeability and the Kynurenine Pathway in Neurological Disorders. *Cell. Mol. Neurobiol.* **2024**, *44*, 64. [CrossRef] [PubMed]
3. Lassale, C.; Batty, G.D.; Baghdadli, A.; Jacka, F.; Sánchez-Villegas, A.; Kivimäki, M.; Akbaraly, T. Healthy Dietary Indices and Risk of Depressive Outcomes: A Systematic Review and Meta-Analysis of Observational Studies. *Mol. Psychiatry* **2019**, *24*, 965–986. [CrossRef] [PubMed]
4. Marx, W.; Lane, M.; Hockey, M.; Aslam, H.; Berk, M.; Walder, K.; Borsini, A.; Firth, J.; Pariante, C.M.; Berding, K.; et al. Diet and depression: Exploring the biological mechanisms of action. *Mol. Psychiatry* **2021**, *26*, 134–150. [CrossRef]
5. Tsai, S.-J. Role of Interleukin 8 in Depression and Other Psychiatric Disorders. *Prog. Neuropsychopharmacol. Biol. Psychiatry* **2021**, *106*, 110173. [CrossRef]
6. Niranjan, R. The Role of Inflammatory and Oxidative Stress Mechanisms in the Pathogenesis of Parkinson's Disease: Focus on Astrocytes. *Mol. Neurobiol.* **2014**, *49*, 28–38. [CrossRef]
7. Franceschi, C.; Garagnani, P.; Parini, P.; Giuliani, C.; Santoro, A. Inflammaging: A New Immune–Metabolic Viewpoint for Age-Related Diseases. *Nat. Rev. Endocrinol.* **2018**, *14*, 576–590. [CrossRef]
8. Hale, J.M.; Schneider, D.C.; Mehta, N.K.; Myrskylä, M. Cognitive impairment in the U.S.: Lifetime risk, age at onset, and years impaired. *SSM Popul. Health* **2020**, *11*, 100577. [CrossRef]
9. Song, S.; Stern, Y.; Gu, Y. Modifiable lifestyle factors and cognitive reserve: A systematic review of current evidence. *Ageing Res. Rev.* **2022**, *74*, 101551. [CrossRef]
10. Xiao, L.; Zan, G.; Qin, J.; Wei, X.; Lu, G.; Li, X.; Zhang, H.; Zou, Y.; Yang, L.; He, M.; et al. Combined exposure to multiple metals and cognitive function in older adults. *Ecotoxicol. Environ. Saf.* **2021**, *222*, 112465. [CrossRef]
11. Song, S.; Liu, N.; Wang, G.; Wang, Y.; Zhang, X.; Zhao, X.; Chang, H.; Yu, Z.; Liu, X. Sex Specificity in the Mixed Effects of Blood Heavy Metals and Cognitive Function on Elderly: Evidence from NHANES. *Nutrients* **2023**, *15*, 2874. [CrossRef] [PubMed]
12. Flieger, J.; Forma, A.; Flieger, W.; Flieger, M.; Gawlik, P.J.; Dzierżyński, E.; Maciejewski, R.; Teresiński, G.; Baj, J. Carotenoid Supplementation for Alleviating the Symptoms of Alzheimer's Disease. *Int. J. Mol. Sci.* **2024**, *25*, 8982. [CrossRef] [PubMed]
13. Leonard, M.; Dickerson, B.; Estes, L.; Gonzalez, D.E.; Jenkins, V.; Johnson, S.; Xing, D.; Yoo, C.; Ko, J.; Purpura, M.; et al. Acute and Repeated Ashwagandha Supplementation Improves Markers of Cognitive Function and Mood. *Nutrients* **2024**, *16*, 1813. [CrossRef] [PubMed]
14. Sood, A.; Kumar, A.; Dhawan, D.K.; Sandhir, R. Propensity of *Withania somnifera* to Attenuate Behavioural, Biochemical, and Histological Alterations in Experimental Model of Stroke. *Cell. Mol. Neurobiol.* **2016**, *36*, 1123–1138. [CrossRef]
15. D'Cruz, M.; Andrade, C. Potential clinical applications of Ashwagandha (*Withania somnifera*) in medicine and neuropsychiatry. *Expert Rev. Clin. Pharmacol.* **2022**, *15*, 1067–1080. [CrossRef]
16. Gomez Afonso, A.; Fernandez-Lazaro, D.; Adams, D.P.; Monserda-Vilaro, A.; Fernandez-Lazaro, C.I. Effects of *Withania somnifera* (Ashwagandha) on Hematological and Biochemical Markers, Hormonal Behavior, and Oxidant Response in Healthy Adults: A Systematic Review. *Curr. Nutr. Rep.* **2023**, *12*, 465–477. [CrossRef]

17. Alanazi, H.H.; Elfaki, E. The immunomodulatory role of *Withania somnifera* (L.) dunal in inflammatory diseases. *Front. Pharmacol.* **2023**, *14*, 1084757. [CrossRef]
18. Dar, N.J.; Ahmad, M. Neurodegenerative diseases and *Withania somnifera* (L.): An update. *J. Ethnopharmacol.* **2020**, *256*, 112769. [CrossRef] [PubMed]
19. Durg, S.; Dhadde, S.B.; Vandal, R.; Shivakumar, B.S.; Charan, C.S. *Withania somnifera* (Ashwagandha) in neurobehavioural disorders induced by brain oxidative stress in rodents: A systematic review and meta-analysis. *J. Pharm. Pharmacol.* **2015**, *67*, 879–899. [CrossRef]
20. Skv, M.; Abraham, S.M.; Eshwari, O.; Golla, K.; Jhelum, P.; Maity, S.; Komal, P. Tremendous Fidelity of Vitamin D3 in Age-related Neurological Disorders. *Mol. Neurobiol.* **2024**, *61*, 7211–7238. [CrossRef]
21. AlGhamdi, S.A. Effectiveness of Vitamin D on Neurological and Mental Disorders. *Diseases* **2024**, *12*, 131. [CrossRef] [PubMed]
22. Riccardi, C.; Perrone, L.; Napolitano, F.; Sampaolo, S.; Melone, M.A.B. Understanding the Biological Activities of Vitamin D in Type 1 Neurofibromatosis: New Insights into Disease Pathogenesis and Therapeutic Design. *Cancers* **2020**, *12*, 2965. [CrossRef]
23. Clarke, R.; Shipley, M.; Lewington, S.; Youngman, L.; Collins, R.; Marmot, M.; Peto, R. Underestimation of risk associations due to regression dilution in long-term follow-up of prospective studies. *Am. J. Epidemiol.* **1999**, *150*, 341–353. [CrossRef]
24. Grant, W.B. Follow-Up Period Affects the Association between Serum 25-Hydroxyvitamin D Concentration and Incidence of Dementia, Alzheimer's Disease, and Cognitive Impairment. *Nutrients* **2024**, *16*, 3211. [CrossRef] [PubMed]
25. Singh, B.; Parsaik, A.K.; Mielke, M.M.; Erwin, P.J.; Knopman, D.S.; Petersen, R.C.; Roberts, R.O. Association of mediterranean diet with mild cognitive impairment and Alzheimer's disease: A systematic review and meta-analysis. *J. Alzheimers Dis.* **2014**, *39*, 271–282. [CrossRef]
26. Féart, C.; Samieri, C.; Rondeau, V.; Amieva, H.; Portet, F.; Dartigues, J.F.; Scarmeas, N.; Barberger-Gateau, P. Adherence to a Mediterranean diet, cognitive decline, and risk of dementia. *JAMA* **2009**, *302*, 638–648. [CrossRef] [PubMed]
27. Gardener, S.; Gu, Y.; Rainey-Smith, S.R.; Keogh, J.B.; Clifton, P.M.; Mathieson, S.L.; Taddei, K.; Mondal, A.; Ward, V.K.; Scarmeas, N.; et al. Adherence to a Mediterranean diet and Alzheimer's disease risk in an Australian population. *Transl. Psychiatry* **2012**, *2*, e164. [CrossRef] [PubMed]
28. Scarmeas, N.; Luchsinger, J.A.; Mayeux, R.; Stern, Y. Mediterranean diet and Alzheimer disease mortality. *Neurology* **2007**, *69*, 1084–1093. [CrossRef]
29. Scarmeas, N.; Stern, Y.; Mayeux, R.; Luchsinger, J.A. Mediterranean diet, Alzheimer disease, and vascular mediation. *Arch. Neurol.* **2006**, *63*, 1709–1717. [CrossRef]
30. Scarmeas, N.; Stern, Y.; Mayeux, R.; Manly, J.J.; Schupf, N.; Luchsinger, J.A. Mediterranean diet and mild cognitive impairment. *Arch. Neurol.* **2009**, *66*, 216–225. [CrossRef]
31. Scarmeas, N.; Stern, Y.; Tang, M.X.; Mayeux, R.; Luchsinger, J.A. Mediterranean diet and risk for Alzheimer's disease. *Ann. Neurol.* **2006**, *59*, 912–921. [CrossRef] [PubMed]
32. Real, H.; Dias, R.R.; Graça, P. Mediterranean Diet Conceptual Model and Future Trends of Its Use in Portugal. *Health Promot. Int.* **2021**, *36*, 548–560. [CrossRef] [PubMed]
33. Bavaro, A.R.; Tarantini, A.; Bruno, A.; Logrieco, A.F.; Gallo, A.; Mita, G.; Valerio, F.; Bleve, G.; Cardinali, A. Functional foods in Mediterranean diet: Exploring the functional features of vegetable case-studies obtained also by biotechnological approaches. *Aging Clin. Exp. Res.* **2024**, *36*, 208. [CrossRef]
34. Perrone, L.; Grant, W.B. Observational and ecological studies of dietary advanced glycation end products in national diets and Alzheimer's disease incidence and prevalence. *J. Alzheimers Dis.* **2015**, *45*, 965–979. [CrossRef]
35. Dominguez, L.J.; Veronese, N.; Parisi, A.; Seminara, F.; Vernuccio, L.; Catanese, G.; Barbagallo, M. Mediterranean Diet and Lifestyle in Persons with Mild to Moderate Alzheimer's Disease. *Nutrients* **2024**, *16*, 3421. [CrossRef]
36. Sbai, O.; Torrisi, F.; Fabrizio, F.P.; Rabbeni, G.; Perrone, L. Effect of the Mediterranean Diet (MeDi) on the Progression of Retinal Disease: A Narrative Review. *Nutrients* **2024**, *16*, 3169. [CrossRef] [PubMed]
37. French, S. *Visual Impairment and Work: Experiences of Visually Impaired People*, 1st ed.; Routledge: London, UK, 2017.
38. Vasconcelos, A.R.; Dos Santos, N.B.; Scavone, C.; Demarchi Munhoz, C. Nrf2/ARE Pathway Modulation by Dietary Energy Regulation in Neurological Disorders. *Front. Pharmacol.* **2019**, *10*, 33. [CrossRef]
39. Cryan, J.F.; Dinan, T.G. Mind-altering microorganisms: The impact of the gut microbiota on brain and behaviour. *Nat. Rev. Neurosci.* **2012**, *13*, 701–712. [CrossRef]
40. Perrone, L.; Valente, M. The emerging role of metabolism in brain-heart axis: New challenge for the therapy and prevention of Alzheimer Disease. May Thioredoxin Interacting Protein (TXNIP) play a role? *Biomolecules* **2021**, *11*, 1652. [CrossRef]
41. Dinan, T.G.; Cryan, J.F. The microbiome-gut-brain axis in health and disease. *Gastroenterol. Clin. N. Am.* **2017**, *46*, 77–89. [CrossRef]
42. Chatterjee, A.; Kumar, S.; Roy-Sarkar, S.; Halder, R.; Kumari, R.; Banerjee, S.; Sarkar, B. Dietary polyphenols represent a phytotherapeutic alternative for gut dysbiosis associated neurodegeneration: A systematic review. *J. Nutr. Biochem.* **2024**, *129*, 109622. [CrossRef] [PubMed]
43. Hrnciarova, J.; Kubelkova, K.; Bostik, V.; Rychlik, I.; Karasova, D.; Babak, V.; Datkova, M.; Simackova, K.; Macela, A. Modulation of Gut Microbiome and Autism Symptoms of ASD Children Supplemented with Biological Response Modifier: A Randomized, Double-Blinded, Placebo-Controlled Pilot Study. *Nutrients* **2024**, *16*, 1988. [CrossRef] [PubMed]
44. Venigalla, H.; Mekala, H.M.; Hassan, M.; Ahmed, R.; Zain, H.; Dar, S.; Veliz, S. An update on biomarkers in psychiatric disorders–are we aware use in our clinical practice. *Ment. Health Fam. Med.* **2017**, *13*, 471–479.

45. Fiorentino, M.; Sapone, A.; Senger, S.; Camhi, S.S.; Kadzielski, S.M.; Buie, T.M.; Kelly, D.L.; Cascella, N.; Fasano, A. Blood–brain barrier and intestinal epithelial barrier alterations in autism spectrum disorders. *Mol. Autism* **2016**, *7*, 49. [CrossRef] [PubMed]
46. Louis, P. Does the human gut microbiota contribute to the etiology of autism spectrum disorders? *Dig. Dis. Sci.* **2012**, *57*, 1987–1989. [CrossRef]
47. Sivamaruthi, B.S.; Suganthy, N.; Kesika, P.; Chaiyasut, C. The Role of Microbiome, Dietary Supplements, and Probiotics in Autism Spectrum Disorder. *Int. J. Environ. Res. Public Health* **2020**, *17*, 2647. [CrossRef]
48. Shanmugamprema, D.; Muthuswamy, K.; Subramaniam, S. Emerging Perspectives: The Interplay of Taste Perception and Oral Microbiota Composition in Dietary Preferences and Obesity. *Nutr. Res. Rev.* **2024**, *21*, 1–10. [CrossRef]
49. Rasul, T.; Frederiksen, J.L. Link between overweight/obese in children and youngsters and occurrence of multiple sclerosis. *J. Neurol.* **2018**, *265*, 2755–2763. [CrossRef]
50. Gordon-Lipkin, E.; Banwell, B. An update on multiple sclerosis in children: Diagnosis, therapies, and prospects for the future. *Expert. Rev. Clin. Immunol.* **2017**, *13*, 975–989. [CrossRef]
51. Papetti, L.; Panella, E.; Monte, G.; Ferilli, M.A.N.; Tarantino, S.; Checchi, M.P.; Valeriani, M. Pediatric Onset Multiple Sclerosis and Obesity: Defining the Silhouette of Disease Features in Overweight Patients. *Nutrients* **2023**, *15*, 4880. [CrossRef]
52. Jinnah, H.A.; Albanese, A.; Bhatia, K.P.; Cardoso, F.; Da Prat, G.; de Koning, T.J.; Espay, A.J.; Fung, V.; Garcia-Ruiz, P.J.; Gershanik, O.; et al. Treatable inherited rare movement disorders. *Mov. Disord.* **2018**, *33*, 21–35. [CrossRef] [PubMed]
53. Briglia, M.; Allia, F.; Avola, R.; Signorini, C.; Cardile, V.; Romano, G.L.; Giurdanella, G.; Malaguarnera, R.; Bellomo, M.; Graziano, A.C.E. Diet and Nutrients in Rare Neurological Disorders: Biological, Biochemical, and Pathophysiological Evidence. *Nutrients* **2024**, *16*, 3114. [CrossRef] [PubMed]

nutrients

MDPI

Article

Mediterranean Diet and Lifestyle in Persons with Mild to Moderate Alzheimer's Disease

Ligia J. Dominguez [1,*,†], Nicola Veronese [2,†], Angela Parisi [2], Flavia Seminara [2], Laura Vernuccio [2], Giuseppina Catanese [2] and Mario Barbagallo [2]

1 Department of Medicine and Surgery, University Kore of Enna, 94100 Enna, Italy
2 Geriatric Unit, Department of Internal Medicine and Geriatrics, University of Palermo, 90144 Palermo, Italy; nicola.veronese@unipa.it (N.V.); angelaparisi89@libero.it (A.P.); semifla94@gmail.com (F.S.); lvernuccio@virgilio.it (L.V.); giuseppina.catanese@policlinico.pa.it (G.C.); mario.barbagallo@unipa.it (M.B.)
* Correspondence: ligia.dominguez@unikore.it
† These authors contributed equally to this work.

Abstract: Due to the continuous aging of the population and consequent increase in dementia, focus on its prevention is of growing importance for public health. Since effective pharmacological treatments are not yet available, other determinants of cognitive decline have become fundamental. Several studies have indicated that the Mediterranean diet (MedDiet) is associated with reduced incident cognitive decline and dementia, but few studies have been conducted in persons already diagnosed with Alzheimer's disease (AD). We age-matched 73 patients with mild–moderate AD with 73 controls (mean age for the whole group = 76.5 ± 6.5; 67.5% women). The cases had a significantly lower adherence to the MedDiet and lower physical activity vs. controls, where only one participant (1.4%) had a high adherence to the MedDiet among cases compared to 5.5% among controls, while 52.5% of the cases had a moderate adherence to the MedDiet vs. 82.2% in controls. In multivariate analysis, only the presence of AD was significantly associated with a lower adherence to the MedDiet vs. controls. Other factors examined (gender, age, physical activity level, multimorbidity, and polypharmacy) were not significantly associated with adherence to the MedDiet. Thus, AD patients had a low adherence to the MedDiet and very low physical activity. Public health strategies aimed at promoting the Mediterranean diet and physical activity for older people should be a priority.

Keywords: Alzheimer's disease; dementia; mediterranean diet; aging; malnutrition; physical activity

1. Introduction

The world is continuously aging. Currently, most humans around the world can expect to live beyond the age of 60 [1]. Consequently, the prevention and treatment of age-related chronic diseases, such as neurodegenerative diseases, are of growing importance for public health [2,3]. It is estimated that the incidence of dementia almost doubles every 5 years after the age of 65. By the time people reach their nineties, almost one-third of adults will meet the diagnostic criteria for dementia. In 2019, it was estimated that 57 million people had dementia worldwide, and the projection for 2050 has increased to 153 million [4]. The most diagnosed forms/types of dementia are Alzheimer's disease (AD), the vascular form, and the mixed form. By causing progressive cognitive and functional deterioration, they represent major causes of disability [5].

Aging is one of the greatest risk factors for developing dementia, as with other chronic conditions [6]. By the time adults are in their thirties, fundamental cognitive abilities, such as episodic memory, reasoning, processing speed, and spatial visualization, begin to decline [7]. Regardless of other factors, "biological" aging is in itself accompanied by a systemic pro-inflammatory response and modifications in the central nervous system, including brain atrophy, white matter degradation, and neuropathological protein accumulation [8], which increases susceptibility to the development of neurodegenerative diseases. However, further investigations are needed to determine which additional factors or pathophysiological mechanisms lead some persons to develop a neurodegenerative disease while others will remain "resilient" to any form of dementia [9].

Other factors may contribute to the onset of cognitive decline, including genetic predisposition and socioeconomic and environmental influences such as diet and physical activity [10]. Since effective pharmacological treatments are not yet available to treat cognitive deterioration and dementia [11], prevention becomes increasingly fundamental [12]. Certain nutritional regimes and specific foods, especially if associated with a healthy lifestyle that includes physical activity and regular circadian rhythm, have long been studied as possible modifiable factors capable of delaying the onset and severity of cognitive decline [12]. Several studies have indicated the Mediterranean diet (MedDiet), or rather the "Mediterranean lifestyle", as having significant features associated with a reduced incidence of various non-communicable diseases, including dementia [12–16].

Nutritional problems such as weight loss, lack of appetite, and sarcopenia are frequent among patients suffering from cognitive decline, especially in the more advanced stages of the disease [17]. These correlate with relevant adverse events, including institutionalization, morbidity, and mortality [18]. Some studies have also shown that in patients with early-stage AD or mild cognitive decline, nutritional alterations increase the risk of disease progression to a more severe stage [19,20] and of behavioral and psychiatric symptoms [21]. In the early stages of the disease, memory and concentration deficits can affect planning, shopping, and preparing meals. As cognitive deterioration progresses, eating adequately becomes an increasingly difficult challenge: sensory alterations (e.g., reduction in smell and taste), loss of appetite (senile anorexia), difficulty in communicating discomfort or sensation (e.g., hunger, thirst, tiredness, pain, constipation), behavioral disorders (e.g., aggression, apathy, depression, wandering), and in the last stages, the possible onset of dysphagia. These factors increase the risk of malnutrition, complications, and disease progression [17].

Based on this background, we aimed to evaluate the adherence to the MedDiet and lifestyle of persons affected by mild to moderate AD and its association with the severity of the disease as well as with anthropometric parameters, self-efficacy, comorbidity, polypharmacy, and physical activity in comparison with age-matched control participants without cognitive issues. We also evaluated the consumption of specific foods that have been shown to have neuroprotective properties.

2. Participants and Methods

2.1. Participants

In the present cross-sectional study, older men and women undergoing an evaluation at the Cognitive Disorders and Dementia (CDCD) ambulatory clinics of the Geriatrics Section of the "Azienda Ospedaliera Universitaria Policlinico 'Paolo Giaccone'" in Palermo, Italy, were consecutively enrolled from 1 February 2023 to 1 August 2024. All patients lived at home in an urban setting with their relatives or caregivers, and none of them lived alone or were institutionalized. Inclusion criteria were (1) age > 60 years; (2) diagnosis of AD (mild to moderate), according to DSM-5 criteria [22]. In brief, this includes significant cognitive decline from a previous level of performance in one or more cognitive domains referred by the patient, a knowledgeable informant, or the clinician; a substantial impairment in cognitive performance in standardized neuropsychological testing; the cognitive deficits interfere with independence in everyday activities; exclusion of delirium or other mental disorder. In addition, all the participants included retained a minimum degree of autonomy and self-sufficiency; therefore, they were able to carry out regular physical activity (even a simple daily walk) and could consume different types of food as they were free from problems that prevented their intake (e.g., dysphagia, total edentulism, etc.). On the contrary, we excluded patients without a complete evaluation of functional and cognitive status or a cognitive decline due to a primary psychiatric disorder such as schizophrenia or bipolar disorder. The participants in the study were recruited based on the voluntary acceptance by the patient or caregiver to be part of the study and interviewed via a questionnaire that included various parameters (anthropometric, nutritional, and geriatric assessments) as described below. The participants in the control group were identified among outpatients attending the osteoporosis ambulatory clinic of the same University Hospital; in particular, autonomous and self-sufficient persons not affected by cognitive decline were enrolled. Written informed consent was obtained from all participants involved in the study, and in case of the patient's inability, the legally authorized delegate provided informed consent. The study was conducted in accordance with the Declaration of Helsinki on the ethical principles for medical research involving human subjects [23].

All data analyzed in the present study were obtained as part of routine evaluation, diagnosis, and treatment. In accordance with the current Italian law (Gazzetta Ufficiale della Repubblica Italiana, Serie Generale n. 76 del 31 May 2008), we acknowledge our Institution's Ethical Committee (Comitato Etico Palermo 1 A.O.U.P. 'P. Giaccone') (protocol n. 22 del 3 September 2024) about this observational research regarding usual clinical practice by sending an official letter.

2.2. Outcome: Adherence to the Mediterranean Diet

The degree of adherence to the MedDiet was assessed through the MEDAS (*Mediterranean Diet Adherence Screener*) [24], a questionnaire initially developed during the PREDIMED study (*Prevención con Dieta Mediterránea*) [25] and consisting of 14 items, which are quick and easy to administer. Briefly, each question was given a score of "0" or "1". One point was awarded for the consumption of olive oil as the main source of cooking fats (question 1); 4 or more tablespoons of olive oil per day (question 2); 2 or more portions of vegetables per day (question 3); 3 or more servings of fruit per day (question 4); less than 1 serving of red meat or sausage per day (question 5); less than one serving of animal fats per day (question 6); less than 1 carbonated or sweetened drink per day (question 7); 7 or more glasses of wine per week (question 8); more than 2 portions of legumes per week (question no. 9); 3 or more portions per week of fish or seafood (question 10); less than 3 times a week of commercial pastry (question no. 11); 3 or more times a week of dried fruit (question no. 12); chicken, turkey, or rabbit instead of beef, pork, and cured meats (question n.13); and 2 or more times a week of boiled vegetables, pasta, or rice seasoned and sautéed (question no. 14). The MEDAS score ranges from 0 to 14 points. Even if no definitive cut-offs exist, we used, for the purposes of this work, a score less than 6 to indicate low adherence to the MedDiet, a score between 7 and 10 as intermediate adherence, and a score between 11 and 14 considered as high adherence [25].

2.3. Demographics

Other than age and sex, we included physical activity level carried out in the last seven days through the IPAQ SHORT FORM (*International Physical Activity Questionnaire*) [26], which differentiates between heavy activity (e.g., heavy lifting, aerobics, digging, or fast bicycling...), moderate activity (lifting light weights, doubles tennis, cycling at a moderate pace...), or simply walking (of at least 10 min), as well as time spent sitting on weekdays (e.g., sitting at a desk, reading, visiting friends, lying down to watch television, or sitting), indicating for each duration and weekly frequency; smoking status, categorized in no, previous, or current.

2.4. Multidimensional Evaluation

A multidimensional evaluation was performed on all patients, investigating the ability to perform activities of daily living (ADL), which defines the level of dependence/independence in six activities (bathing, feeding, toileting, dressing, transferring in and out of bed or chair, urine, and bowel continence) ranging from zero to six [27,28]; Instrumental Activities of Daily Living (IADL) considering eight activities, which are more demanding cognitively and physically with respect to ADL, i.e., using the telephone, managing finances, shopping, taking medications, preparing meals, using transportation, doing housework, and washing [29,30]; the global degree of cognitive decline assessed with the Mini-Mental State Examination (MMSE) considering seven different cognitive areas and 30 different questions divided into orientation over time, orientation in space, word recording, attention and calculation, evocation, language, and constructive praxis. Multimorbidity was considered when a participant had a diagnosis of two or more medical conditions [31] commonly present in older people, such as heart attack, hypertension, hypercholesterolemia, stroke, diabetes, chronic lung disease, arthritis, asthma, osteoporosis, gastric or duodenal ulcer, cancer, hip fracture, and cataracts. Polypharmacy is defined as the regular use of over five medications [32].

2.5. Nutritional Evaluation

Weight and height were recorded by trained residents in geriatric medicine. Body mass index (BMI) was recorded classifying participants into underweight (BMI < 19 kg/m^2), normal weight (BMI between 19.1 and 25 kg/m^2), overweight (BMI between 25.1 and 30 kg/m^2), and obese (BMI > 30 kg/m^2). We also recorded data about the dietary intakes of some specific nutrients that have been associated with beneficial or harmful effects on cognition, such as alcohol drinking, spices, tea, coffee, cocoa, red fruits, and salt [33].

2.6. Statistical Analysis

We used mean and standard deviation (SD) to describe continuous variables and percentages for categorical variables. Baseline characteristics of the study participants were compared between cases and controls, using an independent T-test for continuous variables and Chi-squared for categorical variables. We categorized the adherence to the MedDiet as moderate-to-high vs. low since only one case reported a high adherence to the MedDiet. The association between the adherence to the MedDiet and factors significantly different between cases and controls ($p < 0.05$) was explored using a logistic regression analysis. The factors potentially associated with the outcome of interest were based on the previous literature about risk factors of poor adherence to the MedDiet [34]. The results are reported as odds ratios (ORs) and their 95% confidence intervals (CIs). Finally, in people affected by AD, we explored the association between MEDAS and the domains of multidimensional evaluation, using simple correlation analysis and reporting data as rho and p-values due to a non-parametric distribution of the MEDAS. All statistical analyses were carried out using SPSS software version 26.0, were two-tailed, and a p-value < 0.05 was considered statistically significant.

3. Results

In this study, we matched 73 patients with mild to moderate AD (cases) with 73 controls for age. The mean age of the whole group was 76.5 ± 6.5, and 67.5% were women. As shown in Table 1, the cases did not differ in terms of age ($p = 0.69$), but they were more frequently males. The cases reported significantly lower moderate ($p = 0.03$) and low ($p < 0.0001$) physical activity levels compared to the controls. The cases reported an important impairment in ADL (mean = 3.4) and IADL (mean = 1.9) as well as in MMSE (mean = 16.7/30). As expected, multimorbidity and polypharmacy were more frequent among the cases compared to the controls (Table 1).

About nutritional information, the cases reported less frequently the use of spices ($p = 0.04$) and coffee ($p < 0.0001$) but more frequently the consumption of red fruits and salt. As shown in Table 1, the cases reported significantly lower adherence to the MedDiet in the mean of 1.4 points. In Figure 1, data from the differences in the degree of adherence to the MedDiet between the cases and controls are shown: very few (1.4%) had high adherence to the MedDiet among the cases compared to 5.5% among the controls; 52.5% of the cases had a moderate adherence to the MedDiet compared to 82.2% in the control group ($p < 0.0001$ for these comparisons).

Table 1. Descriptive characteristics in cases (affected by dementia) and controls.

Parameter	Cases (n = 73)	Controls (n = 73)	*p*-Value
Demographics			
Age (mean, SD)	76.6 (4.7)	76.3 (3.9)	0.69
Males (%)	42.5	20.5	0.004
Moderate-to-high physical activity level (%)	1.4	13.7	0.03
Low physical activity level (%)	38.4	93.2	<0.0001
Smoking status, no (%)	82.2	86.3	0.34
Smoking status, current (%)	5.5	6.8	
Smoking status, previous (%)	12.3	6.9	
Multidimensional evaluation			
ADL (mean, SD)	3.4 (1.8)	6	-
IADL (mean, SD)	1.9 (2.1)	8 for women, 6 for men	-
MMSE (mean, SD)	16.7 (6.0)	30	-
Multimorbidity (%)	68.5	9.6	<0.0001
Polypharmacy (%)	74.0	28.8	<0.0001
Nutritional evaluation			
BMI (mean, SD)	26.5 (4.0)	26.1 (4.4)	0.54
Alcohol drinking, non-red wine (%)	5.5	0	0.12
Spices more than 3 times in a week (%)	32.9	49.3	0.04
Tea consumption more than 3 times in a week (%)	24.7	34.2	0.20
Cocoa, more than 3 times in a week (%)	23.3	13.7	0.21
Red fruits, more than 3 times in a week (%)	52.1	11.0	<0.0001
Coffee, yes/no (%)	72.6	100	<0.0001
Salt, 3 spoons/day (%)	4.1	0	<0.0001
MEDAS (mean, SD)	6.7 (1.4)	8.1 (1.5)	<0.0001

ADL: activities of daily living; BMI: body mass index; IADL: instrumental activities of daily living; MEDAS: Mediterranean Diet Adherence Screener; MMSE: mini-mental state examination; SD: standard deviation.

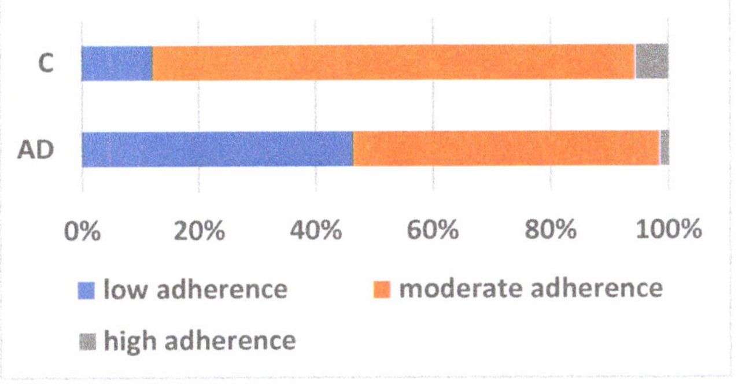

Figure 1. Adherence to Mediterranean diet in cases and controls. AD: Alzheimer's disease; C: controls.

Table 2 shows the multivariate analysis, taking as an outcome the moderate-to-high adherence to the MedDiet and as factors all the factors significantly different among the cases and controls. Only the presence of AD was associated with a lower adherence to the MedDiet compared to the controls (OR = 0.222; 95% CI: 0.058–0.848; $p = 0.028$), whilst the other factors examined (gender, age, physical activity level, multimorbidity, polypharmacy, the consumption of red fruits, more than 3 times in a week, coffee consumption, and the use of salt) reported a p-value > 0.20.

Finally, we analyzed the correlations between the adherence to the MedDiet and the domains of multidimensional evaluation. Considering these domains, only IADL (rho = 0.334, $p < 0.001$) and MMSE scores (rho = 0.331, $p < 0.001$) were associated with MEDAS, as graphically reported in Figure 2.

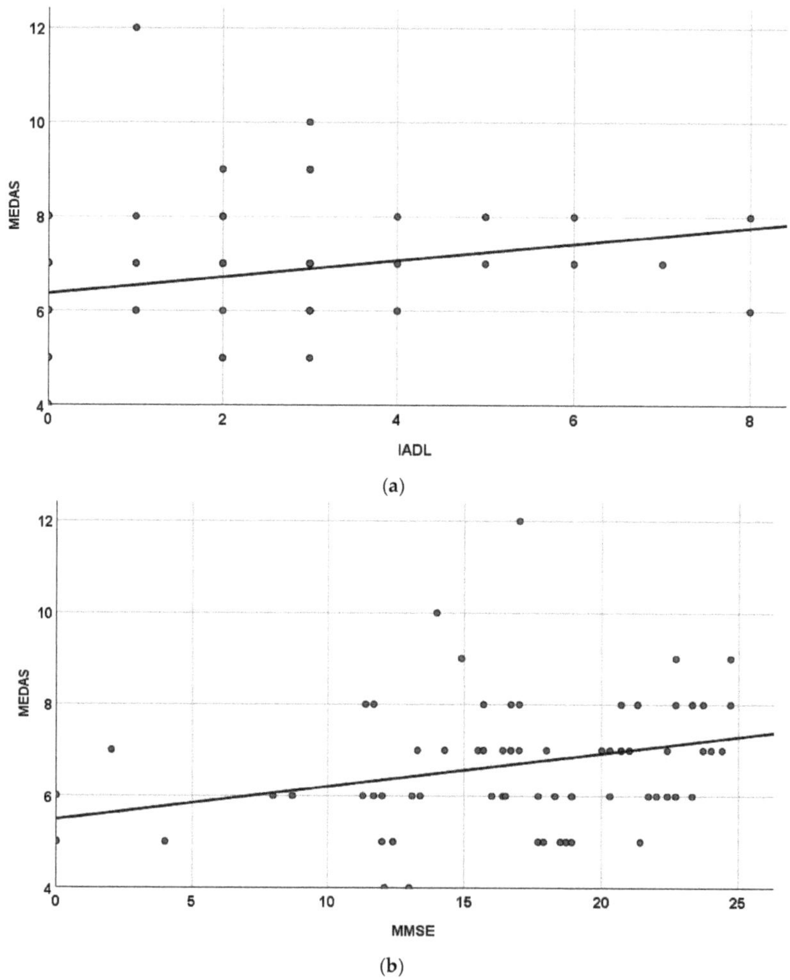

Figure 2. Correlations between adherence to the Mediterranean diet and instrumental activities of daily living (**a**) and with mini-mental state examination (**b**). IADL: Instrumental Activities of Daily Living); MEDAS: Mediterranean Diet Adherence Screener; MMSE: Mini-Mental State Examination.

Table 2. Factors associated with moderate-to-high adherence to Mediterranean diet.

Parameter	Odds Ratio	95% Low CI	95% High CI	*p* -Value
Presence of Alzheimer's Disease	0.222	0.058	0.848	0.028
Males	0.916	0.368	2.281	0.850
Age	0.929	0.845	1.022	0.129
Low physical activity level	0.905	0.345	2.375	0.839
Multimorbidity	0.762	0.238	2.443	0.648
Polypharmacy	0.930	0.297	2.917	0.902
Red fruits, more than 3 times per week	0.855	0.350	2.090	0.731
Coffee consumption	1.475	0.494	4.403	0.486
Salt, 2 spoons/day	0.587	0.232	10.484	0.260
Salt, 3 spoons/day	10.358	0.101	18.185	0.817

Data are reported as odds ratio (OR) with their 95% confidence intervals (CIs) and *p*-values, among factors statistically significant between cases (presence of dementia) and controls.

4. Discussion

In the present study, among 146 participants examined, those with a diagnosis of mild to moderate AD had significantly lower adherence to the MedDiet pattern assessed with the MEDAS questionnaire compared to those participants who were cognitively competent. Practically none of the participants with dementia had a high adherence to this high-quality dietary pattern, while most of the controls (>80%) had an intermediate good adherence in this Mediterranean population. Likewise, the level of physical activity carried out in the last seven days assessed with the IPAQ SHORT FORM was also remarkably low among dementia patients, with almost half (45%) not carrying out any physical activity vs. almost all (93%) of cognitively competent patients habitually carrying out at least light physical activity and 14% moderate physical activity. Few participants from both groups used to smoke or had a habitual consumption of alcoholic beverages (other than red wine); both parameters were comparable for both groups. The consumption of certain foods considered neuroprotective resulted slightly different between the two groups, with cases reporting less frequently the use of spices and coffee but more frequently consumption of red fruits and salt. Nevertheless, all participants had a low average consumption of coffee (two cups per day), below the general recommendation of 3–5 cups per day [35]. We found significant differences between the two groups regarding comorbidity and pharmacological history, with most participants with dementia having more than three pathologies and taking over five medicaments per day vs. lower numbers for controls. In multivariate analysis only the presence of AD was significantly associated with a lower adherence to the MedDiet compared to controls. Other factors examined (gender, age, physical activity level, multimorbidity, and polypharmacy) were not significant. There was a positive significant association between adherence to the MedDiet and cognition assessed with the MMSE score as well as with self-autonomy as an ability to perform IADL.

Although there are several studies specifically investigating the role of the MedDiet in the prevention of cognitive decline and dementia that have found that this high-quality dietary pattern seems to be protective for the development of dementia [12–16], this is one of the few studies exploring the adherence to the MedDiet in patients with a clinical diagnosis of AD. Several systematic reviews and meta-analyses have explored the association of adherence to the MedDiet with cognitive decline and/or incident dementia [13], and most of them have shown an inverse association of a higher adherence to the MedDiet with a slower cognitive decline and/or reduced progression to AD. There are also conflicting results regarding the efficacy of the MedDiet for age-related cognitive function [36,37]. In an umbrella review, out of 11 meta-analyses, 3 reported convincing evidence of benefit, 2 were highly suggestive of benefit, 3 were suggestive of benefit, 1 reported weak evidence, and only 2 reported no evidence [38]. Disparity of neuropsychological assessment methods used appeared to be a plausible contributor to the lack of consensus among study findings. Other diets very similar to MedDiet, such as the Mediterranean dietary approaches to stop hypertension intervention for neurodegenerative delay (MIND) diet, have shown positive results [39–42]. A large meta-analysis of three cohort studies among over 200,000 participants reported that increased adherence to the MIND diet was significantly associated with a decreased risk of dementia [39]. Likewise, a US cohort study of older adults with a baseline mean age of 84.2 years showed that the MIND and the MedDiet patterns, in particular the consumption of green leafy vegetables, were inversely associated with amyloid beta load, phosphorylated tau tangles, and global AD pathology at post mortem analyses [40]. A recent 6-month pilot randomized controlled trial (RCT) was conducted among 93 participants with prodromal AD in four European countries with 3 intervention arms: (1) multimodal lifestyle intervention (nutritional guidance, exercise, cognitive training, vascular/metabolic risk management, and social stimulation); (2) multimodal lifestyle intervention + medical food product; and (3) regular health advice (control group). Adherence to dietary advice was assessed with the Healthy Diet Index and MEDAS. Dietary quality in the intervention groups improved in this population with prodromal AD. Nutrient intakes remained unchanged in the intervention groups, while the control group showed a decreasing nutrient density [43]. These preliminary results suggest that dietary intervention as part of multimodal lifestyle interventions is feasible in this type of patient and that further studies are needed to confirm these encouraging results. There is also evidence of an inverse significant association between adherence to the MedDiet and other relevant mental health conditions, in particular depression [44].

We have considered some foods for which there is some evidence of neuroprotection or neurodamage that were available in the questionnaire. For example, red fruits (berries), which are rich in potent antioxidants such as flavonoids and anthocyanins, have shown neuroprotective effects, but most of the evidence comes from experimental studies in animal models or cellular cultures [45]. There are few intervention studies in humans. A systematic review of studies exploring anthocyanin consumption (i.e., berry juices) and cognitive outcomes in humans found that 6 of 7 studies reported improvements in single or multiple cognitive outcomes. However, there were important methodological limitations because most were small trials with high heterogeneity [46]. Coffee is currently considered helpful in many aspects [47]. However, in terms of neuroprotection, there is no consistency in the literature on the association of the positive effects of coffee on long-term cognition [48,49]. A recent meta-analysis indicated that limited (1–4 cups/day) daily coffee consumption reduces the risk of AD, whereas excessive consumption (>4 cups/day) might increase the risk [49]. A recent meta-analysis reported a linear association between tea intake and risk of dementia, with a significantly decreased risk of dementia for each 1 cup per day increase in tea consumption [50]. Alcohol consumption has contradictory evidence; a small amount, less for women, is reported to be protective [51], but some studies that inhibit any consumption of alcohol state that it is especially harmful to the brain [52]. Preliminary evidence showed some neuroprotective effects of cocoa consumption [53] but there are still debated results about its potential use in the prevention of cognitive decline and dementia [54]. Likewise, there is no consensus on the indication of the consumption of dietary spices for neuroprotection [55]. A recent Mendelian randomization study exploring the causal relationship between dietary salt intake and dementia risk found strong evidence of this association [56]. We found a significantly lower consumption of spices and coffee in AD patients vs. controls and a higher consumption of red fruits and salt in these patients. However, none of them were significantly associated with the multivariate analysis.

There is also evidence showing that combining a healthy diet with a healthy lifestyle is associated with a reduced risk of developing dementia [3,41,57–59]. In this regard, a recent remarkable study using data from 586 deceased participants of the Rush Memory and Aging Project who were followed for up to 24 years with data on cognitive testing and lifestyle factors collected close to death and a complete neuropathologic autoptic evaluation found that a higher lifestyle score was associated with better global cognitive functioning close to death; in multivariable-adjusted models, a one-point increase in lifestyle score was significantly associated with higher global cognitive scores. Even when including common dementia-related brain pathologies in the multivariable-adjusted models the strength and significance of the association was maintained. A higher lifestyle score was associated with lower beta-amyloid load in the brain. Thus, a healthy lifestyle may afford a cognitive reserve in older adults, contributing to the maintenance of cognitive abilities independently of the presence of dementia neuropathologies [41].

There are fewer studies considering the role of nutrition in the initial stages of AD in clinical settings. Due to the lack of an efficacious pharmacological treatment, this evidence is relevant in order to target modifiable nutritional factors that potentially may help prevent or delay further cognitive deterioration. Former evidence showed that malnourished patients with AD display a more rapid cognitive deterioration and progression within one year when compared with well-nourished patients [60,61]. This has been confirmed by a recent study showing that participants with a less healthy diet and worse nutritional status were associated with a higher risk of cognitive decline clinical progression in a sample of patients with subjective cognitive disease, mild cognitive impairment (MCI), and AD, that is, across the complete AD spectrum [19]. In addition, this study showed that a lower fat-free mass, an important indicator of malnutrition [62], was associated with a higher risk of cognitive deterioration, with a comparable effect extent as BMI. Previous findings from the same group showed that patients with subjective cognitive decline (considered one of the earliest noticeable manifestations of AD and related dementias) reporting the lowest consumption of vegetables had the worst cognitive alterations [63]. Soysal et al., in a cross-sectional study, compared the nutritional status and micronutrient levels in outpatients with different types of dementia (AD, frontotemporal dementia, Lewy body dementia, vascular dementia, and normal pressure hydrocephalus) and found a prevalence of malnutrition of 17.2% and a risk of malnutrition in 43.2% of patients, according to the mini nutritional assessment [64].

A Korean study reported that among persons with an early diagnosis of AD, a reduction in cortical thickness was associated with an unhealthy and inadequate nutritional intake [65]. The course of dementia is very variable; diverse factors contribute to the uneven progression of cognitive, physical, and functional decline, with nutrition playing a crucial role [18]. Appropriate nutrition is essential for the proper functioning and repair of body systems, including the brain and the nervous system [12,66]. Inadequate nutrition is a frequent condition among older people and even more in residents of nursing homes [67], who are frequently affected by cognitive decline and dementia. It is confirmed that poor nutritional status is associated with adverse health outcomes such as increased hospitalizations, morbidity, and mortality [68]. Nutritional status seems to impact the prognosis of the progression of functional performance, cognitive decline [19], and behavioral disturbances [21]. A recent retrospective study [69] of the "PROtein enriched MEDiterranean diet to combat undernutrition and promote healthy neuroCOGnitive ageing" (PROMED-COG) project evaluated the association between undernutrition and cognitive decline and incident dementia in 9071 older adults (age range between 42 and 101 years) from three Italian population-based studies. The authors found a 14.3% prevalence of undernutrition at baseline, which was higher among women and in older participants, ranging from 3.5% in those aged <60 years to 28.8% in those aged >85 years. Undernutrition was associated with both incident cognitive decline and incident dementia over a median follow-up of 8.3 and 8.6 years, respectively [69]. These results highlight the importance of early identification of inadequate nutritional status because its management may be an important nonpharmacologic strategy to counteract neurodegeneration. These results highlight the need for an early detection of malnutrition in order to take action even in the early stages of the disease [70,71]. Other studies have shown an association between malnutrition, in particular weight loss, and cognitive decline in persons with dementia [72–74]. There is evidence that weight loss may precede the onset of AD [75–77], and older adults with lower BMI had worse cognitive deterioration and an increased risk of incident dementia [78,79]. Other studies have confirmed that low baseline BMI was associated with clinical progression after two years of follow-up in patients with MCI [80,81]. In our study, participants with AD and cognitively normal controls had a mean normal to slightly increased BMI, but persons with dementia had significantly lower adherence scores to the MedDiet, probably reflecting a poor-quality diet, which, besides BMI, as already discussed, may have important consequences from a cognitive and functional point of view. As a matter of fact, there was a significant correlation between adherence to the MedDiet and self-autonomy, as well as with cognitive decline. A study including MCI patients suggested that lower adherence to the MedDiet was associated with a greater risk of progression to AD [16]. The maintenance of functional status is a major goal in geriatric medicine because it is an essential component of the older adult's health and quality of life. Functional loss is a common pathway to many chronic diseases associated with aging, and this is particularly true in patients with dementia [82]. Maintenance of functional status depends on several factors, some of which are modifiable, preventable, and reversible, including poor nutritional status [68] or poor-quality diet and lifestyle. Therefore, identifying these modifiable determinants in old age, especially for people with dementia, is crucial to help improve the quality of life and reduce disability and dependence [83].

According to widely recognized recommendations, malnutrition screening should be performed in all people with a diagnosis of dementia, administering food according to the person's needs and preferences [17]. Appetite stimulants and oral nutritional supplements to correct cognitive impairment are not recommended. Conversely, ESPEN guidelines for nutrition and hydration in geriatrics recommend the use of oral nutritional supplements if needed in order to prevent protein–energy undernutrition in older adults [68].

We observed a remarkably reduced physical activity in patients with AD. The low levels of physical activity seen in patients with AD could be related to the progression of the disease that limits physical activity. It has been shown that exercise can help cognitive function in all phases of life, probably due to its effects on improving blood circulation and vascular function by reducing hypertension and increasing nitric oxide production, thus promoting neuronal brain plasticity and reducing neuroinflammation [84]. Data support exercise training to improve cognitive function in healthy older adults, while evidence in AD patients has not been strongly conclusive. This means that exercise seems to be useful for the management of AD at initial stages and for the prevention of incident dementia, but it is not yet defined, and no certain recommendations can be made about the specific exercise recommendations for AD prevention or treatment. More well-designed research is still needed to be able to recommend it with certainty. However, there is evidence that people who regularly exercise at a moderate to vigorous intensity on several days during the week have larger brains compared to those who do not exercise [85]. In addition, both physical and cognitive activity seem clearly related to better brain and cognitive resilience markers across cohorts with differing educational, racial, and disease statuses, supporting their potential neuroprotective effects [86], while studies in experimental animals have suggested that myokines secreted during exercise, such as irisin, may have neuroprotective actions [87].

According to the latest review by the *Lancet Commission* on dementia [3], there is evidence reporting that harmonized care and being actively engaged in diverse types of activities may contribute to reducing depression and neuropsychiatric symptoms and improve well-being in people with dementia, which may also have relevant benefits for caregivers. The success of such interventions has been related to personalizing the programs according to individuals' interests, abilities, and preferences, and involving the family and carers. Few studies have evaluated the cost-effectiveness of these types of programs; hence, further investigations are needed in order to evaluate the requirements and feasibility of their implementation [3]. Conversely, RCTs of exercise in persons with dementia have been so far negative for mental health domain improvements in either care homes or communities [88,89].

A recent international collaborative guideline from European scientific societies and other stakeholders recommends considering physical activity for the primary prevention of dementia. The role of physical activity in slowing the progression from MCI to dementia is still uncertain; the greatest supporting evidence is that of mind–body interventions; the guideline recommends that exercise may be used for maintaining the functional ability and cognition in people with moderate dementia. Even if the scientific evidence on the beneficial effects of physical activity and exercise in the maintenance of cognitive functions in persons with normal cognition, MCI, or dementia comes from still inconclusive studies with very low or low certainty of evidence, the guideline recommends their application due to their beneficial effects on nearly all facets of health [90].

Some of the mechanisms that may help explain the potential neuroprotective effects of the MedDiet and exercise include the reduction in neuroinflammation and oxidative stress, reduced insulin resistance, and improved vascular function and cerebral blood flow. Neuroinflammation has demonstrated a crucial role in the pathogenesis of AD [43]. Neuroinflammation is characterized by a primary immune reaction to brain injury mediated by key pro-inflammatory cytokines involving the activation and priming of glial cells, during which microglia exert macrophage-like functions such as vital surveillance, scavenging, antigen presentation, and cell repair [91]. The released cytokines increase blood–brain barrier permeability [92], which increases their own concentrations in the brain and intensifies microglia's pro-inflammatory responses [93]. Several components of MedDiet exert antioxidant and anti-inflammatory actions that may contribute to attenuating the neuroinflammatory state [94,95]. Various polyphenols (e.g., tyrosol, anthocyanins, and isoflavones) have been found in extra-virgin olive oil, vegetables, fruits, nuts, and legumes and have demonstrated anti-inflammatory properties by modulating various pathways involved in oxidative stress, inflammatory mediators, and promoting cell survival mechanisms [96]. Moreover, the gut–brain axis has recently been proposed as a key player influencing multiple states or diseases, being susceptible to modification by factors such as diet [97]. Likewise, although physical activity and exercise are considered an essential strategy for the prevention and treatment of AD, their specific mechanisms for ameliorating AD are still not fully understood. Nevertheless, there is evidence highlighting their role in reducing neuroinflammatory responses even in the early stages [98] as well as a prominent role in the prevention of hypertension, diabetes, and the resulting cardiovascular damage [99], all recognized risk factors for developing AD (Figure 3).

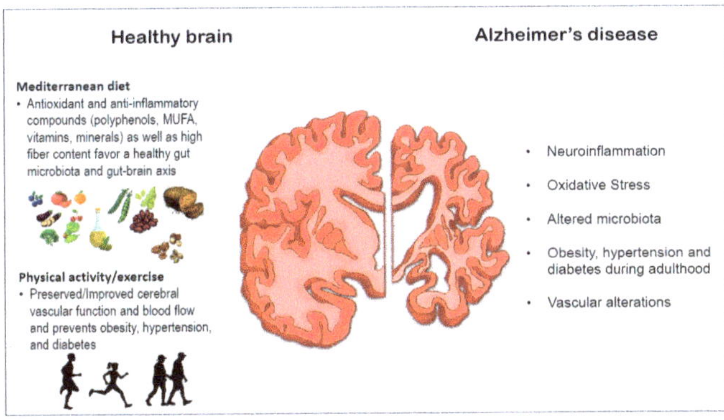

Figure 3. Some mechanisms that may help explain the neuroprotective effects of the Mediterranean diet and physical activity/exercise in the prevention of AD. MUFA: mono-unsaturated fatty acids.

The findings of our study should be interpreted within its limitations. First, it was conducted on a relatively small group of patients. When separating the groups by sex, the numbers were small and did not consent analyses comparing data from men and women; therefore, future studies are needed to include a larger number of participants from both sexes. Second, any cross-sectional study biases due to sampling, length-time bias, and residual confounding cannot be excluded, as well as the inability to establish a clear temporal association between exposure and outcome, preventing any conclusion of causality. Third, unfortunately, we did not have complete data on other crucial variables that may have an impact on cognitive decline and dementia, such as socioeconomic status, educational level, loneliness, and sleep quality/disorders. We also did not have information on other nutritional components that may be relevant to dementia risk, such as fiber, sweetened beverages, or ultra-processed food. Furthermore, age-matched case–control studies have several limitations, as indicated in two studies [100,101]. First, matching does not eliminate confounding; it may introduce selection bias if age is associated with exposure. Second, matching for age may distort dose–response relationships between age and outcomes. Finally, residual confounding requires complex adjustments and challenging interpretation. The strength of our study is that it is one of the few studies exploring the adherence of the high-quality Mediterranean diet and physical activity among patients who already have a diagnosis of Alzheimer's disease.

5. Conclusions

Alzheimer's disease patients had low adherence to the Mediterranean diet and very low levels of physical activity, reflecting unfavorable diet quality and lifestyle. Further attention is needed in these already compromised patients because of the possible implications of the increased risk of malnutrition and faster disease progression.

Author Contributions: Conceptualization, L.J.D., N.V. and A.P.; methodology, L.J.D., N.V., A.P. and F.S.; validation, L.J.D., N.V. and M.B.; formal analysis, N.V.; data curation, A.P. and F.S.; writing—original draft preparation, L.J.D. and N.V.; writing—review and editing, L.J.D., N.V., A.P., F.S., L.V., G.C. and M.B.; supervision, L.J.D., N.V. and M.B. All authors have read and agreed to the published version of the manuscript.

Funding: This research received no external funding.

Institutional Review Board Statement: In agreement with the current Italian law (Gazzetta Ufficiale della Repubblica Italiana, Serie Generale n. 76 del 31 May 2008), we informed the Ethical Committee of our Institution (protocol n. 22 del 3 September 2024) of this observational research regarding normal critical practice by sending a formal letter.

Informed Consent Statement: Informed consent was obtained from all subjects involved in the study.

Data Availability Statement: The data and the databases are available upon reasonable request to the corresponding author.

Conflicts of Interest: The authors declare no conflicts of interest.

References

1. Foreman, K.J.; Marquez, N.; Dolgert, A.; Fukutaki, K.; Fullman, N.; McGaughey, M.; Pletcher, M.A.; Smith, A.E.; Tang, K.; Yuan, C.W.; et al. Forecasting life expectancy, years of life lost, and all-cause and cause-specific mortality for 250 causes of death: Reference and alternative scenarios for 2016–40 for 195 countries and territories. *Lancet* **2018**, *392*, 2052–2090. [CrossRef] [PubMed]
2. Akselrod, S.; Bloomfield, A.; Marmot, M.; Moran, A.E.; Nishtar, S.; Placella, E.; on behalf of the NCD Expert Editorial Group. Mobilising society to implement solutions for non-communicable diseases. *BMJ* **2019**, *365*, l360. [CrossRef]
3. Livingston, G.; Huntley, J.; Liu, K.Y.; Costafreda, S.G.; Selbaek, G.; Alladi, S.; Ames, D.; Banerjee, S.; Burns, A.; Brayne, C.; et al. Dementia prevention, intervention, and care: 2024 report of the Lancet standing Commission. *Lancet* **2024**, *404*, 572–628. [CrossRef] [PubMed]
4. Collaborators, G.B.D.D.F. Estimation of the global prevalence of dementia in 2019 and forecasted prevalence in 2050: An analysis for the Global Burden of Disease Study 2019. *Lancet Public Health* **2022**, *7*, e105–e125. [CrossRef]
5. Malik, R.; Kalra, S.; Bhatia, S.; Harrasi, A.A.; Singh, G.; Mohan, S.; Makeen, H.A.; Albratty, M.; Meraya, A.; Bahar, B.; et al. Overview of therapeutic targets in management of dementia. *Biomed. Pharmacother.* **2022**, *152*, 113168. [CrossRef] [PubMed]
6. Gonzales, M.M.; Garbarino, V.R.; Pollet, E.; Palavicini, J.P.; Kellogg, D.L., Jr.; Kraig, E.; Orr, M.E. Biological aging processes underlying cognitive decline and neurodegenerative disease. *J. Clin. Investig.* **2022**, *132*, e158453. [CrossRef]
7. Anderton, B.H. Ageing of the brain. *Mech. Ageing Dev.* **2002**, *123*, 811–817. [CrossRef]
8. Yankner, B.A.; Lu, T.; Loerch, P. The aging brain. *Annu. Rev. Pathol.* **2008**, *3*, 41–66. [CrossRef]
9. Minhas, P.S.; Latif-Hernandez, A.; McReynolds, M.R.; Durairaj, A.S.; Wang, Q.; Rubin, A.; Joshi, A.U.; He, J.Q.; Gauba, E.; Liu, L.; et al. Restoring metabolism of myeloid cells reverses cognitive decline in ageing. *Nature* **2021**, *590*, 122–128. [CrossRef]
10. Armstrong, R.A. Risk factors for Alzheimer's disease. *Folia Neuropathol.* **2019**, *57*, 87–105. [CrossRef]
11. Profyri, E.; Leung, P.; Huntley, J.; Orgeta, V. Effectiveness of treatments for people living with severe dementia: A systematic review and meta-analysis of randomised controlled clinical trials. *Ageing Res. Rev.* **2022**, *82*, 101758. [CrossRef] [PubMed]

12. Scarmeas, N.; Anastasiou, C.A.; Yannakoulia, M. Nutrition and prevention of cognitive impairment. *Lancet Neurol.* **2018**, *17*, 1006–1015. [CrossRef] [PubMed]
13. Dominguez, L.J.; Veronese, N.; Vernuccio, L.; Catanese, G.; Inzerillo, F.; Salemi, G.; Barbagallo, M. Nutrition, Physical Activity, and Other Lifestyle Factors in the Prevention of Cognitive Decline and Dementia. *Nutrients* **2021**, *13*, 4080. [CrossRef] [PubMed]
14. Scarmeas, N.; Stern, Y.; Mayeux, R.; Manly, J.J.; Schupf, N.; Luchsinger, J.A. Mediterranean diet and mild cognitive impairment. *Arch. Neurol.* **2009**, *66*, 216–225. [CrossRef]
15. Shannon, O.M.; Ranson, J.M.; Gregory, S.; Macpherson, H.; Milte, C.; Lentjes, M.; Mulligan, A.; McEvoy, C.; Griffiths, A.; Matu, J.; et al. Mediterranean diet adherence is associated with lower dementia risk, independent of genetic predisposition: Findings from the UK Biobank prospective cohort study. *BMC Med.* **2023**, *21*, 81. [CrossRef]
16. Singh, B.; Parsaik, A.K.; Mielke, M.M.; Erwin, P.J.; Knopman, D.S.; Petersen, R.C.; Roberts, R.O. Association of mediterranean diet with mild cognitive impairment and Alzheimer's disease: A systematic review and meta-analysis. *J. Alzheimers Dis.* **2014**, *39*, 271–282. [CrossRef]
17. Volkert, D.; Chourdakis, M.; Faxen-Irving, G.; Fruhwald, T.; Landi, F.; Suominen, M.H.; Vandewoude, M.; Wirth, R.; Schneider, S.M. ESPEN guidelines on nutrition in dementia. *Clin. Nutr.* **2015**, *34*, 1052–1073. [CrossRef]
18. Sanders, C.L.; Wengreen, H.J.; Schwartz, S.; Behrens, S.J.; Corcoran, C.; Lyketsos, C.G.; Tschanz, J.T.; Cache County, I. Nutritional Status is Associated with Severe Dementia and Mortality: The Cache County Dementia Progression Study. *Alzheimer Dis. Assoc. Disord.* **2018**, *32*, 298–304. [CrossRef]
19. Doorduijn, A.S.; de van der Schueren, M.A.E.; van de Rest, O.; de Leeuw, F.A.; Hendriksen, H.M.A.; Teunissen, C.E.; Scheltens, P.; van der Flier, W.M.; Visser, M. Nutritional Status Is Associated with Clinical Progression in Alzheimer's Disease: The NUDAD Project. *J. Am. Med. Dir. Assoc.* **2023**, *24*, 638–644. [CrossRef]
20. Blasko, I.; Hinterberger, M.; Kemmler, G.; Jungwirth, S.; Krampla, W.; Leitha, T.; Heinz Tragl, K.; Fischer, P. Conversion from mild cognitive impairment to dementia: Influence of folic acid and vitamin B12 use in the VITA cohort. *J. Nutr. Health Aging* **2012**, *16*, 687–694. [CrossRef]
21. Kimura, A.; Sugimoto, T.; Kitamori, K.; Saji, N.; Niida, S.; Toba, K.; Sakurai, T. Malnutrition is Associated with Behavioral and Psychiatric Symptoms of Dementia in Older Women with Mild Cognitive Impairment and Early-Stage Alzheimer's Disease. *Nutrients* **2019**, *11*, 1951. [CrossRef] [PubMed]
22. Sachdev, P.S.; Blacker, D.; Blazer, D.G.; Ganguli, M.; Jeste, D.V.; Paulsen, J.S.; Petersen, R.C. Classifying neurocognitive disorders: The DSM-5 approach. *Nat. Rev. Neurol.* **2014**, *10*, 634–642. [CrossRef] [PubMed]
23. World Medical, A. World Medical Association Declaration of Helsinki. Ethical principles for medical research involving human subjects. *Bull. World Health Organ.* **2001**, *79*, 373–374.
24. Schröder, H.; Fitó, M.; Estruch, R.; Martínez-González, M.A.; Corella, D.; Salas-Salvadó, J.; Lamuela-Raventós, R.; Ros, E.; Salaverría, I.; Fiol, M. A short screener is valid for assessing Mediterranean diet adherence among older Spanish men and women. *J. Nutr.* **2011**, *141*, 1140–1145. [CrossRef]
25. Estruch, R.; Ros, E.; Salas-Salvadó, J.; Covas, M.-I.; Corella, D.; Arós, F.; Gómez-Gracia, E.; Ruiz-Gutiérrez, V.; Fiol, M.; Lapetra, J. Primary prevention of cardiovascular disease with a Mediterranean diet supplemented with extra-virgin olive oil or nuts. *N. Engl. J. Med.* **2018**, *378*, e34. [CrossRef]
26. Craig, C.L.; Marshall, A.L.; Sjöström, M.; Bauman, A.E.; Booth, M.L.; Ainsworth, B.E.; Pratt, M.; Ekelund, U.; Yngve, A.; Sallis, J.F. International physical activity questionnaire: 12-country reliability and validity. *Med. Sci. Sports Exerc.* **2003**, *35*, 1381–1395. [CrossRef]
27. Wallace, M.; Shelkey, M. Katz index of independence in activities of daily living (ADL). *Urol. Nurs.* **2007**, *27*, 93–94.
28. Katz, S.; Ford, A.B.; Moskowitz, R.W.; Jackson, B.A.; Jaffe, M.W. Studies of Illness in the Aged. The Index of Adl: A Standardized Measure of Biological and Psychosocial Function. *JAMA* **1963**, *185*, 914–919. [CrossRef]
29. Graf, C. The Lawton instrumental activities of daily living scale. *AJN Am. J. Nurs.* **2008**, *108*, 52–62. [CrossRef]
30. Lawton, M.P.; Brody, E.M. Assessment of older people: Self-maintaining and instrumental activities of daily living. *Gerontologist* **1969**, *9*, 179–186. [CrossRef]
31. Willadsen, T.G.; Bebe, A.; Køster-Rasmussen, R.; Jarbøl, D.E.; Guassora, A.D.; Waldorff, F.B.; Reventlow, S.; Olivarius, N.d.F. The role of diseases, risk factors and symptoms in the definition of multimorbidity—A systematic review. *Scand. J. Prim. Health Care* **2016**, *34*, 112–121. [CrossRef] [PubMed]
32. Gnjidic, D.; Hilmer, S.N.; Blyth, F.M.; Naganathan, V.; Waite, L.; Seibel, M.J.; McLachlan, A.J.; Cumming, R.G.; Handelsman, D.J.; Le Couteur, D.G. Polypharmacy cutoff and outcomes: Five or more medicines were used to identify community-dwelling older men at risk of different adverse outcomes. *J. Clin. Epidemiol.* **2012**, *65*, 989–995. [CrossRef] [PubMed]
33. Hill, E.; Goodwill, A.M.; Gorelik, A.; Szoeke, C. Diet and biomarkers of Alzheimer's disease: A systematic review and meta-analysis. *Neurobiol. Aging* **2019**, *76*, 45–52. [CrossRef] [PubMed]
34. Hu, E.A.; Toledo, E.; Diez-Espino, J.; Estruch, R.; Corella, D.; Salas-Salvado, J.; Vinyoles, E.; Gomez-Gracia, E.; Aros, F.; Fiol, M. Lifestyles and risk factors associated with adherence to the Mediterranean diet: A baseline assessment of the PREDIMED trial. *PLoS ONE* **2013**, *8*, e60166. [CrossRef] [PubMed]
35. van Dam, R.M.; Hu, F.B.; Willett, W.C. Coffee, Caffeine, and Health. *N. Engl. J. Med.* **2020**, *383*, 369–378. [CrossRef]
36. Knight, A.; Bryan, J.; Murphy, K. The Mediterranean diet and age-related cognitive functioning: A systematic review of study findings and neuropsychological assessment methodology. *Nutr. Neurosci.* **2017**, *20*, 449–468. [CrossRef]

37. Radd-Vagenas, S.; Duffy, S.L.; Naismith, S.L.; Brew, B.J.; Flood, V.M.; Fiatarone Singh, M.A. Effect of the Mediterranean diet on cognition and brain morphology and function: A systematic review of randomized controlled trials. *Am. J. Clin. Nutr.* **2018**, *107*, 389–404. [CrossRef]
38. Dinu, M.; Pagliai, G.; Casini, A.; Sofi, F. Mediterranean diet and multiple health outcomes: An umbrella review of meta-analyses of observational studies and randomised trials. *Eur. J. Clin. Nutr.* **2018**, *72*, 30–43. [CrossRef]
39. Chen, H.; Dhana, K.; Huang, Y.; Huang, L.; Tao, Y.; Liu, X.; Melo van Lent, D.; Zheng, Y.; Ascherio, A.; Willett, W.; et al. Association of the Mediterranean Dietary Approaches to Stop Hypertension Intervention for Neurodegenerative Delay (MIND) Diet with the Risk of Dementia. *JAMA Psychiatry* **2023**, *80*, 630–638. [CrossRef]
40. Agarwal, P.; Leurgans, S.E.; Agrawal, S.; Aggarwal, N.T.; Cherian, L.J.; James, B.D.; Dhana, K.; Barnes, L.L.; Bennett, D.A.; Schneider, J.A. Association of Mediterranean-DASH Intervention for Neurodegenerative Delay and Mediterranean Diets with Alzheimer Disease Pathology. *Neurology* **2023**, *100*, e2259–e2268. [CrossRef]
41. Dhana, K.; Agarwal, P.; James, B.D.; Leurgans, S.E.; Rajan, K.B.; Aggarwal, N.T.; Barnes, L.L.; Bennett, D.A.; Schneider, J.A. Healthy Lifestyle and Cognition in Older Adults with Common Neuropathologies of Dementia. *JAMA Neurol.* **2024**, *81*, 233–239. [CrossRef] [PubMed]
42. van den Brink, A.C.; Brouwer-Brolsma, E.M.; Berendsen, A.A.M.; van de Rest, O. The Mediterranean, Dietary Approaches to Stop Hypertension (DASH), and Mediterranean-DASH Intervention for Neurodegenerative Delay (MIND) Diets Are Associated with Less Cognitive Decline and a Lower Risk of Alzheimer's Disease—A Review. *Adv. Nutr.* **2019**, *10*, 1040–1065. [CrossRef] [PubMed]
43. Levak, N.; Lehtisalo, J.; Thunborg, C.; Westman, E.; Andersen, P.; Andrieu, S.; Broersen, L.M.; Coley, N.; Hartmann, T.; Irving, G.F.; et al. Nutrition guidance within a multimodal intervention improves diet quality in prodromal Alzheimer's disease: Multimodal Preventive Trial for Alzheimer's Disease (MIND-AD(mini)). *Alzheimers Res. Ther.* **2024**, *16*, 147. [CrossRef] [PubMed]
44. Shafiei, F.; Salari-Moghaddam, A.; Larijani, B.; Esmaillzadeh, A. Mediterranean diet and depression: Reanalysis of a meta-analysis. *Nutr. Rev.* **2023**, *81*, 889–890. [CrossRef] [PubMed]
45. Gildawie, K.R.; Galli, R.L.; Shukitt-Hale, B.; Carey, A.N. Protective Effects of Foods Containing Flavonoids on Age-Related Cognitive Decline. *Curr. Nutr. Rep.* **2018**, *7*, 39–48. [CrossRef]
46. Kent, K.; Charlton, K.E.; Netzel, M.; Fanning, K. Food-based anthocyanin intake and cognitive outcomes in human intervention trials: A systematic review. *J. Hum. Nutr. Diet.* **2017**, *30*, 260–274. [CrossRef]
47. Barrea, L.; Pugliese, G.; Frias-Toral, E.; El Ghoch, M.; Castellucci, B.; Chapela, S.P.; Carignano, M.L.A.; Laudisio, D.; Savastano, S.; Colao, A.; et al. Coffee consumption, health benefits and side effects: A narrative review and update for dietitians and nutritionists. *Crit. Rev. Food Sci. Nutr.* **2023**, *63*, 1238–1261. [CrossRef]
48. Santos, C.; Costa, J.; Santos, J.; Vaz-Carneiro, A.; Lunet, N. Caffeine intake and dementia: Systematic review and meta-analysis. *J. Alzheimers Dis.* **2010**, *20* (Suppl. 1), S187–S204. [CrossRef]
49. Nila, I.S.; Villagra Moran, V.M.; Khan, Z.A.; Hong, Y. Effect of Daily Coffee Consumption on the Risk of Alzheimer's Disease: A Systematic Review and Meta-Analysis. *J. Lifestyle Med.* **2023**, *13*, 83–89. [CrossRef]
50. Li, F.; Liu, X.; Jiang, B.; Li, X.; Wang, Y.; Chen, X.; Su, Y.; Wang, X.; Luo, J.; Chen, L.; et al. Tea, coffee, and caffeine intake and risk of dementia and Alzheimer's disease: A systematic review and meta-analysis of cohort studies. *Food Funct.* **2024**, *15*, 8330–8344. [CrossRef]
51. Xie, C.; Feng, Y. Alcohol consumption and risk of Alzheimer's disease: A dose-response meta-analysis. *Geriatr. Gerontol. Int.* **2022**, *22*, 278–285. [CrossRef] [PubMed]
52. Shafiee, A.; Jafarabady, K.; Rafiei, M.A.; Beiky, M.; Seighali, N.; Golpayegani, G.; Jalali, M.; Soltani Abhari, F.; Arabzadeh Bahri, R.; Safari, O.; et al. Effect of alcohol on Brain-Derived Neurotrophic Factor (BDNF) blood levels: A systematic review and meta-analysis. *Sci. Rep.* **2023**, *13*, 17554. [CrossRef]
53. Zeli, C.; Lombardo, M.; Storz, M.A.; Ottaviani, M.; Rizzo, G. Chocolate and Cocoa-Derived Biomolecules for Brain Cognition during Ageing. *Antioxidants* **2022**, *11*, 1353. [CrossRef] [PubMed]
54. Baker, L.D.; Manson, J.E.; Rapp, S.R.; Sesso, H.D.; Gaussoin, S.A.; Shumaker, S.A.; Espeland, M.A. Effects of cocoa extract and a multivitamin on cognitive function: A randomized clinical trial. *Alzheimers Dement.* **2023**, *19*, 1308–1319. [CrossRef] [PubMed]
55. Seibel, R.; Schneider, R.H.; Gottlieb, M.G.V. Effects of Spices (Saffron, Rosemary, Cinnamon, Turmeric and Ginger) in Alzheimer's Disease. *Curr. Alzheimer Res.* **2021**, *18*, 347–357. [CrossRef] [PubMed]
56. Shi, K.; Yu, Y.; Li, Z.; Hou, M.; Li, X. Causal relationship between dietary salt intake and dementia risk: Mendelian randomization study. *Genes. Nutr.* **2024**, *19*, 6. [CrossRef] [PubMed]
57. Lourida, I.; Hannon, E.; Littlejohns, T.J.; Langa, K.M.; Hypponen, E.; Kuzma, E.; Llewellyn, D.J. Association of Lifestyle and Genetic Risk With Incidence of Dementia. *JAMA* **2019**, *322*, 430–437. [CrossRef]
58. Licher, S.; Ahmad, S.; Karamujic-Comic, H.; Voortman, T.; Leening, M.J.G.; Ikram, M.A.; Ikram, M.K. Genetic predisposition, modifiable-risk-factor profile and long-term dementia risk in the general population. *Nat. Med.* **2019**, *25*, 1364–1369. [CrossRef]
59. Grande, G.; Qiu, C.; Fratiglioni, L. Prevention of dementia in an ageing world: Evidence and biological rationale. *Ageing Res. Rev.* **2020**, *64*, 101045. [CrossRef]
60. Vellas, B.; Lauque, S.; Gillette-Guyonnet, S.; Andrieu, S.; Cortes, F.; Nourhashemi, F.; Cantet, C.; Ousset, P.J.; Grandjean, H.; Group, R.F. Impact of nutritional status on the evolution of Alzheimer's disease and on response to acetylcholinesterase inhibitor treatment. *J. Nutr. Health Aging* **2005**, *9*, 75–80.

61. Guerin, O.; Soto, M.E.; Brocker, P.; Robert, P.H.; Benoit, M.; Vellas, B.; Group, R.F. Nutritional status assessment during Alzheimer's disease: Results after one year (the REAL French Study Group). *J. Nutr. Health Aging* **2005**, *9*, 81–84. [PubMed]
62. Jensen, G.L.; Cederholm, T.; Correia, M.; Gonzalez, M.C.; Fukushima, R.; Higashiguchi, T.; de Baptista, G.A.; Barazzoni, R.; Blaauw, R.; Coats, A.J.S.; et al. GLIM Criteria for the Diagnosis of Malnutrition: A Consensus Report from the Global Clinical Nutrition Community. *JPEN J. Parenter. Enteral Nutr.* **2019**, *43*, 32–40. [CrossRef] [PubMed]
63. Wesselman, L.M.P.; Doorduijn, A.S.; de Leeuw, F.A.; Verfaillie, S.C.J.; van Leeuwenstijn-Koopman, M.; Slot, R.E.R.; Kester, M.I.; Prins, N.D.; van de Rest, O.; de van der Schueren, M.A.E.; et al. Dietary Patterns Are Related to Clinical Characteristics in Memory Clinic Patients with Subjective Cognitive Decline: The SCIENCe Project. *Nutrients* **2019**, *11*, 1057. [CrossRef] [PubMed]
64. Soysal, P.; Dokuzlar, O.; Erken, N.; Dost Gunay, F.S.; Isik, A.T. The Relationship between Dementia Subtypes and Nutritional Parameters in Older Adults. *J. Am. Med. Dir. Assoc.* **2020**, *21*, 1430–1435. [CrossRef] [PubMed]
65. Jang, J.W.; Kim, Y.; Choi, Y.H.; Lee, J.M.; Yoon, B.; Park, K.W.; Kim, S.E.; Kim, H.J.; Yoon, S.J.; Jeong, J.H.; et al. Association of Nutritional Status with Cognitive Stage in the Elderly Korean Population: The Korean Brain Aging Study for the Early Diagnosis and Prediction of Alzheimer's Disease. *J. Clin. Neurol.* **2019**, *15*, 292–300. [CrossRef]
66. Morley, J.E. Nutrition and the brain. *Clin. Geriatr. Med.* **2010**, *26*, 89–98. [CrossRef]
67. Favaro-Moreira, N.C.; Krausch-Hofmann, S.; Matthys, C.; Vereecken, C.; Vanhauwaert, E.; Declercq, A.; Bekkering, G.E.; Duyck, J. Risk Factors for Malnutrition in Older Adults: A Systematic Review of the Literature Based on Longitudinal Data. *Adv. Nutr.* **2016**, *7*, 507–522. [CrossRef]
68. Volkert, D.; Beck, A.M.; Cederholm, T.; Cereda, E.; Cruz-Jentoft, A.; Goisser, S.; de Groot, L.; Grosshauser, F.; Kiesswetter, E.; Norman, K.; et al. Management of Malnutrition in Older Patients-Current Approaches, Evidence and Open Questions. *J. Clin. Med.* **2019**, *8*, 974. [CrossRef]
69. Noale, M.; Prinelli, F.; Conti, S.; Sergi, G.; Maggi, S.; Brennan, L.; de Groot, L.C.; Volkert, D.; McEvoy, C.T.; Trevisan, C.; et al. Undernutrition, cognitive decline and dementia: The collaborative PROMED-COG pooled cohorts study. *Clin. Nutr.* **2024**, *43*, 2372–2380. [CrossRef]
70. Testad, I.; Kajander, M.; Froiland, C.T.; Corbett, A.; Gjestsen, M.T.; Anderson, J.G. Nutritional Interventions for Persons With Early-Stage Dementia or Alzheimer's Disease: An Integrative Review. *Res. Gerontol. Nurs.* **2019**, *12*, 259–268. [CrossRef]
71. Borders, J.C.; Blanke, S.; Johnson, S.; Gilmore-Bykovskyi, A.; Rogus-Pulia, N. Efficacy of Mealtime Interventions for Malnutrition and Oral Intake in Persons With Dementia: A Systematic Review. *Alzheimer Dis. Assoc. Disord.* **2020**, *34*, 366–379. [CrossRef] [PubMed]
72. Gomez-Gomez, M.E.; Zapico, S.C. Frailty, Cognitive Decline, Neurodegenerative Diseases and Nutrition Interventions. *Int. J. Mol. Sci.* **2019**, *20*, 2842. [CrossRef] [PubMed]
73. Soto, M.E.; Secher, M.; Gillette-Guyonnet, S.; Abellan van Kan, G.; Andrieu, S.; Nourhashemi, F.; Rolland, Y.; Vellas, B. Weight loss and rapid cognitive decline in community-dwelling patients with Alzheimer's disease. *J. Alzheimers Dis.* **2012**, *28*, 647–654. [CrossRef] [PubMed]
74. White, H.K.; McConnell, E.S.; Bales, C.W.; Kuchibhatla, M. A 6-month observational study of the relationship between weight loss and behavioral symptoms in institutionalized Alzheimer's disease subjects. *J. Am. Med. Dir. Assoc.* **2004**, *5*, 89–97. [CrossRef] [PubMed]
75. Johnson, D.K.; Wilkins, C.H.; Morris, J.C. Accelerated weight loss may precede diagnosis in Alzheimer disease. *Arch. Neurol.* **2006**, *63*, 1312–1317. [CrossRef]
76. Stewart, R.; Masaki, K.; Xue, Q.L.; Peila, R.; Petrovitch, H.; White, L.R.; Launer, L.J. A 32-year prospective study of change in body weight and incident dementia: The Honolulu-Asia Aging Study. *Arch. Neurol.* **2005**, *62*, 55–60. [CrossRef]
77. Alhurani, R.E.; Vassilaki, M.; Aakre, J.A.; Mielke, M.M.; Kremers, W.K.; Machulda, M.M.; Geda, Y.E.; Knopman, D.S.; Petersen, R.C.; Roberts, R.O. Decline in Weight and Incident Mild Cognitive Impairment: Mayo Clinic Study of Aging. *JAMA Neurol.* **2016**, *73*, 439–446. [CrossRef]
78. Atti, A.R.; Palmer, K.; Volpato, S.; Winblad, B.; De Ronchi, D.; Fratiglioni, L. Late-life body mass index and dementia incidence: Nine-year follow-up data from the Kungsholmen Project. *J. Am. Geriatr. Soc.* **2008**, *56*, 111–116. [CrossRef]
79. Nourhashemi, F.; Deschamps, V.; Larrieu, S.; Letenneur, L.; Dartigues, J.F.; Barberger-Gateau, P.; Quid, P.s.P.A. Body mass index and incidence of dementia: The PAQUID study. *Neurology* **2003**, *60*, 117–119. [CrossRef]
80. Cronk, B.B.; Johnson, D.K.; Burns, J.M.; Alzheimer's Disease Neuroimaging, I. Body mass index and cognitive decline in mild cognitive impairment. *Alzheimer Dis. Assoc. Disord.* **2010**, *24*, 126–130. [CrossRef]
81. Sobow, T.; Fendler, W.; Magierski, R. Body mass index and mild cognitive impairment-to-dementia progression in 24 months: A prospective study. *Eur. J. Clin. Nutr.* **2014**, *68*, 1216–1219. [CrossRef] [PubMed]
82. Colon-Emeric, C.S.; Whitson, H.E.; Pavon, J.; Hoenig, H. Functional decline in older adults. *Am. Fam. Physician* **2013**, *88*, 388–394. [PubMed]
83. Lisko, I.; Kulmala, J.; Annetorp, M.; Ngandu, T.; Mangialasche, F.; Kivipelto, M. How can dementia and disability be prevented in older adults: Where are we today and where are we going? *J. Intern. Med.* **2021**, *289*, 807–830. [CrossRef] [PubMed]
84. Huuha, A.M.; Norevik, C.S.; Moreira, J.B.N.; Kobro-Flatmoen, A.; Scrimgeour, N.; Kivipelto, M.; Van Praag, H.; Ziaei, M.; Sando, S.B.; Wisloff, U.; et al. Can exercise training teach us how to treat Alzheimer's disease? *Ageing Res. Rev.* **2022**, *75*, 101559. [CrossRef]

85. Raji, C.A.; Meysami, S.; Hashemi, S.; Garg, S.; Akbari, N.; Ahmed, G.; Chodakiewitz, Y.G.; Nguyen, T.D.; Niotis, K.; Merrill, D.A.; et al. Exercise-Related Physical Activity Relates to Brain Volumes in 10,125 Individuals. *J. Alzheimers Dis.* **2024**, *97*, 829–839. [CrossRef] [PubMed]
86. Casaletto, K.B.; Renteria, M.A.; Pa, J.; Tom, S.E.; Harrati, A.; Armstrong, N.M.; Rajan, K.B.; Mungas, D.; Walters, S.; Kramer, J.; et al. Late-Life Physical and Cognitive Activities Independently Contribute to Brain and Cognitive Resilience. *J. Alzheimers Dis.* **2020**, *74*, 363–376. [CrossRef]
87. de Freitas, G.B.; Lourenco, M.V.; De Felice, F.G. Protective actions of exercise-related FNDC5/Irisin in memory and Alzheimer's disease. *J. Neurochem.* **2020**, *155*, 602–611. [CrossRef]
88. Sanders, L.M.J.; Hortobagyi, T.; Karssemeijer, E.G.A.; Van der Zee, E.A.; Scherder, E.J.A.; van Heuvelen, M.J.G. Effects of low- and high-intensity physical exercise on physical and cognitive function in older persons with dementia: A randomized controlled trial. *Alzheimers Res. Ther.* **2020**, *12*, 28. [CrossRef]
89. Lamb, S.E.; Sheehan, B.; Atherton, N.; Nichols, V.; Collins, H.; Mistry, D.; Dosanjh, S.; Slowther, A.M.; Khan, I.; Petrou, S.; et al. Dementia And Physical Activity (DAPA) trial of moderate to high intensity exercise training for people with dementia: Randomised controlled trial. *BMJ* **2018**, *361*, k1675. [CrossRef]
90. Veronese, N.; Soysal, P.; Demurtas, J.; Solmi, M.; Bruyere, O.; Christodoulou, N.; Ramalho, R.; Fusar-Poli, P.; Lappas, A.S.; Pinto, D.; et al. Physical activity and exercise for the prevention and management of mild cognitive impairment and dementia: A collaborative international guideline. *Eur. Geriatr. Med.* **2023**, *14*, 925–952. [CrossRef]
91. DiSabato, D.J.; Quan, N.; Godbout, J.P. Neuroinflammation: The devil is in the details. *J. Neurochem.* **2016**, *139* (Suppl. S2), 136–153. [CrossRef] [PubMed]
92. Lyman, M.; Lloyd, D.G.; Ji, X.; Vizcaychipi, M.P.; Ma, D. Neuroinflammation: The role and consequences. *Neurosci. Res.* **2014**, *79*, 1–12. [CrossRef] [PubMed]
93. Traub, J.; Frey, A.; Stork, S. Chronic Neuroinflammation and Cognitive Decline in Patients with Cardiac Disease: Evidence, Relevance, and Therapeutic Implications. *Life* **2023**, *13*, 329. [CrossRef] [PubMed]
94. Kip, E.; Parr-Brownlie, L.C. Healthy lifestyles and wellbeing reduce neuroinflammation and prevent neurodegenerative and psychiatric disorders. *Front. Neurosci.* **2023**, *17*, 1092537. [CrossRef]
95. Dominguez, L.J.; Di Bella, G.; Veronese, N.; Barbagallo, M. Impact of Mediterranean Diet on Chronic Non-Communicable Diseases and Longevity. *Nutrients* **2021**, *13*, 2028. [CrossRef]
96. Grabska-Kobylecka, I.; Szpakowski, P.; Krol, A.; Ksiazek-Winiarek, D.; Kobylecki, A.; Glabinski, A.; Nowak, D. Polyphenols and Their Impact on the Prevention of Neurodegenerative Diseases and Development. *Nutrients* **2023**, *15*, 3454. [CrossRef]
97. Novakovic, M.; Rout, A.; Kingsley, T.; Kirchoff, R.; Singh, A.; Verma, V.; Kant, R.; Chaudhary, R. Role of gut microbiota in cardiovascular diseases. *World J. Cardiol.* **2020**, *12*, 110–122. [CrossRef]
98. Wang, M.; Zhang, H.; Liang, J.; Huang, J.; Chen, N. Exercise suppresses neuroinflammation for alleviating Alzheimer's disease. *J. Neuroinflammation* **2023**, *20*, 76. [CrossRef]
99. Rijal, A.; Nielsen, E.E.; Hemmingsen, B.; Neupane, D.; Gaede, P.H.; Olsen, M.H.; Jakobsen, J.C. Adding exercise to usual care in patients with hypertension, type 2 diabetes mellitus and/or cardiovascular disease: A protocol for a systematic review with meta-analysis and trial sequential analysis. *Syst. Rev.* **2019**, *8*, 330. [CrossRef]
100. Pearce, N. Analysis of matched case-control studies. *BMJ* **2016**, *352*, i969. [CrossRef]
101. Mansournia, M.A.; Jewell, N.P.; Greenland, S. Case-control matching: Effects, misconceptions, and recommendations. *Eur. J. Epidemiol.* **2018**, *33*, 5–14. [CrossRef] [PubMed]

nutrients

MDPI

Article

Follow-Up Period Affects the Association between Serum 25-Hydroxyvitamin D Concentration and Incidence of Dementia, Alzheimer's Disease, and Cognitive Impairment

William B. Grant

Sunlight, Nutrition, and Health Research Center, 1745 Pacific Ave., Suite 504, San Francisco, CA 94109, USA; wbgrant@infionline.net

Abstract: Background/Objectives: Vitamin D's effect on risk health outcomes is often evaluated using prospective cohort studies. For vitamin D, risk ratios (RRs) are based on health outcomes with respect to serum 25-hydroxyvitamin D [25(OH)D] concentrations measured at time of enrollment. Serum 25(OH)D concentrations vary over time, thereby diluting the effect of 25(OH)D for long follow-up periods. Inverse relationships between RR and follow-up period have been reported for all-cause mortality rate and cancer incidence rates. Here, the effect for neurological outcomes is evaluated. Methods: The analysis examines how follow-up period affected results from nine cohort studies of all-cause dementia, six studies of Alzheimer's disease, and nine for cognitive impairment with respect to vitamin D deficiency. Results: For all-cause dementia, Alzheimer's disease, and cognitive impairment, respectively, the linear regression fits are RR = 2.9 − 0.14 × years, r = 0.73, p = 0.02; RR = 2.9 − 0.14 × years, r = 0.69, p = 0.13; and RR = 1.8 − 0.066 × years, r = 0.72, p = 0.03. The regression fit to RR for the shortest follow-up period for each outcome is considered the best estimate of vitamin D deficiency's effect on risk. Those values are approximately twice that found by averaging all RRs without considering the effect of follow-up period. Conclusions: Vitamin D's effect on risk of neurological conditions is inversely correlated with mean follow-up period in prospective cohort studies. This effect should be considered in the design and analysis of such studies. Additional studies should also be conducted regarding raising serum 25(OH)D concentrations to reduce risk of brain function decline.

Keywords: Alzheimer's disease; cognitive impairment; dementia; follow-up period; neurological conditions; risk; vitamin D deficiency

1. Introduction

Prospective cohort studies are often used to ascertain how lifestyle, diet, nutrients, lifestyle, and biomarkers are related to health outcomes. The standard procedure is to enroll participants, obtain values for factors to be studied and those that might affect the outcome, monitor participants for several years, and note changes in the health condition of interest. Because serum 25-hydroxyvitamin D [25(OH)D] concentrations change for various reasons, relying only on the 25(OH)D concentration measured at the time of enrollment is problematic.

Since at least 1999, researchers have known that long-term follow up in prospective studies results in "regression dilution" [1]. Of the articles consulted in preparing this article, only Kuzma and colleagues (2016) [2] cited that article. Since 2011, that same effect has been found in prospective cohort studies regarding serum 25(OH)D and cancer [3] and, since 2012, all-cause mortality rate [4]. An observational study in Norway reported that "depending on the method of adjusting for season, the correlation coefficient between serum 25(OH)D measurements from 1994 and 2008 ranged from 0.42 to 0.52" [5]. A more recent report showed that the analysis of the risk of colorectal cancer with respect to serum 25(OH)D concentration on the basis of prospective cohort studies [6] was incorrect because the researchers had not realized that men had nearly four times the rate of change in relative risk (RR) with respect to follow-up time as women (Figure 1 in [7]).

The article by Zhang and colleagues [8] regarding the association between vitamin D levels and risk of dementia, Alzheimer's disease (AD), and cognitive impairment forms the basis for the present article. The authors used the standard random effects model regarding 17 prospective cohort studies with 486,921 individuals. For dementia with respect to vitamin D deficiency (VDD), RR = 1.42 (95% confidence interval [CI], 1.21–1.65). However, that analysis did not consider the effect of each study's mean follow-up period. The present study plots the RR values for the various health outcomes vs. mean follow-up period. The plots show that RR is highest for the shortest follow-up periods and declines to near 1.00 for follow-up periods near 13 years. This study examines the implications of follow-up period in prospective cohort studies on the estimation of the effect of VDD on risk of the three adverse brain health outcomes. The analysis suggests that using the RR for the prospective studies with the shortest follow-up period results in an estimate of the effect of VDD being about twice as high as from an analysis that ignores the length of follow-up period.

2. Materials and Methods

The sources of the data used in this study were obtained from two meta-analyses [8,9] as well as a search of Google Scholar for additional studies not included in those two meta-analyses. All of the studies in the meta-analyses were added to the data tables in this study. However, studies that did not provide follow-up period, the comparison of serum 25(OH)D concentration between participants with or without adverse brain health, or were based on dietary vitamin D intake were not included in the analyses. One study, (Graf, 2014) [10], was omitted from the dementia analysis due having a very large range of 95% CI.

To evaluate how follow-up period affects risk of dementia with respect to VDD, several sources were used. Much of the data are from Figure 2 from Zhang and colleagues [8] plus the results from Figure 2 in Chen and colleagues [9]. Data for AD were obtained from Figure 3a in Zhang and colleagues [8]. Data for cognitive impairment (CogImp) are from Figure 3b in Zhang and colleagues [8]. Years of follow-up were obtained from Chen and colleagues [9] or from the original studies. Tables 1–5 show the relevant information regarding the data in the cohort studies. For dementia, mean ages of participants at baseline ranged from 53 (SD 17) to 85 (SD 7) years. The mean 25(OH)D concentrations for studies that gave values ranged from 32 (standard deviation [SD] 25) to 69 $\pm$ 19 nmol/L. The 25(OH)D comparisons included <25 versus >50 mol/L, <50 versus >50 mol/L, <50 versus $\geq$75 nmol/L, and so on. The mean follow-up period ranged from 5.6 to 30 years. For CogImp, mean ages of participants at baseline ranged from 67 $\pm$ 5 to 74 (SD 7) years. Mean 25(OH)D concentrations for studies that gave values ranged from 50 (SD 21) to 84 (SD 54) nmol/L. The 25(OH)D comparisons included <25 versus $\geq$50 mol/L, <50 versus $\geq$75 nmol/L, and so on. The mean follow-up period ranged from 4.0 to 13 years.

In the analysis, it is assumed that the only important factor is the mean follow-up period. Though values for various factors could affect the HR, in the analysis it appears that they are smaller than the effect of follow-up period. Studies with mean follow-up period greater than 15 years were omitted because those periods were considered too long to yield meaningful data. One study (Graf, 2014 [10] was omitted from the dementia analysis due to having very large 95% CI range due to the low numbers of participants and those who developed dementia.

Table 1. Data for vitamin D deficiency and risk of dementia or Alzheimer's disease from Figure 2 and Figure 3a in Zhang and colleagues (2024) [8] plus a recent study from the UK Biobank [11].

Country	Mean Age (±SD) (yrs)	N_T	N_D	N_{AD}	Author, yr, Ref.
USA	74 ± 5	1658	171	102	(Littlejohns, 2014) [12]
Germany	84 ± 3	861 F, 473 M	250	209	(van Lent, 2022) [13]
Israel	53 ± 17	2454 F, 1824 M	133		Kiderman, 2023) [14]
UK	64.6	13,486	283	101	(Geng, 2022) [15]
USA	72 ± 7	1663	267	208	(Karakis, 2016) [16]
Norway	78	790 F, 644 M	324		(Asante, 2023) [17]
France	73 ± 5	916	177	124	(Féart, 2017) [18]
Sweden	71	1182 M	250 M	116 M	(Olsson, 2017) [19]
The Netherlands	69 ± 8	3462 F, 2625 M	795	641	(Licher, 2017) [20]
UK	62 ± 3	140,857 F, 128,372 M	7087	3616	(Chen, 2024) [11]
Omitted					
Switzerland	85 ± 7	147 F, 53 M	46		(Graf, 2014) [10]
USA	62	793 B, 859 W	145		(Schneider, 2014) [21]
Finland	Cases: 69 ± 7 Noncases: 56 ± 10	2724 F, 2286 M	100 F, 51 M		(Knekt, 2014) [22]
USA	57 ± 6	13,039	1323		(Fashanu, 2019) [23]
Denmark	58	10,186	418	92	(Afzal, 2014) [24]

Key: B, black; F, female; M, male; N_{AD}, number developing Alzheimer's disease; N_D, number with dementia; N_T, total; SD, standard deviation; W, white.

Table 2. Data for vitamin D deficiency and risk of dementia from Figure 2 in Chen and colleagues [9] and Figure 2 in Zhang and colleagues (2024) [8].

Mean BMI (±SD) (kg/m^2)	Mean 25(OH)D (±SD) (nmol/L)	25(OH)D Comparison (nmol/L)	Mean Follow-Up (yrs)	RR (95% CI)	Author, yr, Ref.
27 ± 5		<25 vs. >50	5.6	2.18 (1.18–4.02)	(Littlejohns, 2014) [12]
27 ± 6	54 ± 24	<25 vs. >50	7	2.38 (1.31–4.23)	Kiderman, 2023) [14]
27 ± 5	63 ± 28		9	1.00 (0.58–1.72)	(Karakis, 2016) [16]
27 ± 3	50 ± 21	<50 vs. >50	10	1.09 (0.64–1.83)	(Asante, 2023) [17]
26 ± 4		<50 vs. >50	11.4	2.12 (1.21–3.71)	(Féart, 2017) [18]
26 ± 3	69 ± 19	<50 vs. ≥75	12	0.86 (0.58–1.30)	(Olsson, 2017) [19]
27 ± 4	49 (IQR 30–69)	<25 vs. >50	13.3	1.22 (0.98–1.54)	(Licher, 2017) [20]
27 ± 4	50 ± 21	<50 vs. >50	13.6	1.25 (1.16–1.34)	(Chen, 2024) [11]
Omitted	from analysis due to long follow-up period				
23 ± 4	32 ± 25	<25 vs. >75	2	2.85 (0.45–17.95)	(Graf, 2014) [10]
27 ± 5, W	64 ± 20 W;	High vs. low tertile	16.6	1.30 (0.62–2.71)	(Schneider, 2014) [21]
30 ± 6, B	43 ± 16 B	High vs. low tertile	16.6	1.81 (0.33–6.50)	(Schneider, 2014) [21]
26 ± 4 F	Cases: 40 ± 20 Noncases: 43 ± 17	High vs. low quartile	17	3.03 (1.37–6.69)	(Knekt, 2014) [22]

Table 2. *Cont.*

Mean BMI (±SD) (kg/m^2)	Mean 25(OH)D (±SD) (nmol/L)	25(OH)D Comparison (nmol/L)	Mean Follow-Up (yrs)	RR (95% CI)	Author, yr, Ref.
26 ± 4 M	Cases: 40 ± 20 Noncases: 43 ± 17	High vs. low quartile	17	1.35 (0.53–3.44)	(Knekt, 2014) [22]
28 ± 5	61 ± 22	<25 vs. >50	20	1.24 (1.05–1.48)	(Fashanu, 2019) [23]
25 ± 3	45 (M) 40 (F)	<25th vs. >50th percentile	30	1.27 (1.01–1.60)	(Afzal, 2014) [24]

Key: 25(OH)D, 25-hydroxyvitamin D; 95% CI, 95% confidence interval; B, black; BMI, body mass index; F, female; IQR, interquartile range; M, male; RR, relative risk; SD, standard deviation; W, white.

Table 3. Data for vitamin D deficiency and risk of Alzheimer's disease from Figure 2 in Chen and colleagues [9] and Figure 3a in Zhang and colleagues (2024) [8].

Mean BMI (±SD) (kg/m^2)	Mean 25(OH)D (±SD) (nmol/L)	25(OH)D Comparison (nmol/L)	Mean Follow-Up (yrs)	RR (95% CI)	Author, yr, Ref.
27 ± 5		<25 vs. >50	5.6	2.20 (1.01–4.80)	(Littlejohns, 2014) [12]
26 ± 4	37 (IQR 25–58)	<25 vs. >50	7	2.28 (1.47–3.53)	(van Lent, 2022) [13]
31 ± 5		<25 vs. >50	8.5	1.72 (1.02–2.91)	(Geng, 2022) [15]
27 ± 5	63 ± 28		9	0.72 (0.37–1.42)	(Karakis, 2016) [16]
26 ± 4		<50 vs. >50	11.4	2.85 (1.36–5.97)	(Féart, 2017) [18]
26 ± 3	69 ± 19	<50 vs. ≥75	12	1.19 (0.67–2.12)	(Olsson, 2017) [19]
27 ± 4	50 ± 21	<50 vs. >50	13.6	1.19 (1.07–1.31)	(Chen, 2024) [11]
Omitted	from analysis due to long follow-up period				

Key: 25(OH)D, 25-hydroxyvitamin D; 95% CI, 95% confidence interval; BMI, body mass index; RR, relative risk; SD, standard deviation.

Tables 4 and 5 give the data associated with the CI studies. The numbers of cognitively normal participants at baseline and the number who developed CI are for those in the 25(OH)D categories used in the HR or OR analyses.

Table 4. Data for vitamin D deficiency and risk of cognitive impairment from Figure 3b in Zhang and colleagues (2024) [8].

Country	Mean Age (±SD) (yrs)	*N*	N_{CI}	Test	Author, yr, Ref.
USA	74 ± 5	1812 F	446, 409	MMSE, TMTB	(Slinin, 2012) [25]
Italy	74 ± 7	1208 F, 719 M	466	MMSE	(Toffanello, 2014) [26]
USA	74 ± 6	806 M	126	MMSE, TMTB	(Slinin, 2010) [27]
Italy	74 ± 7	487 F, 370 M		MMSE	(Llewellyn, 2011) [28]
USA	72 ± 3	1750 F, 832 M	324	BVRT	(Kuzma, 2016 [CHS]) [2]
Chile	67 ± 5	666 F, 289 M	54	MMSE	(Marquez, 2022) [29]
Norway	78	790 F, 644 M	717	MoCA	(Asante, 2023) [17]
Sweden	71	1182 M	80	MMSE	(Olsson, 2017) [19]
The Netherlands	74 ± 6	1010 F, 820 M	346	RAVLT	(Kuzma, 2016 [LASA]) [2]

Key: BVRT, Benton Visual Retention Test; CHS, Cardiovascular Health Study; F, female; LASA, Longitudinal Aging Study Amsterdam; M, male; MMSE, Mini-Mental State Examination; MoCA, Montreal Cognitive Assessment; *N*, number of participants; N_{CI}, number with cognitive impairment; RAVLT, Rey's Auditory Verbal Learning Test; SD, standard deviation; TMTB, Trail Making Test Part B.

Table 5. Data for vitamin D deficiency and risk of cognitive impairment from Figure 3b in Zhang and colleagues (2024) [8].

Assessment (yrs)	Mean BMI (±SD) (kg/m^2)	Mean 25(OH)D (±SD) (nmol/L)	25(OH)D Comparison (nmol/L)	Mean Follow-Up (yrs)	RR (95% CI)	Author, yr, Ref.
2 and 4	26 ± 5, F		<25 vs. ≥75	4.0	1.45 (1.10–1.86)	(Slinin, 2012) [25]
4	27 ± 3	84 ± 54	<50 vs. ≥75	4.4	1.36 (1.04–1.80)	(Toffanello, 2014) [26]
4.6	27 ± 3, M		<50 vs. ≥75	4.6	1.29 (0.91–1.74)	(Slinin, 2010) [27]
3 and 6		52 ± 37	<25 vs. ≥75	5.2	1.64 (1.20–2.05)	(Llewellyn, 2011) [28]
Annual	27 ± 5		<25 vs. ≥50	6.5	1.73 (1.22–2.45)	(Kuzma, 2016 [CHS]) [2]
?	29 ± 5, F 28 ± 4, M	Cases: 58 ± 32 Noncases: 71 ± 38	30–48 vs. >75	9.6	1.25 (0.64–2.85)	(Marquez, 2022) [29]
	27 ± 3	50 ± 21	<50 vs. >50	10	1.06 (0.73–1.44)	(Asante, 2023) [17]
	26 ± 3	69 ± 19	<50 vs. ≥75	12	0.67 (0.31–1.36)	(Olsson, 2017) [19]
Every 3–4	27 ± 4		<25 vs. ≥50	13	1.12 (0.84–1.48)	(Kuzma, 2016 [LASA]) [2]

Key: 25(OH)D, 25-hydroxyvitamin D; 95% CI, 95% confidence interval; BMI, body mass index; F, females; HR, hazard ratio; M, males; RR, relative risk; SD, standard deviation.

These data were used to examine the effect of follow-up period in the risk of dementia, AD, and CogImp as shown in the results section.

3. Results

In the analysis for dementia, omitted were one study with high uncertainty, accounting for only 0.7% of the weight, and three studies with follow-up periods of 17+ years. Two studies were conducted, one with 11 studies and one with 10, omitting Féart and colleagues [18]. The linear fit to the data with 11 studies is RR = 2.8 − 0.12 × years, $r = 0.59$, $p = 0.03$. The linear fit to the data with 10 studies is RR = 2.9 − 0.14 × years, $r = 0.73$, $p = 0.02$ (Figure 1). If the Graf study [10] is included, the regression fit is RR = 3.0 − 0.16 × years, $r = 0.84$, $p = 0.001$. (For 13.3 years, two studies reported RR = 1.22.) Chen and colleagues [9] calculated a pooled RR = 1.39 (95% CI, 1.14–1.47) for low vs. high 25(OH)D concentration based on data from 12 prospective cohort studies. Zhang and colleagues (2024) calculated an estimated pooled RR of 1.42 (95% CI, 1.21–1.65) for low vs. high 25(OH)D concentration based on data from 17 prospective cohort studies. For the shortest follow-up period, 5.6 years, the RR for the analysis with 10 studies is 2.1 (95% CI, 1.04–3.9), 2.6 times higher than the value from Zhang and colleagues, though with much larger 95% CIs.

For the RR of AD versus 25(OH)D concentration as a function of follow-up period, two analyses were conducted. In the analysis with seven studies with less than 15 years of mean follow-up period in Zhang and colleagues (2024) [8] plus Chen and colleagues (2024) [11], the regression fit to the data was RR = 2.5 − 0.08 × years, $r = 0.32$, $p = 0.48$. With Féart and colleagues [18] omitted, the regression fit to the data was RR = 2.9 − 0.14 × years, $r = 0.69$, $p = 0.13$ (Figure 2). The estimated pooled RR in Zhang and colleagues [8] is 1.57 (95% CI, 1.15–2.14). The value in this article for the six studies for the shortest follow-up period, 5.6 years, is 2.12 (95% CI, 1.01–4.13). That estimate is 2.0 times higher than the estimate from Zhang and colleagues but again with higher 95% CI values.

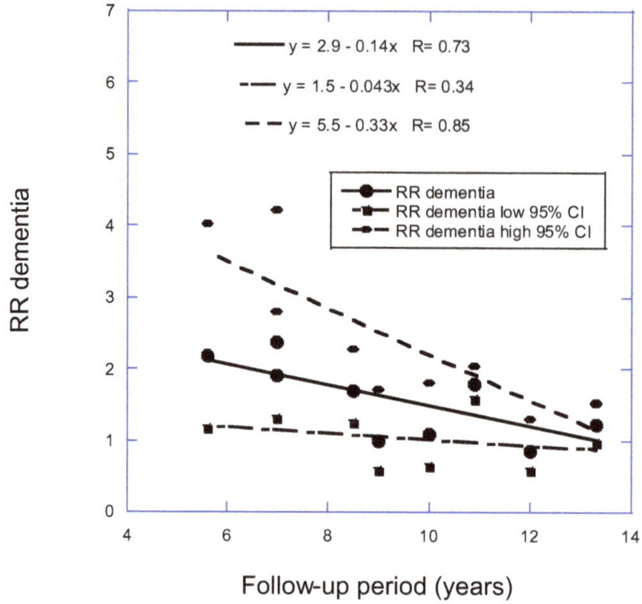

Figure 1. Scatter plot of relative risk (RR) versus low to high 25(OH)D concentration for dementia with respect to mean follow-up period less than 15 years from Figure 2 in Zhang and colleagues (2024) [8] plus Chen and colleagues (2024) [11] but omitting Féart and colleagues (2017) [18]. Key: 95% CI, 95% confidence interval.

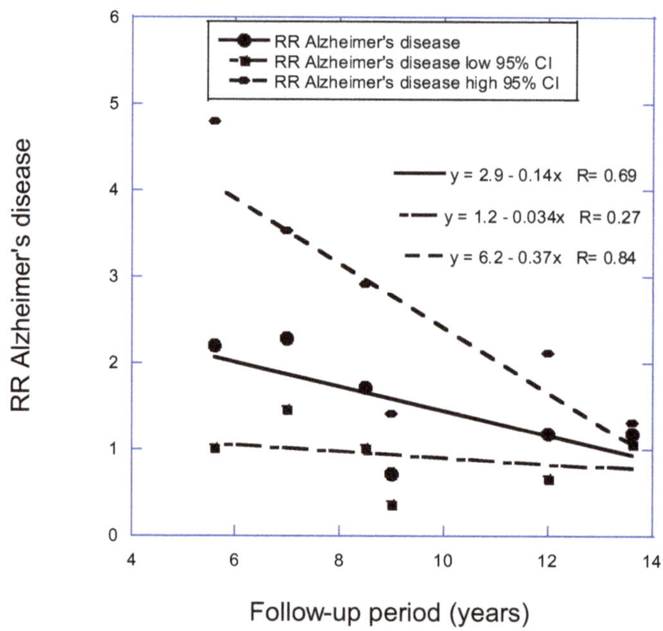

Figure 2. Relative risk (RR) for AD versus low to high 25(OH)D concentration versus mean follow-up period from Figure 3a in Zhang and colleagues (2024) [8] plus Chen and colleagues (2024) [11] but omitting Féart and colleagues (2017) [18]. Key: 95% CI, 95% confidence interval.

The analysis for CI versus 25(OH)D concentration as a function of follow-up period used six of the 10 studies in Figure 3b from Zhang and colleagues (2024) [8], with one study omitted that had very large 95% CI values and three with follow-up times less than 5 years. The regression fit to the data is RR = 2.3 − 0.11 × years, r = 0.88, p = 0.02. (If three studies with mean follow-up period between 4.0 and 4.6 years are added, RR = 1.8 − 0.066 × years, r = 0.72, p = 0.03). Figure 3 is a scatter plot of the data used in the analysis. The estimated pooled RR in Zhang and colleagues (2024) [8] for data from seven prospective cohort studies is 1.34 (95% CI, 1.19–1.52). The estimated pooled RR for AD for low vs. high 25(OH)D concentration from six prospective cohort studies in Chen and colleagues (2018) [9] is 1.28 (95% CI, 1.00–1.67). The value in this article for the six studies for the shortest follow-up period, 4 years, is 1.73 (95% CI, 1.15–2.04). That estimate is 2.1 times higher than the estimate from Zhang and colleagues but again with higher 95% CI values.

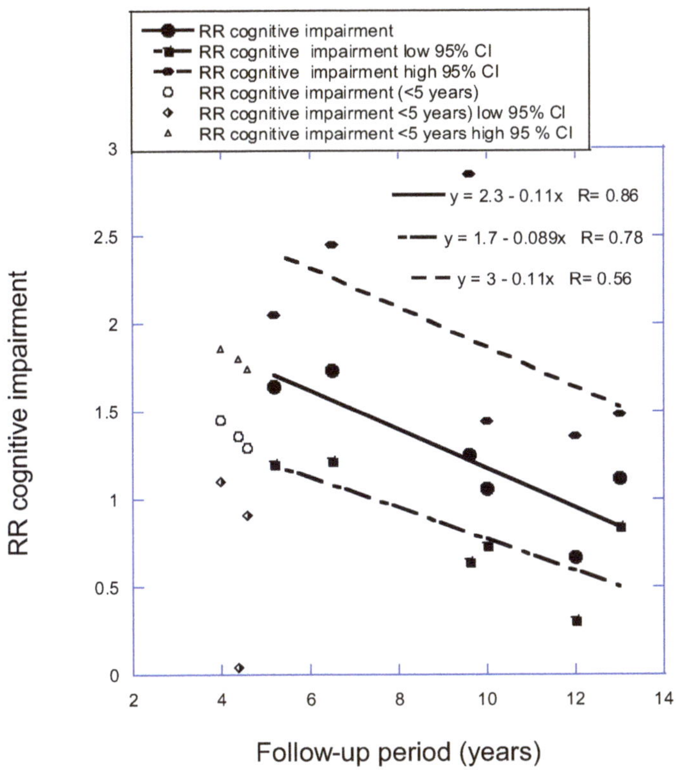

Figure 3. Relative risk (RR) for cognitive impairment versus low to high 25(OH)D concentration with regression fit to follow-up period for six studies with mean follow-up periods from 5 to 13 years from Figure 3b in Zhang and colleagues (2024) [8]. Key: 95% CI, 95% confidence interval; RR, risk ratio.

The implications of the findings regarding follow-up period and risk of dementia, AD, and CogImp are discussed in the next section.

4. Discussion

As shown in the work of Clarke and colleagues [1], values for biological factors change over time. Therefore, apparent health effects related to those factors are reduced in long-term follow-up prospective studies. Serum 25(OH)D concentrations can change for several reasons.

Vitamin D production from solar UVB exposure decreases with age [30]. A recent experimental study reported that vitamin D production from sun exposure decreases by 13% per decade of life [31].

Serum 25(OH)D concentration is generally inversely correlated with body mass index (BMI; kilograms of mass per square meter of body surface area). For example, in the dementia study from Israel [14], mean BMI was 25 ± 4 kg/m^2 for 25(OH)D concentrations > 75 nmol/L, increasing to 29 ± 7 kg/m^2 for 25(OH)D < 25 nmol/L. Thus, if BMI changes, 25(OH)D concentration should also change.

Serum 25(OH)D concentration also is associated with dietary animal product content, especially for fish and meat [32]. If those components of diet change, 25(OH)D will change.

Fortifying food with vitamin D can change 25(OH)D, as it did in Finland, where that approach was approved at the end of 2002 [33]. Measurements of 25(OH)D and dietary assessments of 3650 participants in 1997 at 31 years of age and again in 2012–2013 at 46 years of age determined that fortified foods accounted for most of a 10.6-nmol/L increase in 25(OH)D from 54 ± 19 to 65 ± 19 nmol/L [34].

A 2017 letter to the editor reported changes in daily vitamin D supplementation with 1000 IU or more from data collected in the U.S. National Health and Nutrition Survey [35]. Prevalence for people $\geq$70 years increased from 1.5% (95% CI, 1.1%–2.0%) in 2005–2006 to 8.6% (95% CI, 5.6%–13.1%) in 2007–2008, and up to 38.5% (95% CI, 31.5%–45.7%) in 2013–2014. The Norwegian study noted that 33% of participants had changes in 25(OH)D concentrations over 10 years [17].

These results have implications for long-duration prospective cohort studies with respect to 25(OH)D concentration. One way is to measure the important factors at least every 4 years. That is the approach taken in Harvard University prospective studies of diet and risk of disease, for example, Bernstein and colleagues [36]. An added advantage of that approach is that the latency period between risk factor and health effect can be determined. In that study, the latency period between dietary meat intake and incidence of colorectal cancer was determined to be about 4–8 years. Another way is to perform analyses for various follow-up periods during the study such as for each 3–5 years without remeasuring the values of the biological factors.

The health benefits of vitamin D status may become apparent much more quickly than for diet in the incidence of adverse health outcomes. For example, a vitamin D randomized controlled trial (RCT) was conducted regarding progression from prediabetes to type 2 diabetes mellitus [37]. The vitamin D dose was 4000 IU/day and the median follow-up time was 2.5 years. When the results were reanalyzed, the HR for diabetes for an increase of 25 nmol/L in intratrial 25(OH)D level was 0.75 (95% CI, 0.68–0.82) in the vitamin D treatment arm and 0.90 (95% CI, 0.80–1.02) in the placebo arm.

The shortest follow-up periods for the prospective cohort studies included in this study were 5.6 years for dementia and AD and 4 years for cognitive impairment. It may be the case that raising serum 25(OH)D concentrations through vitamin D supplementation or UVB irradiance could reduce the risk of developing these brain diseases or slow their progression in much shorter time periods.

The evidence that vitamin D reduces risk of AD was reviewed in 2023 in the *Journal of Alzheimer's Disease* [38]. Some important mechanisms include reduced risk of insulin resistance (IR) and inflammation. The mechanisms linking brain insulin/insulin-like growth factor resistance include impaired function of glucose transporter 4, changes in insulin receptor function, energy deficit, increased oxidative stress, and hyperglycemia (see Table 1 in Nguyen and colleagues [39]). A 2019 review discussed vitamin D's role in reducing IR [40]. The mechanisms include maintaining normal levels of reactive oxygen species and ionized calcium, thereby reducing epigenetic changes associated with insulin resistance such as oxidative stress and inflammation.

Therefore, a search was conducted for the effect of vitamin D supplementation regarding health outcomes related to neurodegenerative diseases to ascertain whether supplementation is promising and what time scales are involved. A 3-month study involving elderly people with metabolic disorders showed that supplementation with 2000 IU/day of vitamin D significantly decreased the homeostatic model assessment for insulin and decreased oxidative DNA damage [41]. In addition, supplementation reduced metabolic parameters connected with IR and improved glucose and lipid metabolism.

A 2023 review by Lason and colleagues examined the vitamin D receptor as a potential target for age-related neurodegenerative diseases [42]. The review mentioned a study investigating the effect of vitamin D supplementation involving mild CogImp (MCI) patients [43]. That study included 16 MCI patients, 11 very early AD (VEAD) patients, and 25 healthy control subjects. Patients with 25(OH)D concentrations lower than 75 nmol/L were supplemented with 50,000 IU of vitamin D_3 once a week for 6 weeks, followed by 1500–2000 IU/day for 18 months. In MCI but not VEAD patients, lymphocyte susceptibility to death improved significantly after 6 months. After 18 months, Montreal Cognitive Assessment scores improved in MCI patients but not in VEAD patients. Because MCI is an important risk factor for AD [44,45], this finding supports the role of higher 25(OH)D concentrations in reducing risk of AD.

In addition, that review [42] included Table 1 with information for eight observational and vitamin D supplementation RCTs regarding late-life cognition, dementia, and AD. Three of those studies reported results of interest for this article. An 18-week RCT compared 4000 versus 400 IU/d vitamin D_3 effects on visual memory [46]. Participants in the high-dose group increased mean serum 25(OH)D concentration from 67 $\pm$ 20 to 131 $\pm$ 26 nmol/L, whereas concentration in the low-dose group increased from 61 $\pm$ 22 to 86 $\pm$ 16 nmol/L. Those in the high-dose group with baseline 25(OH)D concentration < 75 nmol/L increased performance in the Pattern Recognition Memory-Delayed task from 86 (SD 14) to 94 (SD 8) (p = 0.005). The change in the low-dose group had p = 0.61. No additional significant differences in cognitive function tests were apparent among the other 11 tests for people with 25(OH)D < 75 nmol/L.

A vitamin D supplementation RCT in AD patients conducted in China reported the best results regarding cognitive function [47]. A total of 105 AD patients who received 800 IU/d of vitamin D increased serum 25(OH)D concentrations from 47 $\pm$ 7 to 57 $\pm$ 4 nmol/L by the end of the year. The 105 participants in the control group decreased 25(OH)D from 49 $\pm$ 3 to 47 $\pm$ 3 nmol/L. The mean BMI in each group was 25 $\pm$ 3 kg/m^2. People in the vitamin D treatment group had modest increases in full-scale IQ, information, digit span, vocabulary, block design, and picture arrangement, whereas participants in the control group had modest-to-large reductions in all those parameters. The p-values for the time and group effects for the vitamin D treatment group compared with the control group were significant to $p < 0.001$ for all but the vocabulary (p = 0.15 for time effect) and block design (p = 0.02). That RCT showed that vitamin D supplementation could significantly improve cognitive function in AD patients. Thus, that intervention study suggests that vitamin D supplementation can rapidly reduce AD risk factors.

As shown in Figures 1–3, RR values increased linearly to the shorter mean follow-up time used for each analysis. However, three studies not included in the regression analysis for CogImp had lower RR than expected from the regression fit to the other six studies. Thus, vitamin D status can affect risk of overall dementia, AD, and CogImp in as little as 5 years. Therefore, any prospective studies of neurodegeneration should measure serum 25(OH)D concentrations at least every 5 years. Harvard has participants in its health studies complete food frequency questionnaires every 4 years [48].

Another measure that should be implemented is to seasonally adjust 25(OH)D concentrations. Many observational studies cited here measured 25(OH)D concentrations at different times of the year and then averaged the values. In the United States, mean adult wintertime serum 25(OH)D concentrations are about 75% of summertime values [49]. In addition, whenever results of meta-analyses of prospective studies are used scientifically or for health policy recommendations, the analyses should be reevaluated with respect to follow-up periods. Also, standardizing 25(OH)D concentration measurements would be helpful since 25(OH)D measured values vary with different assays and instruments. See, for example, Sempos and colleagues (2018) [50].

Low 25(OH)D concentrations have been causally linked to increased risk of AD through Mendelian randomization (MR) studies. MR studies use genetic variants such as alleles of genes involved in the vitamin D pathway to randomize populations. Large-scale vitamin D genome-wide association study (GWAS) datasets are used to determine the relationship between alleles and serum 25(OH)D concentrations [51]. These GWAS data are then used with other large datasets that report health outcomes of interest. With very large datasets, this approach randomizes the effects of other influences on serum 25(OH)D concentration. MR studies are considered capable for causal inference in epidemiological studies [52]. A 2016 MR study found genetically determined serum 25(OH)D concentrations inversely correlated with risk of AD [53]. It included data from an observational study with 17,008 AD cases and 37,154 controls. A 2020 MR study used data from the UK Biobank [51]. It used GWAS data from the International Genomics of Alzheimer's Project and UK Biobank with individuals aged 60 years and over. Six alleles were used in the analysis. For the International Genomics of Alzheimer's Project dataset, the OR for AD per one SD increase in 25(OH)D concentation was 0.64 (95% CI, 0.46–0.89). For the UK Biobank dataset, the OR for AD per one SD increase in 25(OH)D concentation was 0.88 (95% CI, 0.73–1.06). The data from the UK Biobank were based on father's or mother's history of AD or dementia, an indirect measure of risk. A 2022 nonlinear MR analysis of 25(OH)D concentration and incidence of dementia based on UK Biobank data found for serum 25(OH)D concentration of 10 ng/mL compared to 20 ng/mL and adjusted HR = 1.54 (95% CI, 1.21–1.96) [54]. No RCT has demonstrated that vitamin D supplementation reduces risk of AD. That is probably due to the fact that risk appears to be greatest below serum 25(OH)D concentration = 10 ng/mL, and it is difficult to impossible to enroll enough particapants with such low concentrations in most countries.

Thus, one way to reduce risk of AD is to supplement with vitamin D. A recent review outlined the evidence that supplementing with 2000 IU/day of vitamin D might be an appropriate way for many people to avoid VDD [55]. However, supplementation may not be effective for obese people and may not reduce risk of AD as a result of the higher systemic inflammation from visceral adipose tissue. A meta-analysis of 13 RCTs with 1955 overweight and obese subjects with low 25(OH)D concentrations found that vitamin D supplementation did not influence the inflammatory biomarkers C-reactive protein, tumor necrosis factor-α, and interleuken-6 concentrations [56].

5. Conclusions

Vitamin D's effect on risk of neurological conditions is inversely correlated with mean follow-up period in prospective cohort studies. This effect should be considered in the design and analysis of such studies. As shown in the analysis in this article, using the HR or OR for the shortest follow-up period increases the apparent beneficial effect of high vs. low 25(OH)D concentration on risk of three adverse brain outcomes by about a factor of two compared to analyses without considering follow-up period. Additional studies should also be conducted regarding raising serum 25(OH)D concentrations to reduce risk of brain function decline.

Funding: This research received no funding. I have received funding in the past from Bio-Tech Pharmacal Inc. (Fayetteville, AR, USA), a supplier of vitamin D supplements.

Institutional Review Board Statement: Not applicable.

Informed Consent Statement: Not applicable.

Data Availability Statement: The original contributions presented in the study are included in the article, further inquiries can be directed to the corresponding author.

Conflicts of Interest: The author declares no conflict of interest.

References

1. Clarke, R.; Shipley, M.; Lewington, S.; Youngman, L.; Collins, R.; Marmot, M.; Peto, R. Underestimation of risk associations due to regression dilution in long-term follow-up of prospective studies. *Am. J. Epidemiol.* **1999**, *150*, 341–353. [CrossRef] [PubMed]
2. Kuzma, E.; Soni, M.; Littlejohns, T.J.; Ranson, J.M.; van Schoor, N.M.; Deeg, D.J.; Comijs, H.; Chaves, P.H.; Kestenbaum, B.R.; Kuller, L.H.; et al. Vitamin D and Memory Decline: Two Population-Based Prospective Studies. *J. Alzheimers Dis.* **2016**, *50*, 1099–1108. [CrossRef] [PubMed]
3. Grant, W.B. Effect of interval between serum draw and follow-up period on relative risk of cancer incidence with respect to 25-hydroxyvitamin D level: Implications for meta-analyses and setting vitamin D guidelines. *Dermato-Endocrinology* **2011**, *3*, 199–204. [CrossRef] [PubMed]
4. Grant, W.B. Effect of follow-up time on the relation between prediagnostic serum 25-hydroxyvitamin D and all-cause mortality rate. *Dermato-Endocrinology* **2012**, *4*, 198–202. [CrossRef] [PubMed]
5. Jorde, R.; Sneve, M.; Hutchinson, M.; Emaus, N.; Figenschau, Y.; Grimnes, G. Tracking of serum 25-hydroxyvitamin D levels during 14 years in a population-based study and during 12 months in an intervention study. *Am. J. Epidemiol.* **2010**, *171*, 903–908. [CrossRef] [PubMed]
6. McCullough, M.L.; Zoltick, E.S.; Weinstein, S.J.; Fedirko, V.; Wang, M.; Cook, N.R.; Eliassen, A.H.; Zeleniuch-Jacquotte, A.; Agnoli, C.; Albanes, D.; et al. Circulating Vitamin D and Colorectal Cancer Risk: An International Pooling Project of 17 Cohorts. *J. Natl. Cancer Inst.* **2019**, *111*, 158–169. [CrossRef]
7. Munoz, A.; Grant, W.B. Vitamin D and Cancer: An Historical Overview of the Epidemiology and Mechanisms. *Nutrients* **2022**, *14*, 1448. [CrossRef]
8. Zhang, X.X.; Wang, H.R.; Meng, W.; Hu, Y.Z.; Sun, H.M.; Feng, Y.X.; Jia, J.J. Association of Vitamin D Levels with Risk of Cognitive Impairment and Dementia: A Systematic Review and Meta-Analysis of Prospective Studies. *J. Alzheimers Dis.* **2024**, *98*, 373–385. [CrossRef]
9. Chen, H.; Xue, W.; Li, J.; Fu, K.; Shi, H.; Zhang, B.; Teng, W.; Tian, L. 25-Hydroxyvitamin D Levels and the Risk of Dementia and Alzheimer's Disease: A Dose-Response Meta-Analysis. *Front. Aging Neurosci.* **2018**, *10*, 368. [CrossRef]
10. Graf, C.E.; Rossi, C.; Giannelli, S.V.; Nobari, B.H.; Gold, G.; Herrmann, F.R.; Zekry, D. Vitamin D is not associated with cognitive status in a cohort of very old hospitalized patients. *J. Alzheimers Dis.* **2014**, *42* (Suppl. S3), S53–S61. [CrossRef]
11. Chen, L.J.; Sha, S.; Stocker, H.; Brenner, H.; Schottker, B. The associations of serum vitamin D status and vitamin D supplements use with all-cause dementia, Alzheimer's disease, and vascular dementia: A UK Biobank based prospective cohort study. *Am. J. Clin. Nutr.* **2024**, *119*, 1052–1064. [CrossRef] [PubMed]
12. Littlejohns, T.J.; Henley, W.E.; Lang, I.A.; Annweiler, C.; Beauchet, O.; Chaves, P.H.; Fried, L.; Kestenbaum, B.R.; Kuller, L.H.; Langa, K.M.; et al. Vitamin D and the risk of dementia and Alzheimer disease. *Neurology* **2014**, *83*, 920–928. [CrossRef]
13. Melo van Lent, D.; Egert, S.; Wolfsgruber, S.; Kleineidam, L.; Weinhold, L.; Wagner-Thelen, H.; Stoffel-Wagner, B.; Bickel, H.; Wiese, B.; Weyerer, S.; et al. Low Serum Vitamin D Status Is Associated with Incident Alzheimer's Dementia in the Oldest Old. *Nutrients* **2022**, *15*, 61. [CrossRef]
14. Kiderman, D.; Ben-Shabat, N.; Tsur, A.M.; Anis, S.; Watad, A.; Cohen, A.D.; Paz, Z.; Amital, H. Vitamin D Insufficiency is Associated with Higher Incidence of Dementia, a Large Community-Based Retrospective Cohort Study. *J. Geriatr. Psychiatry Neurol.* **2023**, *36*, 511–518. [CrossRef] [PubMed]
15. Geng, T.; Lu, Q.; Wan, Z.; Guo, J.; Liu, L.; Pan, A.; Liu, G. Association of serum 25-hydroxyvitamin D concentrations with risk of dementia among individuals with type 2 diabetes: A cohort study in the UK Biobank. *PLoS Med.* **2022**, *19*, e1003906. [CrossRef] [PubMed]
16. Karakis, I.; Pase, M.P.; Beiser, A.; Booth, S.L.; Jacques, P.F.; Rogers, G.; DeCarli, C.; Vasan, R.S.; Wang, T.J.; Himali, J.J.; et al. Association of Serum Vitamin D with the Risk of Incident Dementia and Subclinical Indices of Brain Aging: The Framingham Heart Study. *J. Alzheimers Dis.* **2016**, *51*, 451–461. [CrossRef]
17. Asante, E.O.; Mai, X.M.; Eldholm, R.S.; Skjellegrind, H.K.; Kolberg, M.; Brumpton, B.M.; Selbaek, G.; Chen, Y.; Sun, Y.Q. Vitamin D Status Over Time and Cognitive Function in Norwegian Older Adults: A Prospective Cohort of the HUNT Study. *J. Nutr. Health Aging* **2023**, *27*, 30–37. [CrossRef] [PubMed]
18. Féart, C.; Helmer, C.; Merle, B.; Herrmann, F.R.; Annweiler, C.; Dartigues, J.F.; Delcourt, C.; Samieri, C. Associations of lower vitamin D concentrations with cognitive decline and long-term risk of dementia and Alzheimer's disease in older adults. *Alzheimers Dement.* **2017**, *13*, 1207–1216. [CrossRef]
19. Olsson, E.; Byberg, L.; Karlstrom, B.; Cederholm, T.; Melhus, H.; Sjogren, P.; Kilander, L. Vitamin D is not associated with incident dementia or cognitive impairment: An 18-y follow-up study in community-living old men. *Am. J. Clin. Nutr.* **2017**, *105*, 936–943. [CrossRef]
20. Licher, S.; de Bruijn, R.; Wolters, F.J.; Zillikens, M.C.; Ikram, M.A.; Ikram, M.K. Vitamin D and the Risk of Dementia: The Rotterdam Study. *J. Alzheimers Dis.* **2017**, *60*, 989–997. [CrossRef]
21. Schneider, A.L.; Lutsey, P.L.; Alonso, A.; Gottesman, R.F.; Sharrett, A.R.; Carson, K.A.; Gross, M.; Post, W.S.; Knopman, D.S.; Mosley, T.H.; et al. Vitamin D and cognitive function and dementia risk in a biracial cohort: The ARIC Brain MRI Study. *Eur. J. Neurol.* **2014**, *21*, 1211-e70. [CrossRef]
22. Knekt, P.; Saaksjarvi, K.; Jarvinen, R.; Marniemi, J.; Mannisto, S.; Kanerva, N.; Heliovaara, M. Serum 25-hydroxyvitamin d concentration and risk of dementia. *Epidemiology* **2014**, *25*, 799–804. [CrossRef]

23. Fashanu, O.E.; Zhao, D.; Schneider, A.L.C.; Rawlings, A.M.; Sharrett, A.R.; Lutsey, P.L.; Gottesman, R.F.; Gross, A.L.; Guallar, E.; Alonso, A.; et al. Mid-life serum Vitamin D concentrations were associated with incident dementia but not late-life neuropsychological performance in the Atherosclerosis Risk in Communities (ARIC) Study. *BMC Neurol.* **2019**, *19*, 244. [CrossRef]
24. Afzal, S.; Bojesen, S.E.; Nordestgaard, B.G. Reduced 25-hydroxyvitamin D and risk of Alzheimer's disease and vascular dementia. *Alzheimers Dement.* **2014**, *10*, 296–302. [CrossRef]
25. Slinin, Y.; Paudel, M.; Taylor, B.C.; Ishani, A.; Rossom, R.; Yaffe, K.; Blackwell, T.; Lui, L.Y.; Hochberg, M.; Ensrud, K.E.; et al. Association between serum 25(OH) vitamin D and the risk of cognitive decline in older women. *J. Gerontol. A Biol. Sci. Med. Sci.* **2012**, *67*, 1092–1098. [CrossRef]
26. Toffanello, E.D.; Coin, A.; Perissinotto, E.; Zambon, S.; Sarti, S.; Veronese, N.; De Rui, M.; Bolzetta, F.; Corti, M.C.; Crepaldi, G.; et al. Vitamin D deficiency predicts cognitive decline in older men and women: The Pro.V.A. Study. *Neurology* **2014**, *83*, 2292–2298. [CrossRef]
27. Slinin, Y.; Paudel, M.L.; Taylor, B.C.; Fink, H.A.; Ishani, A.; Canales, M.T.; Yaffe, K.; Barrett-Connor, E.; Orwoll, E.S.; Shikany, J.M.; et al. 25-Hydroxyvitamin D levels and cognitive performance and decline in elderly men. *Neurology* **2010**, *74*, 33–41. [CrossRef]
28. Llewellyn, D.J.; Lang, I.A.; Langa, K.M.; Melzer, D. Vitamin D and cognitive impairment in the elderly U.S. population. *J. Gerontol. A Biol. Sci. Med. Sci.* **2011**, *66*, 59–65. [CrossRef]
29. Marquez, C.; Angel, B.; Lera, L.; Bendayan, R.; Sanchez, H.; Albala, C. Exploring the Association between Vitamin D and Changes in Cognitive Function in Chilean Older Adults: Evidence from the ALEXANDROS Cohort Study. *J. Pers. Med.* **2022**, *12*, 1078. [CrossRef]
30. MacLaughlin, J.; Holick, M.F. Aging decreases the capacity of human skin to produce vitamin D3. *J. Clin. Investig.* **1985**, *76*, 1536–1538. [CrossRef]
31. Chalcraft, J.R.; Cardinal, L.M.; Wechsler, P.J.; Hollis, B.W.; Gerow, K.G.; Alexander, B.M.; Keith, J.F.; Larson-Meyer, D.E. Vitamin D Synthesis Following a Single Bout of Sun Exposure in Older and Younger Men and Women. *Nutrients* **2020**, *12*, 2237. [CrossRef]
32. Crowe, F.L.; Steur, M.; Allen, N.E.; Appleby, P.N.; Travis, R.C.; Key, T.J. Plasma concentrations of 25-hydroxyvitamin D in meat eaters, fish eaters, vegetarians and vegans: Results from the EPIC-Oxford study. *Public. Health Nutr.* **2011**, *14*, 340–346. [CrossRef]
33. Raulio, S.; Erlund, I.; Mannisto, S.; Sarlio-Lahteenkorva, S.; Sundvall, J.; Tapanainen, H.; Vartiainen, E.; Virtanen, S.M. Successful nutrition policy: Improvement of vitamin D intake and status in Finnish adults over the last decade. *Eur. J. Public Health* **2017**, *27*, 268–273. [CrossRef]
34. Ikonen, H.; Lumme, J.; Seppala, J.; Pesonen, P.; Piltonen, T.; Jarvelin, M.R.; Herzig, K.H.; Miettunen, J.; Niinimaki, M.; Palaniswamy, S.; et al. The determinants and longitudinal changes in vitamin D status in middle-age: A Northern Finland Birth Cohort 1966 study. *Eur. J. Nutr.* **2021**, *60*, 4541–4553. [CrossRef]
35. Rooney, M.R.; Harnack, L.; Michos, E.D.; Ogilvie, R.P.; Sempos, C.T.; Lutsey, P.L. Trends in Use of High-Dose Vitamin D Supplements Exceeding 1000 or 4000 International Units Daily, 1999–2014. *JAMA* **2017**, *317*, 2448–2450. [CrossRef]
36. Bernstein, A.M.; Song, M.; Zhang, X.; Pan, A.; Wang, M.; Fuchs, C.S.; Le, N.; Chan, A.T.; Willett, W.C.; Ogino, S.; et al. Processed and Unprocessed Red Meat and Risk of Colorectal Cancer: Analysis by Tumor Location and Modification by Time. *PLoS ONE* **2015**, *10*, e0135959. [CrossRef]
37. Dawson-Hughes, B.; Staten, M.A.; Knowler, W.C.; Nelson, J.; Vickery, E.M.; LeBlanc, E.S.; Neff, L.M.; Park, J.; Pittas, A.G.; Group, D.d.R. Intratrial Exposure to Vitamin D and New-Onset Diabetes Among Adults With Prediabetes: A Secondary Analysis From the Vitamin D and Type 2 Diabetes (D2d) Study. *Diabetes Care* **2020**, *43*, 2916–2922. [CrossRef]
38. Gezen-Ak, D.; Dursun, E. Vitamin D, a Secosteroid Hormone and Its Multifunctional Receptor, Vitamin D Receptor, in Alzheimer's Type Neurodegeneration. *J. Alzheimer's Dis.* **2023**, *95*, 1273–1299. [CrossRef]
39. Nguyen, T.T.; Ta, Q.T.H.; Nguyen, T.T.D.; Le, T.T.; Vo, V.G. Role of Insulin Resistance in the Alzheimer's Disease Progression. *Neurochem. Res.* **2020**, *45*, 1481–1491. [CrossRef]
40. Szymczak-Pajor, I.; Sliwinska, A. Analysis of Association between Vitamin D Deficiency and Insulin Resistance. *Nutrients* **2019**, *11*, 794. [CrossRef]
41. Wenclewska, S.; Szymczak-Pajor, I.; Drzewoski, J.; Bunk, M.; Sliwinska, A. Vitamin D Supplementation Reduces Both Oxidative DNA Damage and Insulin Resistance in the Elderly with Metabolic Disorders. *Int. J. Mol. Sci.* **2019**, *20*, 2891. [CrossRef]
42. Lason, W.; Jantas, D.; Leskiewicz, M.; Regulska, M.; Basta-Kaim, A. The Vitamin D Receptor as a Potential Target for the Treatment of Age-Related Neurodegenerative Diseases Such as Alzheimer's and Parkinson's Diseases: A Narrative Review. *Cells* **2023**, *12*, 660. [CrossRef]
43. SanMartin, C.D.; Henriquez, M.; Chacon, C.; Ponce, D.P.; Salech, F.; Rogers, N.K.; Behrens, M.I. Vitamin D Increases Abeta140 Plasma Levels and Protects Lymphocytes from Oxidative Death in Mild Cognitive Impairment Patients. *Curr. Alzheimer Res.* **2018**, *15*, 561–569. [CrossRef]
44. Michaud, T.L.; Su, D.; Siahpush, M.; Murman, D.L. The Risk of Incident Mild Cognitive Impairment and Progression to Dementia Considering Mild Cognitive Impairment Subtypes. *Dement. Geriatr. Cogn. Dis. Extra* **2017**, *7*, 15–29. [CrossRef]
45. Davis, M.; O'Connell, T.; Johnson, S.; Cline, S.; Merikle, E.; Martenyi, F.; Simpson, K. Estimating Alzheimer's Disease Progression Rates from Normal Cognition Through Mild Cognitive Impairment and Stages of Dementia. *Curr. Alzheimer Res.* **2018**, *15*, 777–788. [CrossRef]
46. Pettersen, J.A. Does high dose vitamin D supplementation enhance cognition?: A randomized trial in healthy adults. *Exp. Gerontol.* **2017**, *90*, 90–97. [CrossRef]

47. Jia, J.; Hu, J.; Huo, X.; Miao, R.; Zhang, Y.; Ma, F. Effects of vitamin D supplementation on cognitive function and blood Abeta-related biomarkers in older adults with Alzheimer's disease: A randomised, double-blind, placebo-controlled trial. *J. Neurol. Neurosurg. Psychiatry* **2019**, *90*, 1347–1352. [CrossRef]
48. Wang, P.; Song, M.; Eliassen, A.H.; Wang, M.; Fung, T.T.; Clinton, S.K.; Rimm, E.B.; Hu, F.B.; Willett, W.C.; Tabung, F.K.; et al. Optimal dietary patterns for prevention of chronic disease. *Nat. Med.* **2023**, *29*, 719–728. [CrossRef]
49. Kroll, M.H.; Bi, C.; Garber, C.C.; Kaufman, H.W.; Liu, D.; Caston-Balderrama, A.; Zhang, K.; Clarke, N.; Xie, M.; Reitz, R.E.; et al. Temporal relationship between vitamin D status and parathyroid hormone in the United States. *PLoS ONE* **2015**, *10*, e0118108. [CrossRef]
50. Sempos, C.T.; Heijboer, A.C.; Bikle, D.D.; Bollerslev, J.; Bouillon, R.; Brannon, P.M.; DeLuca, H.F.; Jones, G.; Munns, C.F.; Bilezikian, J.P.; et al. Vitamin D assays and the definition of hypovitaminosis D: Results from the First International Conference on Controversies in Vitamin D. *Br. J. Clin. Pharmacol.* **2018**, *84*, 2194–2207. [CrossRef]
51. Wang, L.; Qiao, Y.; Zhang, H.; Zhang, Y.; Hua, J.; Jin, S.; Liu, G. Circulating Vitamin D Levels and Alzheimer's Disease: A Mendelian Randomization Study in the IGAP and UK Biobank. *J. Alzheimers Dis.* **2020**, *73*, 609–618. [CrossRef] [PubMed]
52. Davey Smith, G.; Hemani, G. Mendelian randomization: Genetic anchors for causal inference in epidemiological studies. *Hum. Mol. Genet.* **2014**, *23*, R89–R98. [CrossRef] [PubMed]
53. Mokry, L.E.; Ross, S.; Morris, J.A.; Manousaki, D.; Forgetta, V.; Richards, J.B. Genetically decreased vitamin D and risk of Alzheimer disease. *Neurology* **2016**, *87*, 2567–2574. [CrossRef] [PubMed]
54. Navale, S.S.; Mulugeta, A.; Zhou, A.; Llewellyn, D.J.; Hypponen, E. Vitamin D and brain health: An observational and Mendelian randomization study. *Am. J. Clin. Nutr.* **2022**, *116*, 531–540. [CrossRef] [PubMed]
55. Pludowski, P.; Grant, W.B.; Karras, S.N.; Zittermann, A.; Pilz, S. Vitamin D Supplementation: A Review of the Evidence Arguing for a Daily Dose of 2000 International Units (50 microg) of Vitamin D for Adults in the General Population. *Nutrients* **2024**, *16*, 391. [CrossRef]
56. Jamka, M.; Wozniewicz, M.; Walkowiak, J.; Bogdanski, P.; Jeszka, J.; Stelmach-Mardas, M. The effect of vitamin D supplementation on selected inflammatory biomarkers in obese and overweight subjects: A systematic review with meta-analysis. *Eur. J. Nutr.* **2016**, *55*, 2163–2176. [CrossRef]

nutrients

MDPI

Review

Effect of the Mediterranean Diet (MeDi) on the Progression of Retinal Disease: A Narrative Review

Oualid Sbai [1], Filippo Torrisi [2], Federico Pio Fabrizio [2], Graziella Rabbeni [2] and Lorena Perrone [2,*]

[1] Laboratory of Transmission, Control and Immunobiology of Infections (LTCII), LR11IPT02, Institut Pasteur de Tunis, Tunis 1068, Tunisia; oualid.sbai@pasteur.tn
[2] Faculty of Medicine and Surgery, University KORE of Enna, 94100 Enna, Italy; filippo.torrisi@unikore.it (F.T.); federicopio.fabrizio@unikore.it (F.P.F.); graziella.rabbeni@unikore.it (G.R.)
* Correspondence: lorena.perrone@unikore.it

Abstract: Worldwide, the number of individuals suffering from visual impairment, as well as those affected by blindness, is about 600 million and it will further increase in the coming decades. These diseases also seriously affect the quality of life in working-age individuals. Beyond the characterization of metabolic, genetic, and environmental factors related to ocular pathologies, it is important to verify how lifestyle may participate in the induction of the molecular pathways underlying these diseases. On the other hand, scientific studies are also contributing to investigations as to whether lifestyle could intervene in modulating pathophysiological cellular responses, including the production of metabolites and neurohormonal factors, through the intake of natural compounds capable of interfering with molecular mechanisms that lead to ocular diseases. Nutraceuticals are promising in ameliorating pathophysiological complications of ocular disease such as inflammation and neurodegeneration. Moreover, it is important to characterize the nutritional patterns and/or natural compounds that may be beneficial against certain ocular diseases. The adherence to the Mediterranean diet (MeDi) is proposed as a promising intervention for the prevention and amelioration of several eye diseases. Several characteristic compounds and micronutrients of MeDi, including vitamins, carotenoids, flavonoids, and omega-3 fatty acids, are proposed as adjuvants against several ocular diseases. In this review, we focus on studies that analyze the effects of MeDi in ameliorating diabetic retinopathy, macular degeneration, and glaucoma. The analysis of knowledge in this field is requested in order to provide direction on recommendations for nutritional interventions aimed to prevent and ameliorate ocular diseases.

Keywords: Mediterranean diet; retinopathy; diabetic retinopathy; age-related macular degeneration; glaucoma; Nrf2

1. Introduction

Eye diseases have a serious influence on overall quality of life, health, and the possibility of obtaining an appropriate education, developing one's own capabilities in the working environment, and contributing to sustainable development, and these conditions ultimately affect the economy and well-being across society. Indeed, the effects of visual impairment are not limited to daily routines, while they have a negative effect on both psychological and cognitive development by affecting educational and employment possibilities [1]. Moreover, impaired vision is a predisposing factor for various health disorders, including cardiovascular illness, dementia, cancer, and depression [2].

It has been estimated that about 600 million people have visual impairment and the large majority of these patients can be cured only with high-cost therapies [3]. Notably, 43 million people exhibit partial or complete blindness, while about 500 million patients show poor socio-economic conditions that block them from ameliorating their vision impairment because they cannot afford the cost of reading glasses [3]. Indeed, vision impairment shows a more elevated prevalence in women and people living in rural areas, creating a vicious circle that exacerbates poverty and the low educational level of such people, affecting their ability to improve their socio-economic conditions.

Poor educational and socio-economic conditions are associated with a poor food supply. Interestingly, inadequate dietary patterns are associated with an enhanced risk of developing eye diseases. Considering the high cost needed to cure vision impairment, a good diet can exert preventive action against eye diseases that can lead to blindness. However, in highly developed countries, a good standard of living does not necessarily ensure an amelioration of the quality of life, especially in the quality of aging. Indeed, despite higher levels of education, poor lifestyle choices can promote inadequate nutrition, increasing the risk of chronic and degenerative diseases [4]. The Mediterranean diet (MeDi) is considered a healthy diet that is very efficient for the prevention of several diseases [5]. It is characterized by several healthy components and is not expensive; thus, it is a sustainable resource for the prevention of visual impairment. Herein, we will summarize the effect of MeDi in preventing retinal disorders. We will also focus on the Nrf2 pathway in mediating the advantageous effects promoted via MeDi. To summarize the data contained in this narrative review, we used the following PubMed analysis method to collect published data. (i) We typed in PubMed "Mediterranean diet, diabetic retinopathy", providing 25 articles. We analyzed these articles and the articles in their citations, resulting in the analysis of data derived from 882 articles. (ii) We typed in PubMed "Mediterranean diet, macular degeneration", providing 36 articles. We analyzed these articles and those included in their citations, resulting in the analysis of data derived from 688 articles. (iii) We typed in PubMed "Mediterranean diet, glaucoma", providing 10 articles. We analyzed these articles and those included in their citations, resulting in the analysis of data derived from 623 articles. (iv) We typed in PubMed "Mediterranean diet Nrf2", resulting in 31 publications that we analyzed. (v) We typed in PubMed "retinopathy, Nrf2", resulting in 402 publications that we analyzed.

2. Visual System and Conditions Leading to Retinal Diseases

The visual system consists of the coordinated interaction of visual pathways between the eyes and the brain, and it also involves the tissues associated with the eyes. The cornea and the lens of the eyes direct the light onto the retinal photoreceptors that, in turn, transform the light-induced stimulation into neuronal impulses. Finally, these impulses are translated into tri-dimensional images in the brain.

The architecture of the retina is highly ordered and conserved among all vertebrates. It is constituted by five types of neurons dispersed in three nuclear layers that are separated by two plexiform layers formed by synaptic interactions. Photoreceptors are localized in the outer nuclear layer (ONL); different types of interneurons (bipolar, amacrine, and horizontal cells) are present in the inner nuclear layer (INL); the ganglion cell layer (GCL) contains the retinal ganglion cells (RGCs) and dislocated amacrine cells. Photoreceptors are stimulated by the light, producing electrochemical signals that are transmitted by synapses with bipolar and horizontal cells in the outer plexiform layer (OPL). The inner plexiform layer (IPL) contains the synapses from ganglion cells to amacrine and bipolar cells. The projections of ganglion cell axons constitute the optic nerve, which transmits the signals from the eye to the brain for visual processing. Vision derives from specific patterns of connections from each type of neuron, which leads to the formation of different ganglion cells that show specific sensitivity to different stimulations, including stationary or moving objects, color contrast, and edges. In the retina are present various types of glial cells. The predominant type is the Muller glia that interacts with all neuronal types. The Muller glia modulates neuronal and microglia function through various secreted molecules, such as neurotransmitters [6], and this glial type seems to be involved in retina regeneration approaches [7]. Finally, astrocytes are mostly present in the IPL, together with the microglia, and they modulate retina homeostasis and are involved in promoting inflammation in retinal diseases [8].

Vision can be affected by genetic mutations, age, malnutrition, environment, and lifestyle [9–11]. Eye diseases affecting vision and leading to retinopathy are increasing. Retinal diseases include a range of genetic and non-genetic disorders. Genetic retinal diseases show an incidence of 1 to 3000 individuals [12], and more that 340 genes are implicated in those disorders [13]. Non-genetic retinal diseases can be modulated by genetic factors, but are mostly influenced by environmental factors, lifestyle, infections, and aging. Age-related macular degeneration is one of the most prevalent non-genetic retinal disorders and leads to central vision loss caused by the accumulation of deposits and by retinal pigment damage in the non-vascular variant, or by increased vascular growth in the neovascular variant [14]. Diabetic retinopathy (DR) is also a common retinal disease characterized by alterations of the retinal blood vessels [15], leading to inflammation, gliosis, and neuronal injury in the GCL [16,17]. Glaucoma is characterized by optic nerve structural damage with axonal loss, and RGC apoptosis [18]. Herein, we will focus on the following retinopathies: diabetic retinopathy (DR) [19], age-related macular degeneration [20], and glaucoma [3], as well as the effect of MeDi in preventing/ameliorating these ocular disorders.

3. The Mediterranean Diet (MeDi)

The Mediterranean Diet (MeDi) is not limited to a dietary pattern, while it includes a specific lifestyle. The pattern of lifestyle and diet at the basis of the MeDi originated a long time ago and it is the combination of various cultures that characterized the Mediterranean region: Roman, Greek, Phoenician, Arabic, and other cultures that shared their cultural and nutritional patterns, influencing their lifestyle [21]. In 2010, UNESCO acknowledged MeDi as an Intangible Cultural Heritage of Humanity and developed the model of the food pyramid (Figure 1) in order to communicate the MeDi model to people and health professionals [22].

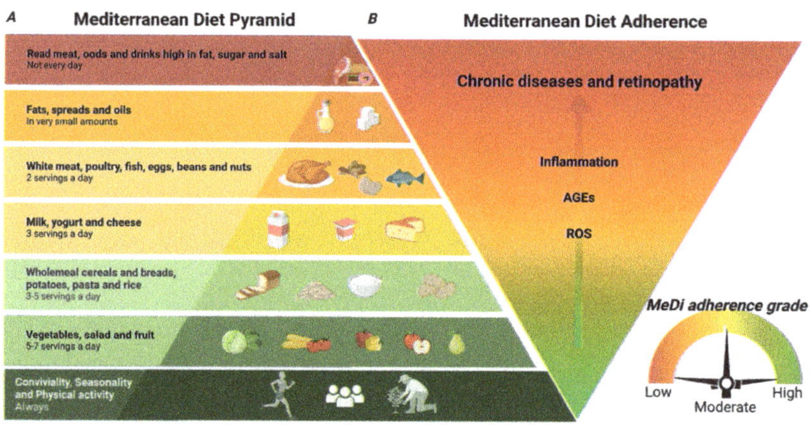

Figure 1. (**A**) **The MeDi pyramid.** The basis of the pyramid includes the Mediterranean lifestyle, with conviviality and daily moderate physical activity. The food categories and the frequency of consumption that represent a high adherence to the MeDi are indicated. (**B**) **MeDi adherence.** Progressive effects induced by lowering the adherence to the MeDi from high (green) to very low (red): starting from ROS production, to increased AGEs formation, followed by inflammation and finally chronic progression of retinopathy (Created with BioRender.com, Licensing and Agreement number GT2769TZYW).

MeDi is constituted not only by a dietary pattern, but also sustainable food production, conviviality, and an active lifestyle, including daily social activity that promotes moderate physical activity as well as an appropriate time for rest [23]. The MeDi dietary pattern includes the daily utilization of fresh vegetables and fruits, nuts and seeds, whole grains, eating legumes several times/week, the utilization of extra virgin oil as source of cooking and seasoning fat, herbs, and spices for flavoring, resulting in low salt consumption, low intake of cakes and desserts, two to three servings/week of fish and seafood, two to four servings of eggs/week, daily consumption of low-fat dairy products (mostly yogurt), the consumption of red meat no more than once a week, drinking water instead of other beverages, and moderate consumption of wine, mostly red wine, during meals [22]. However, there are some differences between MeDi components in different countries of the Mediterranean area that are linked to cultural, economic, and religious differences between the countries [24], as described below. For example, there are differences in alcohol consumption. The Greek MeDi includes general alcohol consumption, while French and Italian MeDi includes the consumption of red wine. In particular, the Lebanese MeDi shows several differences compared to the other Mediterranean countries; it includes dried fruits and burghul, which are traditional food in Lebanon, while red meat, fish, and alcohol are not included in the Lebanese MeDi [25].

4. MeDi Scoring

It has been shown that MeDi decreased the risk of several chronic illness [5]. MeDi scoring has been used in clinical observational investigations aimed at analyzing the impact of the adherence to MeDi on the progression of various diseases. Several scoring systems have been used with the aim of defining MeDi adherence (Figure 2). For the qualitative interpretation of the data, the adherence to MeDi is classified into classes: low, moderate, and high adherence. The first study analyzing the impact of MeDi on the survival of the Greek population was published in 2003 by Trichoupoulo and colleagues [26]. The authors analyzed the adherence to MeDi in a population-based prospective investigation using a food frequency questionnaire and the dietary habits have been classified in a 10-point Mediterranean diet scale (MDS), with a score ranging from 0 to 9, where 9 represents the highest adherence. The scale included the consumption frequency of the most salient foods typical of the Greek diet. In 2006, Panagiotakos et al. defined MedDietScore [27], a five-point scale based on the frequency of consumption of 11 principal constituents of MeDi (whole cereals, vegetables, fruits, legumes, potatoes, fish, poultry, red meat, olive oil, full-fat dairy products, and alcohol). A score of 5 was considered the most adherent to MeDi (low alcohol and red meat consumption/day). Buckland and colleagues created rMed score in 2010 [28], which is an 18-point linear scale containing nine components of the diet. A score of 18 was considered the most adherent to the MeDi. Moreover, the rMed score was divided into low (0–6), medium (7–10), and high (11–18), called tertiles. Schroder et al. defined the Mediterranean Diet Adherence Screener (MEDAS) scoring [29]. MEDAS is composed of twelve questions related to food consumption frequency, and two questions related to food consumption habits. Each question is scored 0 or 1. In particular, one point is assigned for utilizing olive oil as the main kind of fat for cooking, consuming white meat instead of red meat, or for eating (1) 4 or more tablespoons (1 tablespoon = 13.5 g) of olive oil/d (utilized for frying, seasoning, meals consumed far from home, etc.); (2) two or more servings of vegetables/d; (3) three or more servings of fruit/d; (4) <1 portion of red meat or sausages/d; (5) <1 portion of animal fat/d; (6) <1 cup (1 cup = 100 mL) of sweet beverages/d; (7) seven or more portions of red wine/week; (8) three or more portions of pulses/week; (9) three or more portions of fish/week; (10) less than two commercial desserts/week; (11) three or more portions of nuts/week; or (12) two or more portions/week of a traditional plate composed of tomato sauce, garlic, onion, or leeks sautéed in olive oil. When these conditions were not met, 0 points were assigned for the category. Thus, MEDAS scores range from 0 to 14, with 14 corresponding to the highest adherence to MeDi [29]. In 2013, Agnoli et al. described the Italian Mediterranean Index (IMI) [30]. IMI was built on eleven food categories, including six typical Mediterranean food categories (pasta, fish, legumes, Mediterranean vegetables, fruits, and olive oil), four non-Mediterranean food items (sweet beverages, butter potatoes, and red meat), and alcohol. Subjects included in the third tertile of consumption of each characteristic Mediterranean food category were assigned a score of 1, while the others were assigned a score of 0. For non-Mediterranean food items, a score of 1 was assigned for subjects included in the first tertile of consumption and 0 for the others. Concerning alcohol consumption, 1 point was given for the subjects drinking up to 12 g per day, and 0 for abstainers or subjects drinking more than 12 g of alcohol/day. Scores varied between 0 and 11. In 2015, Naja et al. described the Lebanese Mediterranean Index (LMD) [25]. LMD score analyzed the eating frequency of the following categories of foods: vegetables, fruits, dried fruits, legumes, burghul, olive oil, starchy vegetables, eggs, and dairy products. The score was evaluated using a 61-item semi-quantitative questionnaire of food frequency and then compared to the scoring results derived from other studies. The Mediterranean diet scale (MDS) was realized in order to analyze the adherence to the Mediterranean diet in nine European countries (Denmark, Germany, Greece, France, Italy, Spain, the Netherlands, UK, and Sweden) and analyzed nine food categories. Subjects received a score of 1 when their consumption of legumes, vegetables, cereals, fish, and fruits was lower compared to the sex-specific average intake, while they received a score 0 in all the other cases. The opposite scoring was given for the categories of meat and dairy. Concerning alcohol consumption, men drinking 10–50 g/day and women drinking 5–25 g/day received a score of 1, while all the other cases received a score of 0. The ratio of monounsaturated fat (MUFA) and polyunsaturated fat (PUFA) to saturated fat (SFA) was also considered. This scoring ranged between 0 and 9 [25]. Monteagudo and colleagues defined the Mediterranean Diet Scoring System (MDSS) in 2015 [31]. MDSS was calculated from the data obtained using a questionnaire with 129 items divided into 11 food categories (cereals, fruit, vegetables, fish, eggs, meat, fats, commercial foods, sauces, alcohol-free drinks, and alcohol). MDSS was created according to the Mediterranean Diet Pyramid and the recommended eating/drinking frequency of the various categories, and the range of the scores is 0–24 for adults and 0–23 for adolescents (eliminating alcoholic beverages). The points were assigned considering the consumption of food categories in accordance with the recommended servings: a score of 3, 2, or 1 for recommendations calculated in times/meal, times/day, or times/week, respectively. Thus, this scoring provided higher relevance to foods recommended to be consumed at every meal (fruit, vegetables, olive oil, cereals), followed by foods recommended to be consumed daily (dairy products, dried fruit, and nuts), and lastly, foods recommended to be consumed once a week (potatoes, eggs, legumes, white meat, fish, red meat, desserts). In adults, 1 point was assigned for alcohol consumption corresponding to one (women) and two (men) glasses of wine or beer. Sofi and colleagues developed the MeDi-Lite score in 2017 that analyzed nine food groups and compared the results with the MDS score, demonstrating that the higher range score provided an increased sensibility and specificity compared to MDS scoring [32].

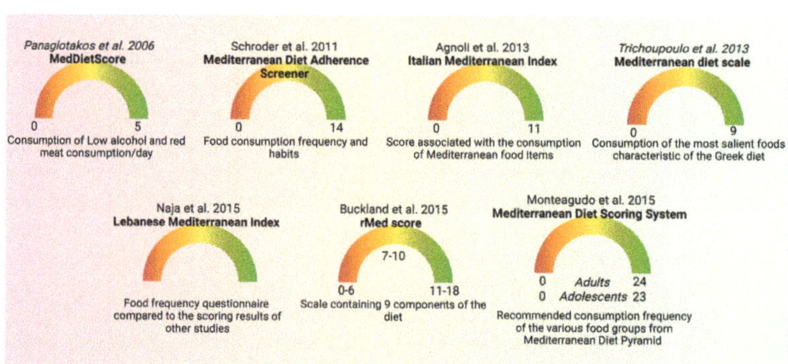

Figure 2. Schematic representation of the various MeDi scoring methods. The different MeDi scoring methods are indicated together with the adherence to the MeDi from lower to higher score (Created with BioRender.com, Licensing and Agreement number SX2769U3H2) [25–31].

These differences in the scoring result in an increased difficulty in comparing the data obtained from different observational, retrospective, and prospective clinical studies (Figure 2). Obeid and colleagues compared the different scoring systems by grouping them in the tertiles that defined the low, medium, and high adherence to the MeDi [33]. Moreover, certain studies provide the scores as consumption of food groups in grams/day, while others used the servings/week. This difference results in an additional complication when analyzing the amount of defined nutrients in different studies, according to published conversion data providing the amount of certain nutrients in defined food. The conversion from servings/week to grams/day has been published in order to solve this complication and extrapolate the content of dietary/nutritional compounds from different clinical studies [34].

5. Natural Molecules Enriched in the MeDi

MeDi is enriched in various nutraceuticals that produce beneficial outcomes for health.

5.1. Phenolic Compounds

Phenolic compounds are plant-derived micronutrients produced and secreted by plants following infection by pathogens or ultraviolet radiation. Polyphenols are classified according to the phenol rings they contain and the association between these rings and carbohydrates or organic acids. Some red fruits, black radish, tea, and onion contain tannins and simple phenols, such as gallic acid, which derives from benzoic acid. Derivatives of cinnamic acid are more common, and frequently, they are glycosylated. This group includes flavonoids, stilbene, and lignans [35]. Flavonoids have been extensively studied and they include flavonols (e.g., quercetin), flavones, flavonones, isoflavonones (e.g., gynestein), and anthocyanins. Resveratrol (RV) is the most studied stilbene and its antioxidant function is well characterized as well as its role in glucose homeostasis [36]. Polyphenols are mostly derived from fruits, vegetables, tea, and red wine (RV). They show an antioxidant activity that exerts a beneficial effect against several chronic diseases. Moreover, they are modified by the gut microbiota, opening the way for the study of their role on the metabolism of the microbiota and the subsequent effect on the human body [37,38]. Blackberries, blueberries, strawberry, kiwi, cherry, apricot, apple, pear, and all other types of fruits, including nuts, contain high levels of polyphenols. Whole grains also contain phenolic compounds, which are lost in refined grains.

5.2. Isoprenoids

This group includes carotenoids, saponins, tocotrienols, tocopherols, and simple terpenes, and they are contained in vegetables and fruits [39]. Carotenoids, in particular β-carotene, lycopene, zeaxanthin, and lutein, exert pro-vitamin A activity and are potent antioxidants [39]. Indeed, they are beneficial against cancer and neurodegenerative diseases, preventing cataracts. They also prevent age-related diseases by enhancing immune system activity [40]. Lycopene modulates the redox signaling that regulates gene expression [41]. Tocopherols and tocotrienols are two isoforms of vitamin E; they are found in plants and seeds. They are potent antioxidants, preventing DNA damage. They also exert anti-inflammatory action and are known to be protective against cancer, cardiovascular diseases, and neurodegeneration [42].

Saponins are also derived from plants and the most known saponins are the ginsenosides, derived from the ginger root. They have anti-cancer activity [43]. Green vegetables show high levels of lutein, β-carotene, and β-cryptoxanthin; carrots and pumpkins show high content of α-carotene; oranges, red bell peppers, broccoli, green vegetables, and potatoes together with carrots contain β-carotene. β-cryptoxanthin is found in tropical fruits like papaya. Tomato, watermelon, and pink grapefruit contain lycopene. Green vegetables including spinach, Brussels sprouts, broccoli, and peas are a source of lutein, while egg yolks and corn contain high levels of zeaxanthin.

5.3. Carbohydrates

Carbohydrates are grouped according to their digestibility in the gastrointestinal tract. Starch and fructans are hydrolyzed and adsorbed in the small intestine. On the contrary, β-glucans cellulose, hemicellulose pectin, and lignin cannot be digested in the small intestine and they are transformed in the large intestine through bacterial fermentation. Several investigations show that β-glucans modulate cholesterol metabolism, are beneficial against colorectal cancer, reduce constipation, and promote the growth of the gut microbiota [44]. Moreover, β-glucans lower the blood level of low-density lipoprotein (LDL) cholesterol particles by enhancing the fecal excretion of bile acids, resulting in an increased transformation of cholesterol in bile acids in the liver [45]. β-glucans ameliorate glycemic rate and the insulin response [46] by promoting insulin signaling in the liver [47]. Pectin is beneficial in promoting lipid and cholesterol metabolism [48], and diminishes intestinal infections by reducing the growth of pathogenic bacteria [49]. Starches that cannot be digested in the small intestine are defined as fibers and are enriched in whole grains. They decrease postprandial glucose and insulin levels [50], and lower cholesterol and triglyceride concentrations [51].

5.4. Proteins

Eggs, fish, meat, and dairy products provide high-quality proteins; beans and peas contain good-quality proteins, and grains are a moderate source of proteins. MeDi is defined according to a low consumption of meat, a moderate intake of fish, and a high consumption of beans and grains. The low intake of meat has been proposed as beneficial against various chronic diseases, including cardiovascular disease, diabetes, and cognitive dysfunction, in particular Alzheimer's disease. Several studies demonstrate that meat contains high levels of Advanced Glycation Endproducts (AGEs), which induce both oxidative stress and inflammatory response [52]. Dietary AGE intake correlates with enhanced incidence of chronic disorders, e.g., Alzheimer's disease [34]; thus, MeDi is also beneficial in preventing chronic illness by providing low AGE content [34]. AGEs participate in the progression of chronic diseases by inducing oxidative stress and by activating the Receptor For Advanced Glycation Endproducts (RAGE), which exerts a major role in ocular diseases characterized by retinopathy, including diabetic retinopathy, age-related macular degeneration, and glaucoma [16,53] (Figure 3).

5.5. Lipids and Fatty Acids

Linoleic acid (LA) is an essential poly unsaturated fatty acid (PUFA), important for the formation of phospholipids that constitute the plasma membranes and for the formation of the lipoprotein particles that regulate cholesterol homeostasis. LA is present in several components of the MeDi, including sunflower oil, grape seed oil, safflower oil, walnuts, salmon, chia seeds, and sardines. LA derivatives constitute the omega-6 fatty acids. In particular, they participate in the formation of high-density lipoprotein (HDL). High dietary LA intake decreases the risk of cardiovascular diseases [54]. Dietary alpha-linoleic acid (LNA) intake is also protective against cardiovascular diseases [55,56]. The conversion products of LNA are present in fish oil. They are eicosapentaenoic acid (EPA) and docosahexaenoic acid (DHA) and are the omega-3 fatty acids. Their dietary intake and fish consumption decrease the risk of cardiovascular diseases [56]. Both LA and LNA undergo several steps of desaturation ($\Delta 6$ and $\Delta 5$ desaturases) and elongation, generating a great number of metabolites, including arachidonic acid (ARA), from LA, and eicosapentaenoic (EPA) and docosahexaenoic (DHA) acids, from LNA. ARA generates the commonly known eicosanoids (ECs) anandamide (AEA) and 2-AG, while eicosapentaenoyl ethanolamine (EPEA) and docosahexaenoyl ethanolamine (DHEA) are produced from EPA and DHA, respectively, recently recognized as weaker ECs. DHEA is also known as synaptamide, a trophic factor that improves cognitive parameters in the nervous system. It is a metabolite generated on the omega-3 arm [57].

Several investigations demonstrate that the ratio between omega-3 and omega-6 fatty acids is relevant for preventing cardiovascular diseases. Indeed, eicosanoids derived from omega-6 PUFA promote inflammation, while eicosanoids derived from omega-3 PUFA have an anti-inflammatory function. Omega-3 fatty acid intake due to the consumption of fatty fish or fish oil exerts a healthy effect by decreasing the risk of cardiovascular diseases, rheumatoid arthritis, cancer, inflammatory bowel disease, and psychiatric and neurodegenerative diseases [58]. Omega-6 PUFA is contained in extra virgin olive oil. PUFAs are contained also in nuts, at high levels in pistachios, and in several vegetable oils derived from safflower, grape, sunflower, wheat germ, pumpkin seeds, sesame, and others [59].

5.6. *Vitamins*

We already described the function of vitamin E as an anti-inflammatory and antioxidant compound.

Vitamins A and D play an essential function in modulating the immune system [60]. The precursors of vitamin A (carotenoids) are pigments contained in vegetables and fruits, and they stimulate the immune system by promoting cell signaling pathways. The MeDi provides high levels of carotenoids with fruit and vegetables (e.g., tomatoes, leafy green vegetables, melons, carrots, bell peppers). Vitamin C modulates other antioxidant systems, including vitamin E, acting as an antioxidant, and can be found in fruits and vegetables.

Shellfish consumption provides high levels of vitamin B12. Rice, seaweed, soybeans, sesame seeds, peanuts, brown rice, and rye bread are great sources of vitamin B1. Spinach, avocados, and apricots provide high levels of vitamin B6 [61]. The vitamin B family is present in milk, cheese, eggs, fish, leafy vegetables, and chicken and exerts both antioxidant and neuroprotective actions [62].

5.7. *Melatonin*

High content of melatonin is present in fish, milk, eggs, seeds, and pistachios, providing neuroprotection and counteracting elevated intraocular pressure (IOP) [63].

5.8. *Saffron*

Saffron is also known as *Curcuma longa* and shows anti-inflammatory and antioxidant activity [64]. It contains curcumin and promotes several health benefits, such as helping to control diabetes, promoting weight loss, and preventing cardiovascular diseases. It inhibits pro-inflammatory cytokine release and modulates the composition of the gut microbiota [65].

5.9. *Taurine*

Taurine is the most copious amino acid in the retina of mammals [66] and is implicated in retinal survival [67]. Dietary taurine can be found in seafood, turkey, and seaweed [66].

5.10. *Palmitoyethanolamide (PEA)*

PEA is an N-acetylethanolamine cell-protective lipid present in various foods and in several living organisms. High concentrations of PEA are present in egg yolk. PEA shows anti-inflammatory and retina-protectant activity [68].

6. MeDi and Stress Response Involved in Retinal Diseases: Focus on Nrf-2 Pathway

Retinal diseases share similar cellular and molecular pathways, involving inflammation, immune response, and neurodegeneration [69]. Notably, the characteristic of retinal disorders is an alteration of the balance between the formation of reactive oxygen and nitrogen species (ROS and RNS, respectively) and the induction of the antioxidant systems, leading to oxidative stress [70]. Indeed, the eye is subjected to both endogenous and environmental damaging factors, resulting in an increased sensitivity to ROS and RNS. Furthermore, age-related diseases characterized by an excess of ROS and RNS production, such as diabetes, are risk factors for retinal diseases [71,72]. Notably, the Nrf-2 pathway seems to be involved in modulating the effect of diet in the prevention of retinopathy (Figure 4). Metabolic disorders can lead to stressful conditions to which the organism must respond to reestablish homeostasis. It is important to identify molecular pathways that signal metabolic changes [73]. The stress response is a set of complex cellular and molecular signals that can be modulated by endogenous and exogenous factors. The correlation between stress and the promotion of chronic degenerative pathologies affecting various organs and systems, including the cardiovascular and nervous systems, has been widely discussed and continues to be a topic of research [74]. Exogenous factors, such as nutrients, can influence this process, while endogenous factors can be grouped into metabolites, hormones, and several molecules that include reactive oxygen species (ROS) and/or reactive nitrogen species (RNS) [75]. Food intake and energetic metabolism can modify the production of endogenous factors by altering the response to stress [76]. The primary molecular pathway that governs the stress response is the nuclear factor erythroid-2-related factor 2 (Nrf2) pathway that plays a crucial role in regulating cellular homeostasis by managing oxidative stress and detoxification [77]. Nrf2 mitigates oxidative damage via the transcriptional activation of antioxidant response elements (AREs), thereby promoting the antioxidant response process. A key domain of Nrf2 is Neh2, which interacts with Kelch-like ECH-associated protein 1 (Keap1). Under normal conditions, this interaction inhibits the transcriptional activity of Nrf2 by ubiquitination and degradation through the proteasome [78]. Disrupted homeostasis promotes excessive production of ROS and RNS that can activate immune cells to release proinflammatory factors. This activation triggers the transcriptional activation of ARE, regulating downstream antioxidant enzymes and various neuroprotective genes that impede oxidative stress and neuroinflammation, thereby blocking the onset of neurological disorders and the subsequent pathological processes [79].

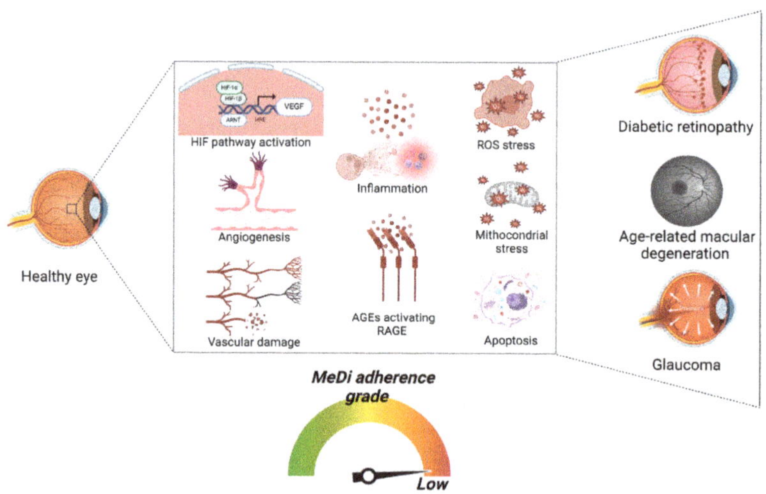

Figure 3. Pathways and cellular alterations implicated in diseases characterized by retinopathy (DR, AMD, and glaucoma) and correlation with low MeDi adherence. Scheme representing the progressive damage and the signaling pathways implicated in retinopathy and associated with low MeDi adherence. Initially, the activation of HIF pathway is observed, leading to VEGF expression, resulting in angiogenesis and subsequent vascular damage. Then, the formation of AGEs stimulates RAGE activation and subsequent inflammation. Ultimately, chronic excessive ROS formation and oxidative stress synergically lead to mitochondrial stress and apoptosis. These pathways are implicated in promoting the risk of DR, AMG, and glaucoma (Created with BioRender.com, Licensing and Agreement number XA2769UB0O).

In the framework of metabolism and nutrition, dietary energy restriction and adhering to the MeDi with the consumption of bioactive nutrients are the most studied approaches for regulating Nrf2 activity [80] (Figure 4). Dietary energy restriction, through either chronic or intermittent calorie reduction, increases Nrf2 activity. This creates an energetic stress in neurons that activates the Nrf2 pathway, leading to numerous health benefits, and increases longevity, promoting the prevention of neurological disorders [81]. Energetic dysfunctions and metabolite production can be correlated in the generation of a state of stress that involves different structures of the nervous system, including the eye [82]. Indeed, when the metabolite content is not sufficient for the energy demand, it increases the risk of retinal neuron death. Mitochondria are the center of energy supplementation and play an essential role in ATP generation and in sustaining redox homeostasis, and they drive the fate as waste of several types of metabolites [83]. Mitochondrial alteration is involved in the pathophysiology of several neurodegenerative diseases of the retina because the retina is highly susceptible to oxidative stress [84].

Nrf2 exerts a major function in promoting mitochondrial quality control and regulating fundamental aspects of mitochondrial function, such as energy production, ROS management, calcium signaling, and the induction of cell death [85]. Additionally, Nrf2 plays an essential function in modulating retinal oxidative stress. The retina shows a high metabolic activity and elevated oxygen consumption. For this reason, the retina is highly subjected to enhanced ROS production [86]. Studies indicate that there is reduced activity of the mitochondrial electron transport chain (ETC) in the aged retina, resulting in ROS augmentation and retinal damage. Impaired mitochondrial function leads to cell death and retinal degeneration. Moreover, the aged retina shows a reduced number of mitochondria, as well as altered mitochondrial activity and morphology compared to the healthy retina. In order to counteract the oxidative stress-induced damage, the retina relies on a crucial antioxidative defense system. Nrf2 is central to modulating the antioxidative stress response, especially against stressors including aging, inflammation, and sunlight exposure [86]. The induction of the Nrf2/Keap1/ARE cascade is considered a key target for neuroprotection in retinal ganglion cells [87]. Sox2 overlapping transcript (Sox2OT), a long non-coding RNA, highly expressed in the human brain, is involved in retinal ganglion cell apoptosis mediated by high glucose-induced reduction. It also induces Nrf2 nuclear translocation, determining HO-1 protein expression [87]. Dysregulation of the Keap1-Nrf2 cascade is also implicated in diabetic retinopathy. Activation of the Nrf2, MAPK, and NFκB signaling pathways effectively alleviates the ocular symptoms of diabetic retinopathy caused by ROS [88]. Moreover, Nrf2 expression in neurons also aids in detoxifying accumulated ROS by blocking mitochondrial complex II, suggesting that Nrf2 can protect neurons from the damage induced by dysfunctional mitochondria [89]. Moreover, Nrf2 plays a vital role in enhancing the antioxidant response, preserving the retina from ROS-induced cell damage.

Several investigations underline the important role of natural products in counteracting oxidative stress by modulating Nrf2 function. Bioactive natural molecules present in food are essential in regulating Nrf2. Flavonoids (hesperidin and quercetin), phenols (curcumin and capsaicin), and terpenes (astaxanthin and lutei) are the major dietary modulators of Nrf2 [81]. Low concentrations of epigallocatechin-3-gallate, the catechin most present in green tea, can induce HO-1 through the ARE/Nrf2 cascade in hippocampal neurons, thereby protecting them against various models of oxidative damage [90]. Similarly, caffeic acid phenethyl ester and ethyl ferulate can protect neurons by inducing HO-1 [91].

One of the most intriguing concepts is that natural compounds, such as phytoestrogens, which are structurally similar to estrogens, can interact with both estrogen receptors, ERα and ERβ, and exhibit weak estrogenic activity [92]. A recent study revealed an epistatic link between the Nrf2-Keap1 cascade and steroid hormone-induced signal transduction, demonstrating the function of the Nrf2-Keap1 pathway in neuronal remodeling through an antioxidant-independent and proteasome-dependent activity [93]. This evidence suggests a close correlation within the hormone–nutrition–stress response axis, mediated by a molecular pathway whose alteration can be associated with various pathologies. Hormones and energy metabolism are closely linked, and alterations in energy metabolism can disrupt mitochondrial homeostasis by affecting stress response systems. In this context, endocrine disruptors can also interfere by altering the hormone homeostasis of target organs. First, they bind hormone receptors and modulate their signaling cascades. Endocrine disruptors can alter sex hormones, thyroid hormones, and insulin [94]. Several research publications underline the key role of Nrf2 in enhancing the thyroid antioxidant defense by inducing the expression of cytoprotective factors that play a key role in modulating the normal thyroid function. These factors include GPx2, GR1, thioredoxin 1, thioredoxin reductase 1, sulfiredoxin 1, and NAD(P)H quinone dehydrogenase (NQO1). Studies have recognized irregularities affecting the reproductive and metabolic systems in animal models subjected to endocrine disruptors. Endocrine disruptors can exert a deleterious effect in ocular diseases. Indeed, cataracts, dry eye, macular degeneration, and diabetic retinopathy occur frequently as a consequence of hormone imbalances [94]. The correlation between many retinopathies and sex has been associated with protective effects of gonadal hormones [95]. Gender differences for retinal diseases have also recently been highlighted [96]; the incidence of central serous chorioretinopathy (CSC) has been found to be higher in young adult males [97]. Notably, retinal diseases are characterized by an overlap of neuronal and endocrine alterations together with aging-induced chronic dysfunction. Thus, retinal diseases deserve great attention for the study of the synergic effect of those alterations.

Despite the fact that the involvement of hormones in the deregulation of molecular pathways is not yet clear, it is known that the Nrf2 pathway can also be subjected to hormonal regulation. Indeed, membrane-associated estrogen receptors (ER)-α36 and G protein-coupled estrogen receptor (GPER) exert an essential function in the estrogen's fast non-genomic activity, such as the induction of cell proliferation [98]. Through these receptors, estrogen promotes fast Nrf2 induction, modulating the metabolic reprogramming in order to enhance cell proliferation, highlighting the effect of estrogen and phytoestrogens in inducing fast Nrf2 activation through membrane-associated estrogen receptors [98]. A study suggested that silibinin, a compound belonging to the flavonolignan family, induces Nrf2-antioxidative pathways in pancreatic β-cells by modulating ERα expression [99]. High glucose and palmitate induce glucolipotoxicity by decreasing the rat pancreatic β-cell line INS-1 viability, whereas preincubation with 5 or 10 μM of silibinin significantly promoted cell viability. In addition, treatment with ERα-selective agonist 4,4′,4″-(4-propyl-[1H]-pyrazole-1,3,5-triyl)trisphenol (PPT) and ERα-selective antagonist methyl-piperidino-pyrazole (MPP) induced an increase and a decrease in the viability of INS-1 cells, respectively. The anti-inflammatory role of estrogen through the Nrf2 pathway was demonstrated, and it was also reported that E2 induced downregulation of proinflammatory protein expression to a much greater extent in wild type mouse embryonic fibroblast MEFs than in Nrf2 knockout (KO) mice [99].

RV can promote epigenetic regulation by inducing the demethylation of the Nrf2 promoter, which was associated with the chemoprotective properties against estrogen-induced breast cancer, which activates the downstream antioxidant genes [100]. The ERβ receptor has also been reported to be linked by S-equol, a gut bacterial metabolite of soy daidzein, inhibiting the interaction of Nrf2 with Keap1 [101]. In support of this mechanism, the induction of estrogen receptor and Nrf2/ARE signaling cascade was also reported by Zhang and colleagues, showing that that S-equol counteracted peroxide-induced endothelial cell apoptosis [102]. Finally, similar molecular pathways involving ERβ and Nrf2 were reported, demonstrating an increase in the expression of the xenobiotic metabolizing enzyme quinone reductase after racemic equol. The complex network that harnesses energy metabolism and food intake in relation to diet and the consumption of bioactive compounds is interesting but at the same time complicated in interpreting the stress mechanisms associated with pathologies closely linked to metabolic disorders. The Nrf2 pathway intersects well with mitochondrial metabolism and with compounds chemically close to estrogen which could reveal interesting perspectives. The challenge to decode this network still remains open [101].

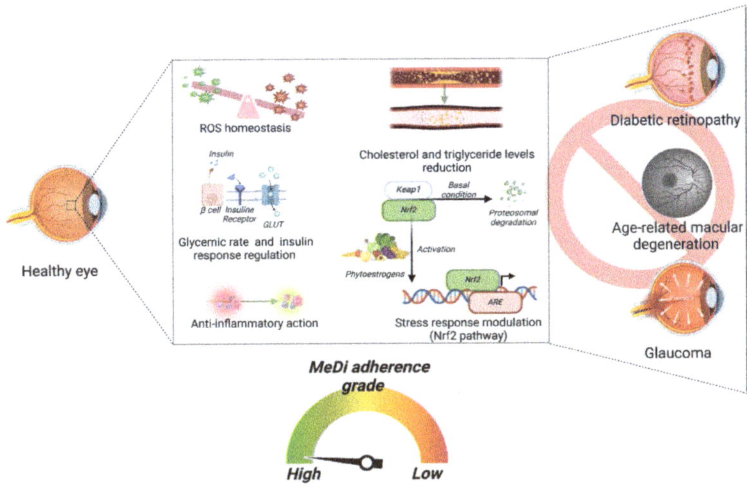

Figure 4. Pathways involved in maintaining a healthy retina and associated with a high adherence to the MeDi. Schematic representation of the conditions and signaling pathways that are induced by a high adherence to the MeDi and that participate in maintaining a healthy retina, lowering the risk of DR, AMD, and glaucoma. High adherence to the MeDi results in: (i) a good glycemic control and insulin response; (ii) anti-inflammatory response; (iii) good ROS homeostasis. Moreover, high adherence to the MeDi promotes: (i) reduction in the cholesterol and triglyceride levels; (ii) vegetable/fruit-derived phytoestrogens that activate the Nrf2 pathway, which activates the stress response. All these conditions and pathways induced by a high adherence to the MeDi participate in lowering the risk of DR, AMD, and glaucoma (Created with BioRender.com, Licensing and Agreement number PK2769UG6I).

7. MeDi's Role in the Prevention and Amelioration of Diabetic Retinopathy

Diabetic retinopathy (DR) is a severe diabetes complication characterized by a progressive, chronic, and irreversible visual impairment due to microvascular abnormalities. DR is one of the major causes of blindness in adults worldwide [103]. Every diabetic patient shows a high risk of developing DR (Figure 5).

DR can be classified in different stages of severity, depending on the morphology and functionality of the retinal vasculature. Two types of DR have been defined: non-proliferative diabetic retinopathy (NPDR) and proliferative diabetic retinopathy (PDR) [104]. Hyperglycemia releases inflammatory responses within the retinal environment that initiate the activation, adhesion, and infiltration of leukocytes, followed by the overexpression of inflammatory cytokines [16,17]. The characteristic of NPDR is the presence of vascular aberrations such as microaneurysms and hemorrhages. NPDR can be classified into mild, moderate, or severe stages based on the presence or absence of retinal bleeding. Patients with NPDR generally show hemorrhages of varying sizes, microaneurysms, exudates, and intra-retinal microvascular abnormalities. PDR is a developed stage of DR and characterized by retinal neovascularization due to diabetes-induced ischemia. PDR presents weak vessels that are prone to bleeding, leading to severe vision loss and even blindness [105]. PDR progression induces serious complications including macular edema, retinal detachment, vitreous hemorrhage, neovascular glaucoma, and irreversible blindness [106].

Many studies strongly suggest that diabetes may be prevented with lifestyle changes. Indeed, diabetes can be delayed or prevented by nutritional intervention based on consuming a low-carbohydrate diet, balanced meals, and eating carbohydrates mostly early in the day. Moreover, during a meal, it is better to first consume protein and vegetables and carbohydrates 30 min later, in order to lower glucose levels. Strong evidence has demonstrated that diabetes can be prevented through energy-restricted diets with routine physical activity. MeDi is considered the best dietary pattern for the prevention of diabetes and has received great attention given its role in improving health and reducing the burden of healthcare costs. MeDi reduces the incidence of diabetic retinopathy for type 2 diabetes patients.

DR development is influenced by hyperglycemia [107] through several pathways: non-enzymatic protein glycation (formation of advanced glycation endproducts (AGEs)), protein kinase C activation, polyol pathway, induction of the hexosamine pathway, accumulation of reactive oxygen species (ROS), and activation of hypoxia-induced factor [108].

MeDi is characterized by whole, nutrient-dense foods and a limited amount of processed and refined foods, which have high sugar content, artificial ingredients, refined carbohydrates, and trans fats.

Compelling evidence has suggested that the risk of DR can be reduced with diet, demonstrating the protective effect of MeDi. Many studies analyzed the effectiveness of MeDi on the incidence of DR. MD is also considered a beneficial diet for type 1 diabetes mellitus (T1DM) patients [109].

The role of diabetes mellitus (DM) and its progression on the development of DR has been widely analyzed, showing that the amelioration of DM and the maintenance of good glycemic control are beneficial against DR by delaying the onset and slowing the progression of DR. Thus, the dietary intervention mostly acts on the prevention and amelioration of DM and in turn is also beneficial for the DR.

Diaz-Lopez and colleagues demonstrated that supplementation with extra virgin olive oil together with a high adherence to MeDi in more than 3600 participants in a prospective 6-year study reduced the risk of DR (40%). The same study revealed that nut oil supplementation resulted in a low and not significant decrease in the risk of developing DR. The composition of olive oil differs depending on the cultivar, altitude, time of harvest, and extraction process. It contains mainly oleic acid, with other fatty acids such as linoleic acid (up to 21%) and palmitic acid (up to 20%). Oleic acid is a monounsaturated fatty acid that decreases the levels of total cholesterol and low-density lipoprotein. Linoleic acid and palmitic acid are polyunsaturated and saturated fatty acids, respectively. The same study revealed that high intake of fruits and vegetables resulted in a reduced risk of DR, suggesting the relevance of flavonoids in inhibiting the molecular pathways involved in DR.

A clinical trial observed that in subjects with type 2 diabetes mellitus (T2DM), the consumption of at least 500 mg/d of dietary LCω3PUFA, easily obtained with two servings/week of oily fish, correlated with a diminished risk of DR [110].

In certain versions of MeDi, milk is substituted with yogurt, kefir, buttermilk, and feta and cottage cheese. Ibsen and colleagues proposed that the substitution of whole-fat yogurt instead of milk among those aged 56–59 lowers the risk of type 2 diabetes, and the replacement of skimmed milk with semi-skimmed milk enhanced the risk among subjects aged 60–64 and 65–72.

The pathogenic progression of DR can be reduced by specific components that are abundant in MeDi. Indeed, polyphenols that are present in several vegetables, seeds, and fruits decrease insulin resistance and secretion, inflammation, and oxidative stress [111].

Díaz-López and colleagues demonstrated a correlation between lower risk of DR and the intake of flavonoid-rich vegetables and fruits [112]. Diabetic retinal microvascular alteration is clinically characterized by microaneurysms, hemorrhages, lipid exudates, and macular edema in T1DM and T2DM patients. High adherence to MeDi resulted in a decreased risk of retinal microvascular dysfunction. Low intake of fibers correlated with a 41% increase in DR in T2DM patients, compared to T2DM patients more adherent to a high-fiber diet. The PREDIMED study included 7447 Spanish participants and was randomized into three groups: one group consumed a highly adherent MeDi supplemented with extra virgin olive oil, the second followed a highly adherent MeDi enriched with mixed nuts, and the control group was subjected to a low-fat diet for a median of 5 years. This Spanish cohort study showed that MeDi enriched in extra virgin olive oil or nut intake significantly decreased the incidence of major cardiovascular events compared to a low-fat diet. However, only MeDi enriched with olive oil protected against DR (60% decrease in DR), while MeDi enriched with nuts resulted in a 37% decrease in DR [112]. In middle-aged and aged T2DM patients, consumption of at least 500 mg/d of dietary LCω3PUFA, obtained with two servings/week of oily fish, correlated with a diminished risk of sight-threatening DR [113].

Oleic acid is an essential component of olive oil. It is a monounsaturated fatty acid. Olive oil shows the presence of polyphenols and vitamins K and E, which can lower oxidative stress, inflammation, and insulin resistance [114].

Nuts contain RV, a polyphenolic compound contained in several plants such as grapes and peanuts, which plays a role in promoting anti-obesity, cardioprotective, neuroprotective, antitumor, antidiabetic, antioxidant, and anti-aging effects, and modulates glucose metabolism. The effects of RV are modulated by various synergistic pathways converging in the control of oxidative stress, cell death, and inflammation. RV modulated apelin gene expression in a rat model of T2DM [115]. The authors found a significant decrease in serum glucose level in rats treated with 5 and 10 mg/kg per day with RV compared with the diabetic control. In agreement, resistin expression in adipose tissue was reduced in RV-treated groups. RV induced heme oxygenase-1 (HO-1) expression through ARE-mediated transcriptional activation of Nrf2, suggesting that RV augmented cellular antioxidant defense capability following the induction of HO-1 through Nrf2-ARE cascade [116]. RV supplementation in diabetic rats resulted in significant amelioration of hyperglycemia, weight loss, increased oxidative markers, superoxide dismutase activity, and inhibition of eNOS activity in the blood and retina [117].

It has been shown that intake of at least 500 mg/d of dietary LCω3PUFA correlated with a decreased risk of sight-threatening DR [113]. Fish intake at least twice a week correlated with a 60% reduction in DR risk [108]. The anti-inflammatory effect of omega-3-fatty acid exerts an essential function in reducing the risk of DR [118]. Fish consumption in Japan, five times higher than in Western countries, reduces the incidence and progression rate of diabetic retinopathy compared to Western populations [119].

Various studies revealed the essential role of micro- and macro-elements in DR. Brazionis and colleagues indicated that plasma carotenoid levels seem to play a role in diabetic retinopathy, independent of established risk factors [120]. Lutein supplementation was shown to delay DR progression within 5 years according to Garcia Medina and colleagues [121].

Tanaka and colleagues demonstrated that fruit intake correlated with a decreased incidence of diabetic retinopathy among patients following a low-fat energy-restricted diet [122]. Fruits are low-glycemic-index foods enriched in fibers that can slow glucose response. Several investigations indicate that adherence to MeDi and high fruit intake are beneficial against the development of diabetic retinopathy. Post and colleagues suggested that fiber supplementation T2DM patients lowered fasting blood glucose and HbA1c [123]. Supplementation with vitamin C diminished the risk of retinopathy [122]. High vitamin C consumption correlated with 40% decreased risk of retinopathy [122] and the intake of vitamin C together with statins diminished the effects of non-proliferative DR more than statins alone [124].

Vitamins have a beneficial effect in lowering the risk of DR. Chatziralli and colleagues reported that vitamin E reduced serum malondialdehyde levels and oxidative stress, suggesting that vitamin E supplementation produced an additional benefit by lowering the risks of developing DR progression [125]. Oral vitamin E treatment (1800 IU daily vitamin E) was effective in normalizing retinal hemodynamic abnormalities and enhancing renal function in T1DM patients [126]. Vitamin C protected against diabetic retinopathy progression. Vitamin C exerts antioxidant and anti-angiogenic actions and enhanced endothelial function [127]. Barba and colleagues showed that PDR patients showed decreased intra-vitreous concentrations of ascorbic acid compared to non-diabetic patients [128]. Vitamin C is an anti-oxidant and modulates oxygen tension and eye-oxidative stress. The lower level of intra-vitreous content of vitamin C is caused by competition between the glucose and the ascorbic acid to bond the GLUT-1 glucose transporter. Rafael Simó and Cristina

Hernández suggested that regular intake of foods enriched in vitamin C, including citrus fruits, together with good glycemic control, was important for preserving the correct intra-retinal levels of ascorbic acid [127]. Park and colleagues observed that the vitreous level of vitamin C in PDR patients were diminished tenfold, which correlated with the degree of macular ischemia, suggesting that vitreous vitamin C depletion can promote macula ischemia in PDR patients [129]. The association of vitamins E and C enhanced the antioxidant effectiveness in the retina [130]. Other vitamins, including vitamins D and B, can exert a pivotal function in decreasing DR risk. The correlation between vitamin D deficiency and retinopathy severity was found in diabetic patients with well-controlled glycemia, suggesting the function of vitamin D in reducing the risk and severity of DR [130]. Vitamin B6 is beneficial against the early death of pericyte cells by maintaining the viability of capillaries, helping to maintain the presence of microvascular cells [114].

Others compounds can decrease the development and progression of DR, like zinc, iron, and manganese copper [114].

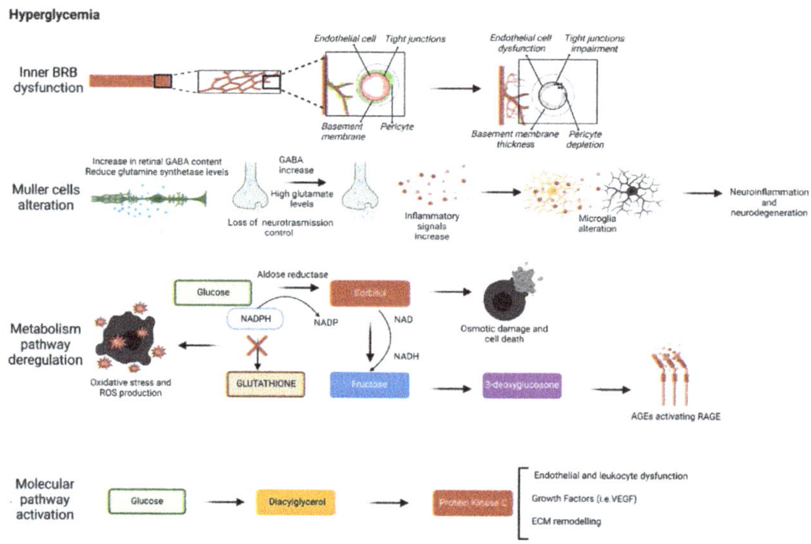

Figure 5. Cellular, metabolic, and signaling alterations in diabetic retinopathy. Four main interconnected mechanisms in retinopathy are shown. Müller cells have numerous functions, including maintaining the proper functioning of the blood–retina barrier. Additionally, they are involved in regulating synaptic neurotransmission and providing neuroprotection. Finally, the most affected processes involve metabolic and genomic pathways linked to hyperglycemia, such as the polyol pathway and the activation of protein kinase C. Created in BioRender.

8. MeDi's Role in the Prevention and Amelioration of Age-Related Macular Degeneration (AMD)

Age-related macular degeneration (AMD) is a retinal degenerative disorder affecting subjects over the age of 55 years. It is a major cause of blindness in industrialized countries, from 196 million affected individuals in 2020 to 288 million by 2040. AMD is a heterogeneous illness, influenced by age, genetics, and environmental factors, such as smoking and diet. Three stages characterize AMD: early, intermediate, and late. The first and second stages, characterized by the absence of symptoms or mild visual symptoms, show macular deposits (drusen) and pigmentary abnormalities [131]. Late AMD can show neovascular and atrophic forms characterized by macular neovascularization and atrophy. The neovascular stage could exist in two subtypes: the non-neovascular (dry) type and the neovascular (wet) type. The most frequent symptoms of age-related macular degeneration are blurry or fuzzy vision, difficulty in recognizing familiar faces, a dark, empty area or blind spot appearing in the center of vision, and the loss of central vision. In the late stage, visual symptoms are serious and irreversible, and include significantly diminished central vision in both eyes.

Many researchers have shown that eating vegetables, fruits, and fish in a Mediterranean-inspired diet is useful in protecting against age-related macular degeneration (AMD). Hogg and colleagues analyzed the correlation between AMD and MeDi in seven European countries [132]. They found that the populations with the highest MD score have the lowest level of advanced AMD [132]. The association between healthy diet, physical activity, and not smoking correlated with 71% lower chance for AMD [133]. In another study, 41,514 participants aged 40 to 70 years and born in Australia or New Zealand who migrated from the United Kingdom, Italy, Greece, or Malta were recruited and assessed for AMD prevalence in a follow-up study analyzing the effect of dietary habits on the onset and progression of AMD. It was demonstrated that dietary factors can regulate AMD risk. Predominant intake of grains, fish, steamed or boiled chicken, vegetables, and nuts correlated with a lower prevalence of advanced AMD, whereas red meat consumption correlated with a higher prevalence of advanced AMD [134]. Merle and colleagues showed that high adherence to the MeDi correlated with a 41% decreased risk of incident advanced AMD in two European population-based prospective cohorts [135,136]. Advanced AMD risk was lowered by 22%, 26%, and 47% in studies by Keenan [131], Merle [136], and Merle [134], respectively.

Drusen size progression decreased by 17% when following the most adherent MeDi compared to subjects following a less adherent MeDi [135]. Analysis of AMD progression showed a significant correlation between highly adherent MeDi and slower enlargement of atrophy [137].

Merle and colleagues studied the correlation between MeDi, AMD, and genetic susceptibility [136]. High adherence to MeDi correlated with a 26% lower risk of progression to advanced AMD [136]. Consuming fish and vegetables reduced the risk of progression of AMD. Genetic variations between different populations are also associated with AMD prevalence [136]. In particular, the prevalence of AMD in different ethnicities and geographic regions should consider genetic variations, in particular the single-nucleotide polymorphism (SNP) Y402H in the complement factor H (CFH) gene [138]. The risk of AMD progression was also significantly decreased among subjects carrying the *CFH* Y402H allele (T), while the individual homozygous for risk allele (CC) showed an enhanced risk of AMD progression [136,139]. European populations showed higher frequencies of risk alleles, correlating with increased progression of AMD, compared to Chinese and Japanese descendants [140]. The relationship between geographic region and prevalence of AMD could be at least partially explained by gene–diet interaction [141]. However, Hogg and colleagues did not find any correlation between AMD progression, MeDi adherence, and the presence of the Y402H allele, probably because a small number of neovascular AMD cases were analyzed [132]. On the contrary, Keenan and colleagues found that patients carrying the rs10922109 allele showed a lower risk of atrophy compared to neovascular AMD when their diet was highly adherent to MeDi [131].

Inflammation markers, including C-reactive protein (CRP), interleukin 6 (IL6), E-selectin, and soluble intercellular adhesion molecule 1 (sICAM-1), exert a function in diabetes development. A positive correlation between serum concentrations of sICAM-1 and E-selectin and diabetes risk has been shown [142]. Notably, the correlation between diet and diabetes is regulated in part through the modulation of the inflammatory response [142], suggesting that the MeDi can modulate the progression of AMD by acting on the inflammatory response.

The beneficial effects of the MeDi are correlated with a decrease in oxidative stress and inflammation, which exert a significant function in AMD [142]. Subjects following a highly adherent MeDi show elevated serum levels of biomarkers considered beneficial against AMD [136]. A high adherence to MeDi is more effective than the consumption of antioxidant and zinc supplementation. Trials examined short- or intermediate-term effects of MeDi on circulating markers of oxidative stress, such as urinary F2-isoprostanes, plasma malondialdehyde, and oxidized LDL. It has been demonstrated that subjects following a highly adherent MeDi showed lower oxidized LDL compared to the control group [143]. Coliij and colleagues proposed that elevated HDL cholesterol levels correlated with augmented risk for AMD [143]. Moreover, it has been shown that healthy habits and a healthy diet associated with supplement assumption are important for the prevention of ADM progression to late stages [136].

Jiang and colleagues demonstrated that a high intake of dietary omega-3 PUFA or fish correlated with a decreased AMD risk [144]. Moreover, a diet enriched in fish and with a low content of linoleic acid reduced the risk of AMD [145]. Interestingly, a high intake of ω-3 fatty acids or fish has no effect in preventing AMD progression in subjects consuming high levels of dietary linoleic acid [145]. Oily fish consumption at least once a week resulted in a lower risk of AMD progression [146]. RV, a bioactive compound present also in nuts, has antioxidant, antithrombotic, and anti-inflammatory properties [147]. It was demonstrated that RV prevented apoptosis of human retinal pigment epithelial (RPE) cells in vitro [148]. Moreover, RV protected RPE cells from autoimmune antibody-promoted apoptosis in vitro [148]. RV prevented oxidative stress-induced RPE degeneration by promoting the activity of superoxide dismutase, glutathione peroxidase, and catalase [149]. Nutritional supplementation with RV exerted beneficial function that resembled the effects induced by anti-VEGF treatment, promoting the anatomical restoration of retinal structure, RPE function, and choroidal blood flow [150].

High dietary intake of lutein correlated with a lower risk of prevalence and incidence of AMD. Lutein is present at elevated concentrations in green leafy vegetables such as spinach, kale, and yellow carrots and also in animal fat. Several studies unveiled the beneficial effect of lutein in lowering AMD risk [151,152]. Lutein is a filter for blue light. For this reason, lutein supplementation has been shown to protect the fovea from blue light-induced damage [153]. Moreover, membrane-bound lutein was demonstrated to exert an ROS scavenger function [154]. Indeed, the unconjugated double bonds in the molecular structure of lutein have a function in ROS quenching. Moreover, lutein decreased lipofuscin accumulation in cultured RPE cells by decreasing oxidative stress [155]. Dietary supplementation with lutein and zeaxanthin for 6 months increased the optical density of macular pigment [150,156–158]. Dietary lutein and zeaxanthin intake decreased the risk of incident early or neovascular AMD over 5 and 10 years [159]. Moreover, lutein, zeaxanthin, eicosapentaenoic acid, and docosahexaenoic acid, which show elevated levels in MeDi, were associated with diminished serum levels of C-reactive protein, suggesting a function in decreasing systemic inflammation in AMD subjects [160].

Vitamins seem to exert a relevant role in lowering AMD risk, since vitamin C is abundant in the retina. Although several studies indicated a correlation between dietary consumption of vitamin C and AMD risk, the function of vitamin C in preventing AMD risk is still controversial. Seddon and colleagues did not find any significant correlation between vitamin C consumption and reduced risk for AMD [161]. In agreement, a more recent study confirmed that vitamin C supplementation did not prevent the risk and the progression of AMD [162]. On the contrary, SanGiovanni and colleagues reported a decreased risk of developing neovascular AMD in subjects consuming elevated levels of dietary β-carotene, vitamin C, and vitamin E [163]. Vitamin E shows an antioxidant activity and RPE show a high concentration of vitamin E, suggesting a protective role of this vitamin in RPE. Vitamin E is composed of four different compounds: α-tocopherol, β-tocopherol, γ-tocopherol, and δ-tocopherol, with α-tocopherol as the most effective scavenger of free radicals. Wiegand and colleagues demonstrated that vitamin E concentrations in the retina were enhanced in response to augmented oxidative stress [164]. Dietary deprivation of vitamin E resulted in an augmented lipofuscin accumulation in the RPE [159]. Moreover, vitamin E deficiency accelerates retinal degenerative damage [165]. Low serum levels of tocopherol were associated with AMD progression [166]. In agreement, an association has been found between fasting α-tocopherol levels and AMD progression to the late stages [167]. Although various studies suggested a protective role of vitamin E in the prevention/amelioration of AMD, other clinical studies did not reveal any significant beneficial effect induced by vitamin E supplementation in AMD prevention or risk reduction [168,169]. Thus, the efficacy of nutrient supplementation in preventing AMD risk is still debated. Although studies show the benefit of specific nutrients, Merle and colleagues demonstrated that none of the nine components of the MeDi, such as vegetables, fruits, legumes, cereals, fish, the MUFA-to-SFA ratio, meat, dairy products, and alcohol intake, significantly correlated with the incidence of advanced AMD, underlining the relevance of assessing dietary patterns rather than single components [134].

9. MeDi's Role in the Prevention and Amelioration of Glaucoma

Worldwide, glaucoma is one of the major causes of irreversible blindness and significant vision impairment due to elevated intraocular pressure (IOP), which is a key modifiable risk factor for preventing the death of retinal ganglion cell [170,171].

To date, the European Glaucoma Society accounts for more than 70 million people with glaucoma diagnosis, with an estimated increase of about 112 million by 2040 [172,173]. It is considered a debilitating and heterogenous neurodegenerative eye disorder and it is characterized by progressive damage in retinal ganglion cells, a bridge between the inner surface of the retina and the optic nerve. Retinal ganglion cell degeneration will lead to the death of smaller nerves around the optic nerve up to a more pronounced blindness [174]. Although peripheral vision is affected first, mainly without the patient realizing it, later stages can destroy in an irreversible way the central visual field [175]. Notably, glaucoma can be divided in two groups, open-angle glaucoma (OAG) and closed-angle glaucoma (CAG), exhibiting characteristic morphological modifications in the optic nerve head and retinal nerve fiber layer. OAG typically occurs with an open drainage angle in the eye, while CAG, which is less common and represents a medical emergency, involves a closed or blocked drainage angle [176]. The disorders of visual function can be related to several risk factors as well as inadequate quantitative and qualitative nutrient supply, onset of genetic alterations, type 2 diabetes, obesity, hypertension, high myopia, environmental factors (pollution, UV rays, cigarette smoking, and particles), old age, and ethnic background (Afro-American or Hispanic) [9,177,178]. Generally, glaucoma is treated with eye drops that decrease the IOP to prevent damage to the optic nerve; however, these treatments, unfortunately, cannot restore the vision loss or cure this disease [179]. For those reasons, the early detection and intervention of modifiable risk factors, such as those linked to MeDi and dietary supplements, could be crucial for reducing the incidence of glaucoma and decelerating its development. Since numerous chronic conditions associated with glaucoma stem from dietary and metabolic disorders, it is clear that nutrition exerts a crucial function in the development, prevention, and treatment of this illness.

However, healthy foods, including fruits, whole grains, olive oil, vegetables, seafood nuts, and beans, which are included in MeDi, have been linked to the prevention of chronic age-related diseases (AREDs). In this context, a systematic review suggested that the consumption of a defined dietary pattern, when compared with a single component or nutrient, could imply potential protective effects by lowering the incidence of OAG (iOAG), although such evidence should be better explored in more detail [180,181]. Thus, in one of the most recent case–control studies, the Rotterdam Study, Vergroesen JE and colleagues demonstrated the correlation between the adherence to Mediterranean-DASH Intervention for Neurodegenerative Delay (MIND) and reduced risk of iOAG. In three independent cohorts from the prospective population, 170 participants developed iOAG in 1991 with follow-up visits every five years. In this program, a high consumption of food rich in nutrients and fiber and low in calories and fat, as well as seafood, strawberries and blueberries kale, collard greens, spinach, cabbage, and so on, with the latter included in green leafy vegetables, showed both anti-inflammatory and protective activities against iOAG. Moreover, the authors also investigated a possible adherence to MeDi or other guidelines, as well as a Dutch diet, and iOAG, but they did not find any remarkable associations. Considering that an

IOP-independent correlation was demonstrated, the authors concluded that the MIND diet tended to be effective in slowing down or halting the progressive neurodegeneration of the optical nerve [180]. Another interesting study came from Moreno-Montañés and colleagues, in a large prospective cohort with more than 10 years of follow-up time (updated with self-reported questionnaires that included lifestyle changes, health-related activities, and medical interventions). In the "Seguimiento Universidad de Navarra" (SUN) Project, the authors assessed the impact of the Mediterranean lifestyle (ML) habits (among no history of smoking, moderate and/or high physical activities, MeDi adherence, body mass index, modest alcohol consumption, and working 40 h per week, to which corresponds the SUN Healthy Lifestyle Score, SHLS, to define the adherence) on the risk of developing glaucoma. As a result, 261 (1.42%) new cases of glaucoma were diagnosed in the largest cohort ever reported with a total of 18,420 participants. They observed a decreased risk of glaucoma in participants with higher SHSL scores (>6) which adhered better to ML, while no significant association was demonstrated regarding MeDi related to each of its components. This work outlined, for the first time, that ML may reduce the incidence of glaucoma as a modifiable and protective risk factor, with a healthy lifestyle system [182]. Although the link between ML and glaucoma remains to be ascertained, a possible explanation could be attributable to the alterations in the nitric oxide (NO)–guanylate cyclase (GC) pathway. As recently reported, MeDi provides L-arginine and nitrate, which act as NO precursors, as well as vitamins, polyphenols, and fatty acids, which potentially boost NO endogenous production, providing both anti-inflammatory and anti-apoptotic properties [183–185]. A previous prospective analysis from the Nurses' Health and the Health Professionals Follow-up Study (63,893 women and 41,094 men, respectively) was reported by Kang and collaborators, who demonstrated that both higher total dietary nitrates, as an exogenous NO source, and vegetable intake were associated with lower IOP and risk of OAG and its subsequent progression. The reason could be due to the elevated concentrations of antioxidants and flavonoids present in these foods, which exerted neuroprotective effects [186]. According to the last piece evidence, Abreu-Reyes and colleagues performed an observational study, then validated it with the Prevention through Mediterranean Diet (PREDIMED), on 100 Spanish Canary Islands patients with the diagnosis of OAG in terms of their adherence to MeDi. Briefly, the authors reported only moderate adherence to MeDi with a high % of participants, approximately 70%, without gender differences [187]. Recently, an extensive review provided by Valero-Vello M and collaborators focused on nutritional hallmarks of foods and oral supplements in a Mediterranean cohort. As a consequence, they did not find any significant correlation between the adherence to MeDi by age and/or gender and the restoration of optic nerve damage in glaucoma patients. Overall, since MeDi plays an important preventive role against progressive eye conditions, the combination with nutritional supplementations as adjuvant factors would allow for high the adherence to healthy diet patterns, thus preventing the vision loss and increasing the quality of life of glaucoma patients [111]. Moreover, Mvitu and colleagues carried out a cross-sectional study that counted 244 Congolese patients affected by type 2 diabetes mellitus (T2 DM, 48% of males; 40% aged $\geq$ 60 years). The assessment of dietary intake was linked to a qualitative-type questionnaire that resumed the frequency of red beans, vegetable, fruit intake, and cataract extraction. Interestingly, they noticed that regular MeDi intake (*Abelmoschus*, *Brassica rapa*, *Musa acuminate*, beans) decreased the risk of blindness, cataracts, and glaucoma in this group of patients; nevertheless, these results focused on a very low rate of vegetable intake in Africa. Particularly, from univariate analysis, red bean intake and consumption equal to or more than three servings of vegetables per day represented independent and protective factors against eye degeneration diseases [188]. To date, further high-quality studies are required to deeply elucidate both molecular mechanisms and healthy benefits of MeDi in the prevention of glaucoma [181].

A large amount of evidence suggests that the effects of nutritional supplements positively impact several ocular dysfunctions, acting as a powerful neuroprotective on the modulation of IOP in preclinical animal models and patients affected by glaucoma [189,190].

For this purpose, a systematic review reported that a high dietary consumption of some micronutrients derived from leafy green vegetables like kale and spinach that are rich in vitamins, minerals and fibers, and contain for instance flavonoids, glutathione and NO, led to reduced levels of reactive oxygen species (ROS) and, consequently, the risk of glaucoma onset in patients affected by OAG. These findings were different from selenium (Se) and iron (Fe), contained in red meats, which would seem to increase the risk of developing glaucoma, although randomized clinical trials (RCTs) will be necessary to confirm these results [191]. Other supplementations as well as blackcurrant, an optimal source of polyphenols, provided a significant improvement in the visual field and ocular blood flow (OBF) in 38 OAG patients, during 24 months of follow-up, but on the other hand, no effects were observed in IOP changes [192]. In addition, Lee J. and colleagues studied the long-term effects on the visual field, particularly at the superior central, following the supplementation of ginkgo biloba extracts, over 12 years of follow-up in a group of 42 patients with normal tension glaucoma (NTG), demonstrating a slower progression in the evaluation of global indices [193]. Alternatively, the supplementation of omega-3 fatty acids, reported by Garcia-Medina et al., did not show any effective treatment in 117 subjects with mild or moderate POAG and IOP over a 2-year follow-up period [194].

Another study considered a possible association between nutritional supplements as well as supplementation with eicosapentaenoic acid (EPA) and docosahexaenoic acid (DHA), two polyunsaturated fatty acids that are typically found in fatty fish, fish oil, and algae, on age-related macular degeneration (AMD) at intermediate and/or advanced stage in a prospective cohort from the Nurses' Health and the Health Professionals Follow-up Study (75,889 women and 38,961 men, respectively). The authors found that an increase in intake of EPA and DHA could slow down the development of visual dysfunction in the intermediate AMD stage [68]. These results confirmed some beneficial effects obtained by previous double-blind, placebo-controlled studies analyzing the anti-inflammatory and neuroprotective effects of EPA and DHA in glaucoma treatment in a dose-dependent manner. Moreover, the study demonstrated that these effects are mediated by peroxisome proliferator-activated receptors PPAR-α, PPAR-γ and PPAR-δ [195]. Further healthy benefits about nutritional supplements, particularly from saffron, came from two studies to evaluate IOP reduction in 22 and 34 OAG patient cohorts with a short-term follow-up, respectively. The first one published by Hecht and colleagues did not show any hypotensive effect deriving from a supplementation of 1g twice/week of saffron to OAG patients [196], whereas the second study subministered 30 mg/day of aqueous saffron dose, after three weeks of conventional timolol and dorzolamide therapy. In these conditions, an ocular hypotensive effect was evident, confirming the anti-inflammatory and neuroprotective role of saffron against glaucomatous optic neuropathy [197]. These results supported the data obtained in vitro by a study reported by Fernández-Albarral JA and colleagues in a mouse model of chronic ocular hypertension (OHT) [198]. Extra virgin olive oil (EVOO) induces beneficial effects because of both anti-inflammatory and antioxidant properties that are provided by the presence in EVOO of over 30 phenolic compounds, as well as oleuropein, verbascoside, tyrosol, hydroxytyrosol, diosmetin, luteolin and rutin, which counteract the pathological pathways that participate to de progression of glaucomatous degeneration [199]. Notably, the first two aforementioned bioactive compounds, oleuropein and verbascoside, exhibited a significant inhibitory effect at low μM concentrations in vitro against human carbonic anhydrase I and II (hCA I and II) isoenzymes [200], which are therapeutic targets against glaucoma and their inhibition is considered a therapeutic strategy against glaucoma [201]. The enzymatic characterization of natural phenolic compounds as well as flavonoids was useful to obtain a clinical amelioration in visual function, minimizing the risk of ophthalmic artery occlusions for those patients with glaucoma [202]. To investigate whether the impact of high dietary fat and sucrose in animal models was able to induce the injury of retinal ganglion cells (RGCs), Chrysostomou and collaborators demonstrated that C57BL/6J mice fed with a short-term high fat/high sucrose diet were more vulnerable to optic nerve damage and showed higher intraocular pressure following the injection of endotoxin-free saline [203]. Similarly, Kong and collaborators tested the effect of diet restriction (DR, with alternate- fasting plan at least for 6 months) in older (18-month) C57BL/6J mice with an inner retinal dysfunction during and after injury caused by IOP. DR treatment resulted in an appreciable functional recovery of retinal neurons at the inner level and enhanced the mitochondrial activity in the retina of older animals when compared with age-matched control mice. The author found that DR decreased ROS levels and oxidation products as well as the levels of oxidative stress markers (heme oxygenase-1, HO-1, and 4 hydroxynonenal, 4-HNE) compared to IOP mice fed with a normal diet [204]. Additionally, Guo X and colleagues tested an every-other-day fasting (EODF)—a form of caloric restriction-, to assess its effects on glaucomatous pathology in EAAC1-/- mice, an animal model with a normal tension glaucoma. They showed that EODF exhibited a neuroprotective function with an improvement of visual impairment in these mice models by ameliorating RGCs and retinal degeneration without modifying IOP [205]. Although the molecular mechanisms induced by dietary restrictions are not yet entirely elucidated, the observed neuroprotective effects seem to be associated with the induction of autophagy and the improvement of retinal ganglion cells survival, indicating that this cytoprotective process could represent an useful therapeutic strategy in glaucoma following the exposure to hypoxic/ischemic stress [206,207]. Surprisingly, a direct correlation was found between hypertensive patients with elevated IOP and higher levels of melatonin in their aqueous humor compared to the normotensive group. This correlation was previously reported in the experimental glaucomatous model (DBA/2J) compared to control mice (C57BL/6J). The authors speculated that the increase of melatonin in the humor was due to hyperactivation of the transient receptor potential vanilloid-type 4 (TRPV4) cation channel that induced higher melatonin levels. These data suggested that IOP promoted an antioxidant protective response by enhancing melatonin concentrations [208]. Indeed, Melatonin exerts a beneficial effect against glaucoma by blocking the oxidative stress-promoted degeneration of the retinal ganglion cells and is proposed as a therapeutic strategy against glaucoma [209–211]. Further evidence about the association between melatonin levels in both aqueous humor and serum and eye disorders in type 2 diabetic patients derived from the study reported by Aydin and colleagues. They hypothesized that the increase in melatonin levels in the eye of glaucoma patients could be due to intraocular concentration and not by melatonin from the pineal gland [212]. In summary, although a large number of studies regarding the long-term advantages and safety of supplements in glaucoma patients seems to be variable, the possibility to find a useful and synergistic combination of different antioxidants or bioactive compounds against sight-threatening or lifelong diseases could be a promising therapeutic option [213]. Moreover, the beneficial effect of MeDi in the prevention of glaucoma is still debated [181].

10. Conclusions

Certainly, healthy life habits prevent the onset and the progression of chronic diseases. Concerning DR, AMD, and glaucoma, several studies underlined the relevance of MeDi for the prevention of these ocular diseases, mostly by preventing oxidative damage and chronic inflammation. Notably, the studies focusing on the effects of supplements are still controversial, suggesting that the adherence to MeDi and the Mediterranean lifestyle exert a major effect in the prevention of retinal diseases and that supplements can be an adjuvant to MeDi, but cannot substitute for healthy dietary habits in the prevention of such diseases. In addition, the studies focusing on the efficacy of MeDi also provide some controversial results, probably due to the different methods used for calculating MeDi scoring. Notably, such studies directly correlate MeDi adherence with the onset and progression of retinal diseases, but there is not a biomarker that clearly associates the effect of the dietary pattern to the disease progression. The absence of a diet-induced or repressed biomarker is a major problem in correlating the effects of dietary patterns with the risk of developing retinal diseases. Herein, we propose to investigate in more detail the effect of MeDi in lowering the serum levels of AGEs and in promoting the activity/expression of key molecules involved in the Nrf2 pathways, in order to have a clear molecular target that will provide a defined measure of the efficacy of MeDi and a molecular correlation with the progression of retinal diseases.

Author Contributions: Conceptualization, L.P.; search and analysis of bibliography, G.R., F.T. and L.P.; writing—original draft preparation, O.S., F.T., F.P.F. and L.P.; writing—review and editing, O.S., F.T., F.P.F., G.R. and L.P.; supervision, L.P. All authors have read and agreed to the published version of the manuscript.

Funding: This research received no external funding.

Conflicts of Interest: The authors declare no conflicts of interest.

References

1. French, S. *Visual Impairment and Work: Experiences of Visually Impaired People*, 1st ed.; Routledge: London, UK, 2017.
2. Bernabei, V.; Morini, V.; Moretti, F.; Marchiori, A.; Ferrari, B.; Dalmonte, E.; De Ronchi, D.; Atti, A.R. Vision and hearing impairments are associated with depressive–anxiety syndrome in Italian elderly. *Aging Ment. Health* **2011**, *15*, 467–474. [CrossRef] [PubMed]
3. Burton, M.J.; Ramke, J.; Marques, A.P.; Bourne, R.R.A.; Congdon, N.; Jones, I.; Ah Tong, B.A.M.; Arunga, S.; Bachani, D.; Bascaran, C.; et al. The Lancet Global Health Commission on Global Eye Health: Vision beyond 2020. *Lancet Glob. Health* **2021**, *9*, e489–e551. [CrossRef]
4. Gropper, S.S. The Role of Nutrition in Chronic Disease. *Nutrients* **2023**, *15*, 664. [CrossRef] [PubMed]
5. Dobroslavska, P.; Silva, M.L.; Vicente, F.; Pereira, P. Mediterranean Dietary Pattern for Healthy and Active Aging: A Narrative Review of an Integrative and Sustainable Approach. *Nutrients* **2024**, *16*, 1725. [CrossRef] [PubMed]
6. Reichenbach, A.; Bringmann, A. New functions of Müller cells. *Glia* **2013**, *61*, 651–678. [CrossRef] [PubMed]
7. Karl, M.O.; Reh, T.A. Regenerative medicine for retinal diseases: Activating endogenous repair mechanisms. *Trends Mol. Med.* **2010**, *16*, 193–202. [CrossRef]
8. Vecino, E.; Rodriguez, F.D.; Ruzafa, N.; Pereiro, X.; Sharma, S.C. Glia-neuron interactions in the mammalian retina. *Prog. Retin. Eye Res.* **2016**, *51*, 1–40. [CrossRef]
9. Falsini, B.; Placidi, G.; De Siena, E.; Savastano, M.C.; Minnella, A.M.; Maceroni, M.; Midena, G.; Ziccardi, L.; Parisi, V.; Bertelli, M.; et al. USH2A-Related Retinitis Pigmentosa: Staging of disease severity and morphofunctional studies. *Diagnostics* **2021**, *11*, 213. [CrossRef]
10. Falsini, B.; Placidi, G.; De Siena, E.; Chiurazzi, P.; Minnella, A.M.; SAvastano, M.C.; Ziccardi, L.; Parosi, V.; Iarossi, G.; Percio, M.; et al. Genetic characteristics of 234 Italian patients with macular and cone/cone-rod dystrophy. *Sci. Rep.* **2022**, *12*, 3774. [CrossRef]
11. Colombo, L.; Maltese, P.E.; Castori, M.; El Shamieh, S.; Zeitz, C.; Audo, I.; Zulian, A.; Marinelli, C.; Benedetti, S.; Costantini, A.; et al. Molecular epidemiology in 591 italian probands with nonsyndromic retinitis pigmentosa and usher syndrome. *Investig. Ophthalmol. Vis. Sci.* **2021**, *62*, 13. [CrossRef]
12. Sohocki, M.M.; Daiger, S.P.; Bowne, S.J.; Rodriquez, J.A.; Northrup, H.; Heckenlively, J.R.; Birch, D.G.; Mintz-Hittner, H.; Ruiz, R.S.; Lewis, R.A.; et al. Prevalence of mutations causing retinitis pigmentosa and other inherited retinopathies. *Hum. Mutat.* **2001**, *17*, 42–51. [CrossRef] [PubMed]
13. Kumaran, N.; Michaelides, M.; Smith, A.J.; Ali, R.R.; Bainbridge, J.W.B. Retinal gene therapy. *Br. Med. Bull.* **2018**, *126*, 13–25. [CrossRef] [PubMed]
14. Fleckenstein, M.; Keenan, T.D.L.; Guymer, R.H.; Chakravarthy, U.; Schmitz-Valckenberg, S.; Klaver, C.C.; Wong, W.T.; Chew, E.Y. Age-related macular degeneration. *Nat. Rev. Dis. Primers* **2021**, *7*, 31. [CrossRef] [PubMed]
15. Saaddine, J.B.; Honeycutt, A.A.; Narayan, K.M.; Zhang, X.; Klein, R.; Boyle, J.P. Projection of diabetic retinopathy and other major eye diseases among people with diabetes mellitus: United States, 2005–2050. *Arch. Ophthalmol.* **2008**, *126*, 1740–1747. [CrossRef] [PubMed]
16. Perrone, L.; Devi, T.S.; Hosoya, K.C.; Terasaki, T.; Singh, L.P. Thioredoxin Interacting Protein (TXNIP) induces inflammation through chromatin modification in retinal capillary endothelial cells under diabetic conditions. *J. Cell Physiol.* **2009**, *221*, 262–272. [CrossRef] [PubMed]
17. Perrone, L.; Devi, T.S.; Hosoya, K.C.; Terasaki, T.; Singh, L.P. Inhibition of TXNIP Expression In Vivo Blocks Early Pathologies of Diabetic Retinopathy. *Cell Death Dis.* **2010**, *1*, e65. [CrossRef]
18. Amerasinghe, N.; Zhang, J.; Thalamuthu, A.; He, M.; Vithana, E.N.; Viswanathan, A.; Wong, T.Y.; Foster, P.J.; Aung, T. The heritability and sibling risk of angle closure in Asians. *Ophthalmology* **2011**, *118*, 480–485. [CrossRef]
19. Lanzetta, P.; Sarao, V.; Scanlon, P.H.; Barratt, J.; Porta, M.; Bandello, F.; Loewenstein, A. Vision Academy. Fundamental principles of an effective diabetic retinopathy screening program. *Acta Diabetol.* **2020**, *57*, 785–798. [CrossRef]
20. Pondorfer, S.G.; Terheyden, J.H.; Heinemann, M.; Wintergerst, M.W.M.; Holz, F.G.; Finger, R.P. Association of vision-related quality of life with visual function in age-related macular degeneration. *Sci. Rep.* **2019**, *9*, 15326. [CrossRef]

21. Dernini, S.; Berry, E.M. Mediterranean Diet: From a Healthy Diet to a Sustainable Dietary Pattern. *Front. Nutr.* **2015**, *2*, 15. [CrossRef]
22. Bach-Faig, A.; Berry, E.M.; Lairon, D.; Reguant, J.; Trichopoulou, A.; Dernini, S.; Medina, F.X.; Battino, M.; Belahsen, R.; Miranda, G.; et al. Mediterranean diet pyramid today. Science and cultural updates. *Public Health Nutr.* **2011**, *14*, 2274–2284. [CrossRef] [PubMed]
23. Real, H.; Dias, R.R.; Graça, P. Mediterranean Diet Conceptual Model and Future Trends of Its Use in Portugal. *Health Promot. Int.* **2021**, *36*, 548–560. [CrossRef] [PubMed]
24. Bach, A.; Serra-Majem, L.; Carrasco, J.L.; Roman, B.; Ngo, J.; Bertomeu, I.; Obrador, B. The use of indexes evaluating the adherence to the Mediterranean diet in epidemiological studies: A review. *Public Health Nutr.* **2006**, *9*, 132–146. [CrossRef] [PubMed]
25. Naja, F.; Hwalla, N.; Itani, L.; Baalbaki, S.; Sibai, A.; Nasreddine, L. A novel Mediterranean diet index from Lebanon: Comparison with Europe. *Eur. J. Nutr.* **2015**, *54*, 1229–1243. [CrossRef] [PubMed]
26. Trichopoulou, A.; Costacou, T.; Bamia, C.; Trichopoulos, D. Adherence to a Mediterranean diet and survival in a Greek population. *N. Engl. J. Med.* **2003**, *348*, 2599–2608. [CrossRef]
27. Panagiotakos, D.B.; Pitsavos, C.; Stefanadis, C. Dietary patterns: A Mediterranean diet score and its relation to clinical and biological markers of cardiovascular disease risk. *Nutr. Metab. Cardiovasc. Dis.* **2006**, *16*, 559–568. [CrossRef]
28. Buckland, G.; Agudo, A.; Luján, L.; Jakszyn, P.; Bueno-de-Mesquita, H.B.; Palli, D.; Boeing, H.; Carneiro, F.; Krogh, V.; Sacerdote, C.; et al. Adherence to a Mediterranean diet and risk of gastric adenocarcinoma within the European Prospective Investigation into Cancer and Nutrition (EPIC) cohort study. *Am. J. Clin. Nutr.* **2010**, *91*, 381–390. [CrossRef]
29. Schröder, H.; Fitó, M.; Estruch, R.; Martinez-Gonzalez, M.A.; Corella, D.; Salas-Salvado, J.; Lamuela-Raventos, R.; Ros, E.; Salaverria, I.; Fiol, M.; et al. A short screener is valid for assessing Mediterranean diet adherence among older Spanish men and women. *J. Nutr.* **2011**, *141*, 1140–1145. [CrossRef]
30. Agnoli, C.; Grioni, S.; Sieri, S.; Palli, S.; Masala, G.; Sacerdote, C.; Vineis, P.; Tumino, R.; Giurdanella, M.C.; Pala, V.; et al. Italian Mediterranean Index and risk of colorectal cancer in the Italian section of the EPIC cohort. *Int. J. Cancer* **2013**, *132*, 1404–1411. [CrossRef]
31. Monteagudo, C.; Mariscal-Arcas, M.; Rivas, A.; LOrenzo-Tovar, M.L.; Tur, J.A.; Olea-Serrano, F. Proposal of a Mediterranean Diet Serving Score. *PLoS ONE* **2015**, *10*, e0128594. [CrossRef]
32. Sofi, F.; Dinu, M.; Pagliai, G.; Marcucci, R.; Casini, A. Validation of a literature-based adherence score to Mediterranean diet: The MEDI-LITE score. *Int. J. Food Sci. Nutr.* **2017**, *68*, 757–762. [CrossRef] [PubMed]
33. Obeid, C.A.; Gubbels, J.S.; Jaalouk, D.; Kremers, S.P.J.; Oenema, A. Adherence to the Mediterranean diet among adults in Mediterranean countries: A systematic literature review. *Eur. J. Nutr.* **2022**, *61*, 3327–3344. [CrossRef] [PubMed]
34. Perrone, L.; Grant, W.B. Observational and ecological studies of dietary advanced glycation end products in national diets and Alzheimer's disease incidence and prevalence. *J. Alzheimers Dis.* **2015**, *45*, 965–979. [CrossRef] [PubMed]
35. Manach, C.; Morand, C.; Gil-Izquierdo, A.; Bouteloup-Demange, C.; Remesy, C. Bioavailability in humans of the flavanones hesperidin and narirutin after the ingestion of two doses of orange juice. *Eur. J. Clin. Nutr.* **2003**, *57*, 235–242. [CrossRef] [PubMed]
36. Borriello, A.; Cucciolla, V.; Della Ragione, F.; Galletti, P. Dietary polyphenols: Focus on resveratrol, a promising agent in the prevention of cardiovascular diseases and control of glucose homeostasis. *Nutr. Metab. Cardiovasc. Dis.* **2010**, *20*, 618–625. [CrossRef] [PubMed]
37. Manach, C.; Williamson, G.; Morand, C.; Scalbert, A.; Remesy, C. Bioavailability and bioefficacy of polyphenols in humans. I. Review of 97 bioavailability studies. *Am. J. Clin. Nutr.* **2005**, *81*, 230S–242S. [CrossRef]
38. Williamson, G.; Manach, C. Bioavailability and bioefficacy of polyphenols in humans. II. Review of 93 intervention studies. *Am. J. Clin. Nutr.* **2005**, *81*, 243S–255S. [CrossRef]
39. Sies, H.; Stahl, W.; Sundquist, A.R. Antioxidant functions of vitamins. Vitamins E and C.; beta-carotene, and other carotenoids. *Ann. N. Y. Acad. Sci.* **1992**, *669*, 7–20. [CrossRef]
40. Fernández-García, E.; Carvajal-Lérida, I.; Jarén-Galán, M.; Garrido-Fernandez, J. Carotenoids bioavailability from foods: From plant pigments to efficient biological activities. *Food Res. Int.* **2012**, *46*, 438–450. [CrossRef]
41. Palozza, P.; Catalano, A.; Simone, R.; Cittadini, A. Lycopene as a guardian of redox signalling. *Acta Biochim. Pol.* **2012**, *59*, 21–25. [CrossRef]
42. Aggarwal, B.B.; Sundaram, C.; Prasad, S.; Kannappan, R. Tocotrienols, the vitamin E of the 21st century: Its potential against cancer and other chronic diseases. *Biochem. Pharmacol.* **2010**, *80*, 1613–1631. [CrossRef] [PubMed]
43. Nag, S.A.; Qin, J.J.; Wang, W.; Wang, M.-H.; Wang, H.; Zhang, R. Ginsenosides as Anticancer Agents: In vitro and in vivo Activities, Structure-Activity Relationships, and Molecular Mechanisms of Action. *Front. Pharmacol.* **2012**, *3*, 25. [CrossRef] [PubMed]
44. Dongowski, G.; Huth, M.; Gebhardt, E.; Flamme, W. Dietary fiber-rich barley products beneficially affect the intestinal tract of rats. *J. Nutr.* **2002**, *132*, 3704–3714. [CrossRef] [PubMed]
45. Drzikova, B.; Dongowski, G.; Gebhardt, E. Dietary fibre-rich oat-based products affect serum lipids, microbiota, formation of short-chain fatty acids and steroids in rats. *Br. J. Nutr.* **2005**, *94*, 1012–1025. [CrossRef] [PubMed]
46. Granfeldt, Y.; Nyberg, L.; Bjorck, I. Muesli with 4 g oat beta-glucans lowers glucose and insulin responses after a bread meal in healthy subjects. *Eur. J. Clin. Nutr.* **2008**, *62*, 600–607. [CrossRef] [PubMed]

47. Choi, J.S.; Kim, H.; Jung, M.H.; Hong, S.; Song, J. Consumption of barley betaglucan ameliorates fatty liver and insulin resistance in mice fed a high-fat diet. *Mol. Nutr. Food Res.* **2010**, *54*, 1004–1013. [CrossRef] [PubMed]
48. Brown, L.; Rosner, B.; Willett, W.W.; Sacks, F.M. Cholesterol-lowering effects of dietary fiber: A meta-analysis. *Am. J. Clin. Nutr.* **1999**, *69*, 30–42. [CrossRef]
49. Olano-Martin, E.; Gibson, G.R.; Rastell, R.A. Comparison of the in vitro bifidogenic properties of pectins and pectic-oligosaccharides. *J. Appl. Microbiol.* **2002**, *93*, 505–511. [CrossRef]
50. Behall, K.M.; Scholfield, D.J.; Hallfrisch, J.G.; Lijeberg-Elmstahl, H.G. Consumption of both resistant starch and beta-glucan improves postprandial plasma glucose and insulin in women. *Diabetes Care* **2006**, *29*, 976–981. [CrossRef]
51. Lattimer, J.M.; Haub, M.D. Effects of dietary fiber and its components on metabolic health. *Nutrients* **2010**, *2*, 1266–1289. [CrossRef]
52. Uribarri, J.; Cai, W.; Peppa, M.; Goodman, S.; Ferrucci, L.; Striker, G.; Vlassara, H. Circulating glycotoxins and dietary advanced glycation endproducts: Two links to inflammatory response, oxidative stress, and aging. *J. Gerontol. A Biol. Sci. Med. Sci.* **2007**, *62*, 427–433. [CrossRef] [PubMed]
53. Mahaling, B.; Low, S.W.Y.; Beck, M.; Kumar, D.; Ahmed, S.; Connor, T.B.; Ahmad, B.; Chaurasia, S.S. Damage-Associated Molecular Patterns (DAMPs) in Retinal Disorders. *Int. J. Mol. Sci.* **2022**, *23*, 2591. [CrossRef] [PubMed]
54. Riemersma, R.A.; Wood, D.A.; Butler, S.; Elton, R.A.; Oliver, M.; Salo, M.; Nikkari, T.; Vartiainen, E.; Puska, P.; Gey, F.; et al. Linoleic acid content in adipose tissue and coronary heart disease. *Br. Med. J. (Clin. Res. Ed.)* **1986**, *292*, 1423–1427. [CrossRef] [PubMed]
55. Djousse, L.; Folsom, A.R.; Province, M.A.; Hunt, S.C.; Ellison, R.C. Dietary linolenic acid and carotid atherosclerosis: The National Heart, Lung, and Blood Institute Family Heart Study. *Am. J. Clin. Nutr.* **2003**, *77*, 819–825. [CrossRef]
56. Wijendran, V.; Hayes, K.C. Dietary n-6 and n-3 fatty acid balance and cardiovascular health. *Annu. Rev. Nutr.* **2004**, *24*, 597–615. [CrossRef]
57. Burdge, G. Alpha-linolenic acid metabolism in men and women: Nutritional and biological implications. *Curr. Opin. Clin. Nutr. Metab. Care* **2004**, *7*, 137–144. [CrossRef]
58. Wall, R.; Ross, R.P.; Fitzgerald, G.F.; Stanton, C. Fatty acids from fish: The antiinflammatory potential of long-chain omega-3 fatty acids. *Nutr. Rev.* **2010**, *68*, 280–289. [CrossRef]
59. Orsavova, J.; Misurcova, L.; Ambrozova, J.V.; Vicha, R.; Micek, J. Fatty Acids Composition of Vegetable Oils and Its Contribution to Dietary Energy Intake and Dependence of Cardiovascular Mortality on Dietary Intake of Fatty Acids. *Int. J. Mol. Sci.* **2015**, *16*, 12871–12890. [CrossRef]
60. Maggini, S.; Pierre, A.; Calder, P.C. Immune function and Micronutrient requirements change over the Life Course. *Nutrients* **2018**, *10*, 1531. [CrossRef]
61. Sharma, Y.; Saxena, S.; Mishra, A.; Saxena, A. Nutrition for diabetic retinopathy: Plummeting the inevitable threat of diabetic vision loss. *Eur. J. Nutr.* **2017**, *56*, 2013–2027. [CrossRef]
62. Cammalleri, M.; Dal Monte, M.; Amato, R.; Bagnoli, P.; Rusciano, D. A dietary combination of forskolin with homotaurine, spearmint and b vitamins protects injured retinal ganglion cells in a rodent model of hypertensive glaucoma. *Nutrients* **2020**, *12*, 1189. [CrossRef] [PubMed]
63. Meng, X.; Li, Y.; Li, S.; Zhou, Y.; Gan, R.Y.; Xu, D.P.; Li, H.B. Dietary sources and bioactivities of melatonin. *Nutrients* **2017**, *9*, 367. [CrossRef] [PubMed]
64. Maccarone, R.; Di Marco, S.; Bisti, S. Saffron supplement maintains morphology and function after exposure to damaging light in mammalian retina. *Investig. Ophthal Vis. Sci.* **2008**, *49*, 1254–1261. [CrossRef] [PubMed]
65. Rashid, M.; Rashid, R.; Saroya, S.; Deverapalli, M.; Brim, H.; Ashktorab, H. Saffron as a Promising Therapy for Inflammatory Bowel Disease. *Nutrients* **2024**, *16*, 2353. [CrossRef]
66. Castelli, V.; Paladini, A.; d'Angelo, M.; Allegretti, M.; Mantelli, F.; Brandolini, L.; Cocchiaro, P.; Cimini, A.; Varrassi, G. Taurine and oxidative stress in retinal health and disease. *CNS Neurosci. Ther.* **2021**, *27*, 403–412. [CrossRef] [PubMed]
67. García-Ayuso, D.; Di Pierdomenico, J.; Hadj-Said, W.; Marie, M.; Agudo-Barriuso, M.; Vidal-Sanz, M.; Picaud, S.; Villegas-Perez, M.P. Taurine depletion causes ipRGC loss and increases light-induced photoreceptor degeneration. *Investig. Ophthalmol. Vis. Sci.* **2018**, *59*, 1396. [CrossRef]
68. Wu, J.; Cho, E.; Giovannucci, E.L.; Rosner, B.A.; Sastry, S.M.; Willett, W.C.; Schaumberg, D.A. Dietary intakes of eicosapentaenoic acid and docosahexaenoic acid and risk of age-related macular degeneration. *Ophthalmology* **2017**, *124*, 634–643. [CrossRef] [PubMed]
69. Zhang, J.; Tuo, J.; Wang, Z.; Zhu, A.; Machalinska, A.; Long, Q. Pathogenesis of Common Ocular Diseases. *J. Ophthalmol.* **2015**, *2015*, 734527. [CrossRef]
70. Pinazo-Durán, M.; Gallego-Pinazo, R.; García-Medina, J.J.; Zanon-Moreno, V.; Nucci, C.; Dolz-Marco, R.; Martinez-Castillo, S.; Galbis-Estrada, C.; Marco-Ramirez, C.; Lopez-Galvez, M.I.; et al. Oxidative stress in aging eyes. *Clin. Interv. Aging* **2014**, *9*, 637–652. [CrossRef]
71. Pinazo-Durán, M.D.; Zanón-Moreno, V.; García-Medina, J.J.; Arevalo, J.F.; Gallego-Pinazo, R.; Nucci, C. Eclectic Ocular Comorbidities and Systemic Diseases with Eye Involvement: A Review. *BioMed Res. Int.* **2016**, *2016*, 6215745. [CrossRef]
72. Nita, M.; Grzybowski, A. The Role of the Reactive Oxygen Species and Oxidative Stress in the Pathomechanism of the Age-Related Ocular Diseases and Other Pathologies of the Anterior and Posterior Eye Segments in Adults. *Oxidative Med. Cell Longev.* **2016**, *2016*, 3164734. [CrossRef] [PubMed]

73. Torrisi, F.; D'Aprile, S.; Denaro, S.; Pavone, A.M.; Alberghina, C.; Zappalà, A.; Giuffrida, R.; Salvatorelli, L.; Broggi, G.; Magro, G.G.; et al. Epigenetics and Metabolism Reprogramming Interplay into Glioblastoma: Novel Insights on Immunosuppressive Mechanisms. *Antioxidants* **2023**, *12*, 220. [CrossRef] [PubMed]
74. Salleh, M.R. Life event, stress and illness. *Malays. J. Med. Sci.* **2008**, *15*, 9–18. [PubMed]
75. Zarkovic, N. Roles and Functions of ROS and RNS in Cellular Physiology and Pathology. *Cells* **2020**, *9*, 767. [CrossRef]
76. Yau, Y.H.C.; Potenza, M.N. Stress and eating behaviors. *Minerva Endocrinol.* **2013**, *38*, 255–267.
77. He, F.; Ru, X.; Wen, T. NRF2, a Transcription Factor for Stress Response and Beyond. *Int. J. Mol. Sci.* **2020**, *21*, 4777. [CrossRef] [PubMed]
78. Baird, L.; Yamamoto, M. The Molecular Mechanisms Regulating the KEAP1-NRF2 Pathway. *Mol. Cell Biol.* **2020**, *40*, e00099–e00200. [CrossRef] [PubMed]
79. Ngo, V.; Duennwald, M.L. Nrf2 and Oxidative Stress: A General Overview of Mechanisms and Implications in Human Disease. *Antioxidants* **2022**, *11*, 2345. [CrossRef]
80. Vasconcelos, A.R.; Dos Santos, N.B.; Scavone, C.; Demarchi Munhoz, C. Nrf2/ARE Pathway Modulation by Dietary Energy Regulation in Neurological Disorders. *Front. Pharmacol.* **2019**, *10*, 33. [CrossRef]
81. Zhao, F.; Ci, X.; Man, X.; Li, J.; Wie, Z.; Zhang, S. Food-Derived Pharmacological Modulators of the Nrf2/ARE Pathway: Their Role in the Treatment of Diseases. *Molecules* **2021**, *26*, 1016. [CrossRef]
82. Zong, Y.; Li, H.; Liao, P.; Chen, L.; Pan, Y.; Zheng, Y.; Zhang, C.; Liu, D.; Zheng, M.; Gao, J. Mitochondrial dysfunction: Mechanisms and advances in therapy. *Signal Transduct. Target. Ther.* **2024**, *9*, 124. [CrossRef] [PubMed]
83. Spinelli, J.B.; Haigis, M.C. The multifaceted contributions of mitochondria to cellular metabolism. *Nat. Cell Biol.* **2018**, *20*, 745–754. [CrossRef] [PubMed]
84. Catalani, E.; Brunetti, K.; Del Quondam, S.; Cervia, D. Targeting Mitochondrial Dysfunction and Oxidative Stress to Prevent the Neurodegeneration of Retinal Ganglion Cells. *Antioxidants* **2023**, *12*, 2011. [CrossRef]
85. Dinkova-Kostova, A.T.; Abramov, A.Y. The emerging role of Nrf2 in mitochondrial function. *Free Radic. Biol. Med.* **2015**, *88*, 179–188. [CrossRef]
86. Zhang, J.; Zhang, T.; Zeng, S.; Zhang, X.; Zhou, F.; Gillies, M.V.; Zhu, L. The Role of Nrf2/sMAF Signalling in Retina Ageing and Retinal Diseases. *Biomedicines* **2023**, *11*, 1512. [CrossRef]
87. Liu, X.-F.; Zhou, D.-D.; Xie, T.; Hao, J.-L.; Malik, T.H.; Lu, C.-B.; Qi, J.; Pant, O.P.; Lu, C.-W. The Nrf2 Signaling in Retinal Ganglion Cells under Oxidative Stress in Ocular Neurodegenerative Diseases. *Int. J. Biol. Sci.* **2018**, *14*, 1090–1098. [CrossRef] [PubMed]
88. Kang, Q.; Yang, C. Oxidative stress and diabetic retinopathy: Molecular mechanisms, pathogenetic role and therapeutic implications. *Redox Biol.* **2020**, *37*, 101799. [CrossRef]
89. Calkins, M.J.; Johnson, D.A.; Townsend, J.A.; Vargas, M.R.; Dowell, J.A.; Williamson, T.P.; Kraft, A.D.; Lee, J.-M.; Jiang, L.; Johnson, J.A. The Nrf2/ARE pathway as a potential therapeutic target in neurodegenerative disease. *Antioxid. Redox Signal.* **2009**, *11*, 497–508. [CrossRef] [PubMed]
90. Scapagnini, G.; Vasto, S.; Abraham, N.G.; Caruso, C.; Zella, D.; Fabio, G. Modulation of Nrf2/ARE pathway by food polyphenols: A nutritional neuroprotective strategy for cognitive and neurodegenerative disorders. *Mol. Neurobiol.* **2011**, *44*, 192–201. [CrossRef]
91. Pérez, R.; Burgos, V.; Marín, V.; Camins, A.; Olloquequi, J.; Gonzales-Chavarria, I.; Ulrich, H.; Wyneke, U.; Luarte, A.; Ortiz, L.; et al. Caffeic Acid Phenethyl Ester (CAPE): Biosynthesis, Derivatives and Formulations with Neuroprotective Activities. *Antioxidants* **2023**, *12*, 1500. [CrossRef]
92. Lecomte, S.; Demay, F.; Ferrière, F.; Pakdel, F. Phytochemicals Targeting Estrogen Receptors: Beneficial Rather Than Adverse Effects? *Int. J. Mol. Sci.* **2017**, *18*, 1381. [CrossRef] [PubMed]
93. Chew, L.Y.; Zhang, H.; He, J.; Yu, F. The Nrf2-Keap1 pathway is activated by steroid hormone signaling to govern neuronal remodeling. *Cell Rep.* **2021**, *36*, 109466. [CrossRef]
94. Pontelli, R.C.N.; Souza, M.C.O.; Fantucci, M.Z.; de Andrade, M.; Rocha, E.M. The role of endocrine disruptors in ocular surface diseases. *Med. Hypotheses* **2019**, *122*, 157–164. [CrossRef] [PubMed]
95. Nuzzi, R.; Scalabrin, S.; Becco, A.; Panzica, G. Gonadal Hormones and Retinal Disorders: A Review. *Front. Endocrinol.* **2018**, *9*, 66. [CrossRef] [PubMed]
96. Tillmann, A.; Ceklic, L.; Dysli, C.; Munk, M.R. Gender differences in retinal diseases: A review. *Clin. Exp. Ophthalmol.* **2024**, *52*, 317–333. [CrossRef]
97. Wang, M.; Ren, Y.; Zheng, H.; Zhang, J.; Zhou, X. Correlative study on retinal microvascular changes and sex hormones in male patients with central serous chorioretinopathy. *J. Men's Health* **2023**, *19*, 71–76.
98. Ishii, T.; Warabi, E. Mechanism of Rapid Nuclear Factor-E2-Related Factor 2 (Nrf2) Activation via Membrane-Associated Estrogen Receptors: Roles of NADPH Oxidase 1, Neutral Sphingomyelinase 2 and Epidermal Growth Factor Receptor (EGFR). *Antioxidants* **2019**, *8*, 69. [CrossRef]
99. Chu, C.; Gao, X.; Li, X.; Zhang, X.; Ma, R.; Jia, Y.; Li, D.; Wang, D.; Xu, F. Involvement of Estrogen Receptor-α in the Activation of Nrf2-Antioxidative Signaling Pathways by Silibinin in Pancreatic β-Cells. *Biomol. Ther.* **2020**, *28*, 163–171. [CrossRef]
100. Ionescu, V.S.; Popa, A.; Alexandru, A.; Manole, E.; Neagu, M.; Pop, S. Dietary Phytoestrogens and Their Metabolites as Epigenetic Modulators with Impact on Human Health. *Antioxidants* **2021**, *10*, 1893. [CrossRef]

101. Liu, T.; Li, N.; Yan, Y.-Q.; Liu, Y.; Xiong, K.; Liu, Y.; Xia, Q.-M.; Zhang, H.; Liu, Z.-D. Recent advances in the anti-aging effects of phytoestrogens on collagen, water content, and oxidative stress. *Phytother. Res.* **2020**, *34*, 435–447. [CrossRef]
102. Zhang, T.; Liang, X.; Shi, L.; Wang, L.; Chen, J.; Kang, C.; Zhu, J.; Mi, M. Estrogen receptor and PI3K/Akt signaling pathway involvement in S-(-)equol-induced activation of Nrf2/ARE in endothelial cells. *PLoS ONE* **2013**, *8*, e79075. [CrossRef] [PubMed]
103. Lin, K.Y.; Hsih, W.H.; Lin, Y.B.; Wen, C.Y.; Chang, T.J.J. Update in the epidemiology, risk factors, screening, and treatment of diabetic retinopathy. *Diabetes Investig.* **2021**, *12*, 1322–1325. [CrossRef] [PubMed]
104. Wong, T.Y.; Cheung, C.M.; Larsen, M.; Sharma, S.; Simo', R. Diabetic retinopathy. *Nat. Rev. Dis. Primers* **2016**, *2*, 16012. [CrossRef] [PubMed]
105. Serikbaeva, A.; Li, Y.; Ma, S.; Yi, D.; Kazlauskas, A. Resilience to diabetic retinopathy. *Prog. Retin. Eye Res.* **2024**, *101*, 101271. [CrossRef]
106. Ahmed, T.S.; Shah, J.; Zhen, Y.N.B.; Chua, J.; Wong, D.W.K.; Nusinovici, S.; Tan, R.; Tan, G.; Schmetterer, L.; Tan, B. Ocular microvascular complications in diabetic retinopathy: Insights from machine learning. *BMJ Open Diabetes Res. Care* **2024**, *12*, e003758. [CrossRef] [PubMed]
107. Kadłubowska, J.; Malaguarnera, L.; Wąż, P.; Zorena, K. Neurodegeneration and Neuroinflammation in Diabetic Retinopathy: Potential Approaches to Delay Neuronal Loss. *Curr. Neuropharmacol.* **2016**, *14*, 831–839. [CrossRef]
108. Dow, C.; Mancini, F.; Rajaobelina, K.; Boutron-Ruault, M.C.; Balkau, B.; Bonnet, F.; Fagherazzi, G. Diet and risk of diabetic retinopathy: A systematic review. *Eur. J. Epidemiol.* **2018**, *33*, 141–156. [CrossRef]
109. Evert, A.B.; Dennison, M.; Gardner, C.D.; Garvey, W.T.; Lau, K.H.K.; MacLeod, J.; Mitri, J.; Pereira, R.F.; Rawlings, K.; Robinson, S.; et al. Nutrition Therapy for Adults With Diabetes or Prediabetes: A Consensus Report. *Diabetes Care* **2019**, *42*, 731–754. [CrossRef]
110. Sala-Vila, A.; Díaz-López, A.; Valls-Pedret, C.; Cofan, M.; Garcia-Layana, A.; Lamuela-Raventos, R.M.; Castaner, O.; Zanon-Moreno, V.; Martinez-Gonzales, M.A.; Toledo, E.; et al. Prevención con Dieta Mediterránea (PREDIMED) Investigators. *JAMA Ophthalmol.* **2016**, *134*, 1142–1149. [CrossRef]
111. Valero-Vello, M.; Peris-Martínez, C.; García-Medina, J.J.; Sanz-Gonzalez, S.M.; Ramirez, A.I.; Fernandez-Albarral, J.A.; Galarreta-Mira, D.; Zanon-Moreno, V.; Casaroli-Marano, R.P.; Pinazo-Duran, M.D. Searching for the antioxidant, anti-inflammatory, and neuroprotective potential of natural food and nutritional supplements for ocular health in the mediterranean population. *Foods* **2021**, *10*, 1231. [CrossRef]
112. Díaz-López, A.; Babio, N.; Martínez-González, M.A.; Corella, D.; Amor, A.J.; Fito', M.; Estruch, R.; Aros, F.; Gomez-Gracia, E.; Fiol, M.; et al. Mediterranean Diet, Retinopathy, Nephropathy, and Microvascular Diabetes Complications: A Post Hoc Analysis of a Randomized Trial. *Diabetes Care* **2015**, *38*, 2134–2141. [CrossRef] [PubMed]
113. Chew, E.Y. Dietary Intake of Omega-3 Fatty Acids From Fish and Risk of Diabetic Retinopathy. *JAMA* **2017**, *317*, 2226–2227. [CrossRef] [PubMed]
114. Bryl, A.; Mrugacz, M.; Falkowski, M.; Zorena, K. A Mediterranean Diet May Be Protective in the Development of Diabetic Retinopathy. *Int. J. Mol. Sci.* **2023**, *24*, 11145. [CrossRef] [PubMed]
115. Zarei, M. Dual antiplatelet therapy may lead to hemorrhagic complications in vitrectomy of proliferative diabetic retinopathy. *Graefes Arch. Clin. Exp. Ophthalmol.* **2019**, *257*, 425–426. [CrossRef]
116. Yuan, D.; Xu, Y.; Xue, L.; Zhang, W.; Gu, L.; Liu, Q. Resveratrol protects against diabetic retinal ganglion cell damage by activating the Nrf2 signaling pathway. *Heliyon* **2024**, *10*, e30786. [CrossRef]
117. Abu-Amero, K.K.; Kondkar, A.A.; Chalam, K.V. Resveratrol and Ophthalmic Diseases. *Nutrients* **2016**, *8*, 200. [CrossRef]
118. Mirmiran, P.; Hosseinpour-Niazi, S.; Naderi, Z.; Bahadoran, Z.; Sadeghi, M.; Azizi, F. Association between interaction and ratio of omega-3 and omega-6 polyunsaturated fatty acid and the metabolic syndrome in adults. *Nutrition* **2012**, *28*, 856–863. [CrossRef]
119. Kawasaki, R.; Tanaka, S.; Tanaka, S.; Yamamoto, T.; Sone, H.; Ohashi, Y.; Akanuma, Y.; Yamada, N.; Yamashita, H.; Japan Diabetes Complications Study Group. Incidence and progression of diabetic retinopathy in Japanese adults with type 2 diabetes: 8 year follow-up study of the Japan Diabetes Complications Study (JDCS). *Diabetologia* **2011**, *54*, 2288–2294. [CrossRef] [PubMed]
120. Brazionis, L.; Rowley, K.; Itsiopoulos, C.; O'Dea, K. Plasma carotenoids and diabetic retinopathy. *Br. J. Nutr.* **2009**, *101*, 270–277. [CrossRef]
121. Garcia-Medina, J.J.; Pinazo-Duran, M.D.; Garcia-Medina, M.; Zanon-Moreno, V.; Pons-Vazquez, S. A 5-year follow-up of antioxidant supplementation in type 2 diabetic retinopathy. *Eur. J. Ophthalmol.* **2011**, *21*, 637–643. [CrossRef]
122. Tanaka, S.; Yoshimura, Y.; Kawasaki, R.; Kamada, C.; Tanaka, S.; Horikawa, C.; Ohashi, Y.; Araki, A.; Ito, H.; Akamura, Y.; et al. Fruit intake and incident diabetic retinopathy with type 2 diabetes. *Epidemiology* **2013**, *24*, 204–211. [CrossRef] [PubMed]
123. Post, R.E.; Mainous, A.G., 3rd; King, D.E.; Simpson, K.N. Dietary fiber for the treatment of type 2 diabetes mellitus: A meta-analysis. *J. Am. Board. Fam. Med.* **2012**, *25*, 16–23. [CrossRef]
124. Bryl, A.; Mrugacz, M.; Mariak, Z. Blood flow in vessels supplying the eye in persons with degenerative myopia. Part II. Blood flow in the central retinal artery. *Klin. Ocz.* **2013**, *115*, 222–225.
125. Chatziralli, I.P.; Theodossiadis, G.; Dimitriadis, P.; Charalambidis, M.; Agorastos, A.; Migkos, Z.; Platogiannis, N.; Moschos, M.M.; Theodossiadis, P.; Kerytopoulos, P. The Effect of Vitamin E on Oxidative Stress Indicated by Serum Malondialdehyde in Insulin-dependent Type 2 Diabetes Mellitus Patients with Retinopathy. *Open Ophthalmol. J.* **2017**, *11*, 51–58. [CrossRef]

126. Bursell, S.E.; Clermont, A.C.; Aiello, L.P.; Aiello, L.M.; Schlossman, D.K.; Feener, E.P.; Laffel, L.; King, G.L. High-dose vitamin E supplementation normalizes retinal blood flow and creatinine clearance in patients with type 1 diabetes. *Diabetes Care* **1999**, *22*, 1245–1251. [CrossRef] [PubMed]
127. Simo, R.; Hernandez, C. What else can we do to prevent diabetic retinopathy? *Diabetologia* **2023**, *66*, 1614–1621. [CrossRef]
128. Barba, I.; Garcia-Ramirez, M.; Hernandez, C.; Alonso, M.A.; Masmiquel, L.; Garcia-Dorado, D.; Simo, R. Metabolic fingerprints of proliferative diabetic retinopathy: An 1H-NMR-based metabonomic approach using vitreous humor. *Investig. Ophthalmol. Vis. Sci.* **2010**, *51*, 4416–4421. [CrossRef] [PubMed]
129. Park, S.W.; Ghim, W.; Oh, S.; Kim, Y.; Park, U.C.; Kang, J.; Yu, H.G. Association of vitreous vitamin C depletion with diabetic macular ischemia in proliferative diabetic retinopathy. *PLoS ONE* **2019**, *14*, e0218433. [CrossRef] [PubMed]
130. Shah, J.; Cheong, Z.Y.; Tan, B.; Wong, D.; Liu, X.; Chua, J. Dietary Intake and Diabetic Retinopathy: A Systematic Review of the Literature. *Nutrients* **2022**, *14*, 5021. [CrossRef]
131. Keenan, T.D.L.; Cukras, C.A.; Chew, E.Y. Age-Related Macular Degeneration: Epidemiology and Clinical Aspects. *Adv. Exp. Med. Biol.* **2021**, *1256*, 1–31.
132. Hogg, R.E.; Woodside, J.V.; McGrath, A.; Young, I.S.; Vioque, J.L.; Chakravarthy, U.; de Jong, P.T.; Rahu, M.; Seland, J.; Soubrane, G.; et al. Mediterranean Diet Score and Its Association with Age-Related Macular Degeneration: The European Eye Study. *Ophthalmology* **2017**, *124*, 82–89. [CrossRef] [PubMed]
133. Mares, J.A.; Voland, R.P.; Sondel, S.A.; Millen, A.E.; Larowe, T.; Moeller, S.M.; Klein, M.L.; Blodi, B.A.; Chappell, R.J.; Tinker, L.; et al. Healthy lifestyles related to subsequent prevalence of age-related macular degeneration. *Arch. Ophthalmol.* **2011**, *129*, 470–480. [CrossRef] [PubMed]
134. Merle, B.M.J.; Colijn, J.M.; Cougnard-Grégoire, A.; de KOnig-Backus, A.P.M.; Delyfer, M.-N.; Kiefte-de Jong, J.C.; Meester-Smoor, M.; Feart, C.; Verzijden, T.; Samieri, C.; et al. Mediterranean Diet and Incidence of Advanced Age-Related Macular Degeneration: The EYE-RISK Consortium. *Ophthalmology* **2019**, *126*, 381–390. [CrossRef] [PubMed]
135. Merle, B.M.J.; Rosner, B.; Seddon, J.M. Genetic Susceptibility, Diet Quality, and Two-Step Progression in Drusen Size. *Investig. Ophthalmol. Vis. Sci.* **2020**, *61*, 17. [CrossRef] [PubMed]
136. Merle, B.M.; Silver, R.E.; Rosner, B.; Seddon, J.M. Adherence to a Mediterranean diet, genetic susceptibility, and progression to advanced macular degeneration: A prospective cohort study. *Am. J. Clin. Nutr.* **2015**, *102*, 1196–1206. [CrossRef] [PubMed]
137. Agrón, E.; Mares, J.; Chew, E.Y.; Keenan, T.D.L.; AREDS2 Research Group. Adherence to a Mediterranean Diet and Geographic Atrophy Enlargement Rate: Age-Related Eye Disease Study 2 Report 29. *Ophthalmol. Retin.* **2022**, *6*, 762–770. [CrossRef]
138. Yiu, G.; Chiu, S.J.; Petrou, P.A.; Stinnett, S.; Sarin, N.; Farsiu, S.; Chew, E.Y.; Wong, W.T.; Toth, C.A. Relationship of central choroidal thickness with age-related macular degeneration status. *Am. J. Ophthalmol.* **2015**, *159*, 617–626. [CrossRef]
139. Zheng, Y.; Lamoureux, E.L.; Lavanya, R.; Wu, R.; Ikram, M.K.; Wang, J.J.; Mitchell, P.; Cheung, N.; Aung, T.; saw, S.-M.; et al. Prevalence and risk factors of diabetic retinopathy in migrant Indians in an urbanized society in Asia: The Singapore Indian eye study. *Ophthalmology* **2012**, *119*, 2119–2124. [CrossRef]
140. Nonyane, B.A.S.; Nitsch, D.; Whittaker, J.C.; Sofat, R.; Smeeth, L.; Chakravarthy, U.; Fletcher, A.E. An ecological correlation study of late age-related macular degeneration and the complement factor H Y402H polymorphism. *Investig. Ophthalmol. Vis. Sci.* **2010**, *51*, 2393–2402. [CrossRef]
141. Wang, J.J.; Buitendijk, G.H.; Rochtchina, E.; Lee, K.E.; Klein, B.E.K.; van Duijn, C.M.; Flood, V.M.; Meuer, S.M.; Attia, J.; Myers, C.; et al. Genetic susceptibility, dietary antioxidants, and long-term incidence of age-related macular degeneration in two populations. *Ophthalmology* **2014**, *121*, 667–675. [CrossRef]
142. Fung, T.T.; McCullough, M.L.; Newby, P.K.; Manson, J.E.; Meigs, J.B.; Rifai, N.; Willett, W.C.; Hu, F.B. Diet-quality scores and plasma concentrations of markers of inflammation and endothelial dysfunction. *Am. J. Clin. Nutr.* **2005**, *82*, 163–173. [CrossRef] [PubMed]
143. Colijn, J.M.; den Hollander, A.I.; Demirkan, A.; Cougnard-Gregoire, A.; Verzijden, T.; Kersten, E.; Meester-Smoor, M.A.; Merle, B.M.J.; Papageorgiou, G.; Ahmad, S.; et al. Increased High-Density Lipoprotein Levels Associated with Age-Related Macular Degeneration: Evidence from the EYE-RISK and European Eye Epidemiology Consortia. *Ophthalmology* **2019**, *126*, 393–406. [CrossRef] [PubMed]
144. Jiang, H.; Shi, X.; Fan, Y.; Wang, D.; Li, B.; Zhou, J.; Pei, C.; Ma, L. Dietary omega-3 polyunsaturated fatty acids and fish intake and risk of age-related macular degeneration. *Clin. Nutr.* **2021**, *40*, 5662–5673. [CrossRef] [PubMed]
145. Seddon, J.M.; Rosner, B.; Sperduto, R.D.; Yannuzzi, L.; Haller, J.A.; Blair, N.P.; Willett, W. Dietary fat and risk for advanced age-related macular degeneration. *Arch. Ophthalmol.* **2001**, *119*, 1191–1199. [CrossRef]
146. Augood, C.; Chakravarthy, U.; Young, I.; Vioque, J.; de Jong, P.T.V.M.; Bentham, G.; Rahu, M.; Seland, J.; Soubrane, G.; Tomazzoli, L.; et al. Oily fish consumption, dietary docosahexaenoic acid and eicosapentaenoic acid intakes, and associations with neovascular age-related macular degeneration. *Am. J. Clin. Nutr.* **2008**, *88*, 398–406. [CrossRef]
147. Athar, M.; Back, J.H.; Tang, X.; Kim, K.H.; Kopelovich, L.; Bickers, D.R.; Kim, A.L. Resveratrol: A review of preclinical studies for human cancer prevention. *Toxicol. Appl. Pharmacol.* **2007**, *224*, 274–283. [CrossRef]
148. Anekonda, T.S.; Adamus, G. Resveratrol prevents antibody-induced apoptotic death of retinal cells through upregulation of Sirt1 and Ku70. *BMC Res. Notes* **2008**, *1*, 122. [CrossRef]
149. Bryl, A.; Falkowski, M.; Zorena, K.; Mrugacz, M. The Role of Resveratrol in Eye Diseases—A Review of the Literature. *Nutrients* **2022**, *14*, 2974. [CrossRef]

150. Richer, S.; Stiles, W.; Ulanski, L.; Carroll, D.; Podella, C. Observation of human retinal remodeling in octogenarians with a resveratrol based nutritional supplement. *Nutrients* **2013**, *5*, 1989–2005. [CrossRef]
151. Kijlstra, A.; Tian, Y.; Kelly, E.R.; Berendschot, T.T.J.M. Lutein: More than just a filter for blue light. *Prog. Retin. Eye Res.* **2012**, *31*, 303–315. [CrossRef]
152. Koushan, K.; Rusovici, R.; Li, W.; Ferguson, L.R.; Chalam, K.V. The role of lutein in eye-related disease. *Nutrients* **2013**, *5*, 1823–1839. [CrossRef] [PubMed]
153. Barker, F.M., 2nd; Snodderly, D.M.; Johnson, E.J.; Schalch, W.; Koepcke, W.; Gerss, J.; Neuringer, M. Nutritional manipulation of primate retinas, V: Effects of lutein, zeaxanthin, and n-3 fatty acids on retinal sensitivity to blue-light-induced damage. *Investig. Ophthalmol. Vis. Sci.* **2011**, *52*, 3934–3942. [CrossRef] [PubMed]
154. Woodall, A.A.; Lee, S.W.; Weesie, R.J.; Jackson, M.J.; Britton, G. Oxidation of carotenoids by free radicals: Relationship between structure and reactivity. *Biochim. Biophys. Acta* **1997**, *1336*, 33–42. [CrossRef]
155. Sundelin, S.P.; Nilsson, S.E. Lipofuscin-formation in retinal pigment epithelial cells is reduced by antioxidants. *Free Radic. Biol. Med.* **2001**, *31*, 217–225. [CrossRef] [PubMed]
156. Schalch, W.; Cohn, W.; Barker, F.M.; Kopcke, W.; Mellerio, J.; Bird, A.C.; Robson, A.G.; Fitzke, F.F.; van Kuijk, F.J.G.M. Xanthophyll accumulation in the human retina during supplementation with lutein or zeaxanthin—The LUXEA (LUtein Xanthophyll Eye Accumulation) study. *Arch. Biochem. Biophys.* **2007**, *458*, 128–135. [CrossRef] [PubMed]
157. Johnson, E.J.; McDonald, K.; Caldarella, S.M.; Chung, H.-Y.; Troen, A.M.; Snodderly, D.M. Cognitive findings of an exploratory trial of docosahexaenoic acid and lutein supplementation in older women. *Nutr. Neurosci.* **2008**, *11*, 75–83. [CrossRef]
158. Weigert, G.; Kaya, S.; Pemp, B.; Sacu, S.; Lasta, M.; Werkmeister, R.M.; Dragostinoff, N.; Simader, C.; Garhofer, G.; Schmidt-Erfurth, U.; et al. Effects of lutein supplementation on macular pigment optical density and visual acuity in patients with age-related macular degeneration. *Investig. Ophthalmol. Vis. Sci.* **2011**, *52*, 8174–8178. [CrossRef]
159. Schmidl, D.; Garhofer, G.; Schmetterer, L. Nutritional supplements in age-related macular degeneration. *Acta Ophthalmol.* **2015**, *93*, 105–121. [CrossRef]
160. Eisenhauer, B.; Natoli, S.; Liew, G.; Flood, V.M. Lutein and Zeaxanthin-Food Sources, Bioavailability and Dietary Variety in Age-Related Macular Degeneration Protection. *Nutrients* **2017**, *9*, 120. [CrossRef]
161. Seddon, J.M.; Ajani, U.A.; Sperduto, R.D.; Hiller, R.; Blair, N.; Burton, T.C.; Farber, M.D.; Gragoudas, E.S.; Haller, J.; Miller, D.T.; et al. Dietary carotenoids, vitamins A.; C.; and E.; and advanced age-related macular degeneration. Eye Disease Case-Control Study Group. *JAMA* **1994**, *272*, 1413–1420. [CrossRef]
162. Evans, J.R.; Lawrenson, J.G. Antioxidant vitamin and mineral supplements for preventing age-related macular degeneration. *Cochrane Database Syst. Rev.* **2017**, *7*, CD000253. [CrossRef] [PubMed]
163. SanGiovanni, J.P.; Chew, E.Y.; Clemons, T.E.; Davis, M.D.; Ferris, F.L., 3rd; Gensler, G.R.; Kurinij, N.; Lindblad, A.S.; Milton, R.C.; Seddon, J.M.; et al. The relationship of dietary lipid intake and age-related macular degeneration in a case-control study: AREDS Report No. 20. *Arch. Ophthalmol.* **2007**, *125*, 671–679. [PubMed]
164. Wiegand, R.D.; Joel, C.D.; Rapp, L.M.; Nielsen, J.C.; Maude, M.B.; Anderson, R.E. Polyunsaturated fatty acids and vitamin E in rat rod outer segments during light damage. *Investig. Ophthalmol. Vis. Sci.* **1986**, *27*, 727–733.
165. Tanito, M.; Yoshida, Y.; Kaidzu, S.; Chen, Z.-H.; Cynshi, O.; Jishage, K.-I.; Niki, E.; Ohira, A. Acceleration of age-related changes in the retina in alpha-tocopherol transfer protein null mice fed a Vitamin E-deficient diet. *Investig. Ophthalmol. Vis. Sci.* **2007**, *48*, 396–404. [CrossRef]
166. West, S.; Vitale, S.; Hallfrisch, J.; Munoz, B.; Muller, D.; Bressler, S.; Bressler, N.M. Are antioxidants or supplements protective for age-related macular degeneration? *Arch. Ophthalmol.* **1994**, *112*, 222–227. [CrossRef]
167. Delcourt, C.; Cristol, J.P.; Tessier, F.; Leger, C.L.; Descomps, B.; Papoz, L. Age-related macular degeneration and antioxidant status in the POLA study. POLA Study Group. Pathologies Oculaires Liees a l'Age. *Arch. Ophthalmol.* **1999**, *117*, 1384–1390. [CrossRef] [PubMed]
168. Taylor, H.R.; Tikellis, G.; Robman, L.D.; McCarty, C.A.; McNeil, J.J. Vitamin E supplementation and macular degeneration: Randomised controlled trial. *BMJ* **2002**, *325*, 11. [CrossRef] [PubMed]
169. Teikari, J.M.; Laatikainen, L.; Rapola, J.M.; Virtamo, J.; Haukka, J.; Liesto, K.; Heinonen, O.P. Retinal vascular changes following supplementation with alpha-tocopherol or beta-carotene. *Acta Ophthalmol. Scand.* **1998**, *76*, 68–73. [CrossRef]
170. Sun, Y.; Chen, A.; Zou, M.; Zhang, Y.; Jin, L.; Li, Y.; Zheng, D.; Jin, G.; Congdon, N. Time trends, associations and prevalence of blindness and vision loss due to glaucoma: An analysis of observational data from the Global Burden of Disease Study 2017. *BMJ Open* **2022**, *12*, e053805. [CrossRef]
171. Heijl, A.; Leske, M.C.; Bengtsson, B.; Hyman, L.; Bengtsson, B.; Hussein, M.; Early Manifest Glaucoma Trial Group. Reduction of intraocular pressure and glaucoma progression: Results from the early manifest glaucoma trial. *Arch. Ophthalmol.* **2002**, *120*, 1268–1279. [CrossRef]
172. Society, E.G. European Glaucoma Society Terminology and Guidelines for Glaucoma, 5th Edition. *Br. J. Ophthalmol.* **2021**, *105*, 1–169.
173. Tham, Y.C.; Li, X.; Wong, T.Y.; Quigley, H.A.; Aung, T.; Cheng, C.Y. Global prevalence of glaucoma and projections of glaucoma burden through 2040: A systematic review and meta-analysis. *Ophthalmology* **2014**, *121*, 2081–2090. [CrossRef] [PubMed]
174. Quigley, H.A. Neuronal death in glaucoma. *Prog. Retin. Eye Res.* **1999**, *18*, 39–57. [CrossRef]
175. Kang, J.M.; Tanna, A.P. Glaucoma. *Med. Clin. North. Am.* **2021**, *105*, 493–510. [CrossRef] [PubMed]

176. Hayashi, K.; Hayashi, H.; Nakao, F.; Hayashi, F. Intraocular lens tilt and decentration after implantation in eyes with glaucoma. *J. Cataract. Refract. Surg.* **1999**, *25*, 1515–1520. [CrossRef]
177. Pérez-De-Arcelus, M.; Toledo, E.; Martínez-González, M.Á.; Martin-Calvo, N.; Fernandez-Montero, A.; Moreno-Montanes, J. Smoking and incidence of glaucoma. The SUN Cohort. *Medicine* **2017**, *96*, e5761. [CrossRef]
178. Kreft, D.; Doblhammer, G.; Guthoff, R.F.; Frech, S. Prevalence, incidence, and risk factors of primary open-angle glaucoma—A cohort study based on longitudinal data from a German public health insurance. *BMC Public Health* **2019**, *19*, 851. [CrossRef]
179. Yadav, K.S.; Rajpurohit, R.; Sharma, S. Glaucoma: Current treatment and impact of advanced drug delivery systems. *Life Sci.* **2019**, *221*, 362–376. [CrossRef]
180. Vergroesen, J.E.; de Crom, T.O.E.; van Duijn, C.M.; Voortman, T.; Klaver, C.C.W.; Ramdas, W.D. MIND diet lowers risk of open-angle glaucoma: The Rotterdam Study. *Eur. J. Nutr.* **2023**, *62*, 477–487. [CrossRef]
181. Wu, Y.; Xie, Y.; Yuan, Y.; Xiong, R.; Hu, Y.; Ning, K.; Ha, J.; Wang, W.; Han, X.; He, M. The Mediterranean Diet and Age-Related Eye Diseases: A Systematic Review. *Nutrients* **2023**, *15*, 2043. [CrossRef]
182. Moreno-Montañés, J.; Gándara, E.; Gutierrez-Ruiz, I.; Moreno-Galarraga, L.; Ruiz-Canela, M.; Bes-Rastrollo, M.; Martinez-Gonzalez, M.A.; Fernandez-Montero, A. Healthy Lifestyle Score and Incidence of Glaucoma: The Sun Project. *Nutrients* **2022**, *14*, 779. [CrossRef] [PubMed]
183. Wareham, L.K.; Buys, E.S.; Sappington, R.M. The nitric oxide-guanylate cyclase pathway and glaucoma. *Nitric Oxide Biol. Chem.* **2018**, *77*, 75–87. [CrossRef] [PubMed]
184. Shannon, O.M.; Stephan, B.C.M.; Minihane, A.M.; Mathers, J.C.; Siervo, M. Nitric Oxide Boosting Effects of the Mediterranean Diet: A Potential Mechanism of Action. *J. Gerontol. Ser. Biol. Sci. Med. Sci.* **2018**, *73*, 902–904. [CrossRef] [PubMed]
185. Reina-Torres, E.; De Ieso, M.L.; Pasquale, L.R.; Madekurozwa, M.; van Batenburg-Sherwood, J.; Overby, D.R.; Stamer, W.D. The vital role for nitric oxide in intraocular pressure homeostasis. *Prog. Retin. Eye Res.* **2021**, *83*, 100922. [CrossRef]
186. Kang, J.H.; Willett, W.C.; Rosner, B.A.; Buys, E.; Wiggs, J.L.; Pasquale, L.R. Association of dietary nitrate intake with primary open-angle glaucoma: A prospective analysis from the nurses' health study and health professionals follow-up study. *JAMA Ophthalmol.* **2016**, *134*, 294–303. [CrossRef] [PubMed]
187. Abreu-Reyes, J.A.; Álvarez-Luis, D.; Arteaga-Hernández, V.; Sanchez-Mendez, M.; Abreu-Gonzalez, R. Mediterranean diet adherence by patients with primary open angle glaucoma Adherencia a la dieta mediterránea en pacientes afectos de glaucoma primario de ángulo abierto. *Arch. Soc. Esp. Oftalmol.* **2017**, *92*, 353–358. [CrossRef]
188. Mvitu, M.; Longo-Mbenza, B.; Tulomba, M.D.; Nge, A. Reduced risk of metabolic syndrome due to regular intake of vegetables rich in antioxidants among African type 2 diabetics. *Diabetes Metab. Syndr. Clin. Res. Rev.* **2010**, *4*, 132–136. [CrossRef]
189. Rusciano, D.; Pezzino, S.; Mutolo, M.G.; Giannotti, R.; Librando, A.; Pescosolido, N. Neuroprotection in Glaucoma: Old and New Promising Treatments. *Adv. Pharmacol. Sci.* **2017**, *2017*, 4320408. [CrossRef]
190. D'Angelo, A.; Vitiello, L.; Lixi, F.; Abbinante, G.; Coppola, A.; Gagliardi, V.; Pellegrino, A.; Giannaccare, G. Optic Nerve Neuroprotection in Glaucoma: A Narrative Review. *J. Clin. Med.* **2024**, *13*, 2214. [CrossRef]
191. Ramdas, W.D. The relation between dietary intake and glaucoma: A systematic review. *Acta Ophthalmol.* **2018**, *96*, 550–556. [CrossRef]
192. Ohguro, H.; Ohguro, I.; Katai, M.T.; Tanaka, S. Two-Year Randomized, Placebo-Controlled Study of Black Currant Anthocyanins on Visual Field in Glaucoma. *Ophthalmologica* **2012**, *228*, 26–35. [CrossRef] [PubMed]
193. Lee, J.; Sohn, S.W.; Kee, C. Effect of Ginkgo Biloba Extract on Visual Field Progression in Normal Tension Glaucoma. *J. Glaucoma* **2013**, *22*, 780–784. [CrossRef] [PubMed]
194. Garcia-Medina, J.J.; Garcia-Medina, M.; Garrido-Fernandez, P.; Galvan-Espinosa, J.; Garcia-Maturana, C.; Zanon-Moreno, V.; Pinazo-Duran, M.D. A Two-Year Follow-up of Oral Antioxidant Supplementation in Primary Open-Angle Glaucoma: An Open-Label, Randomized, Controlled Trial. *Acta Ophthalmol.* **2015**, *93*, 546–554. [CrossRef]
195. Keppel Hesselink, J.M.; Costagliola, C.; Fakhry, J.; Kopsky, D.J. Palmitoylethanolamide, a natural retinoprotectant: Its putative relevance for the treatment of glaucoma and diabetic retinopathy. *J. Ophthalmol.* **2015**, *2015*, 430596. [CrossRef] [PubMed]
196. Hecht, I.; Achiron, A.; Bartov, E.; Maharshak, I.; Mendel, L.; Peer, L.; Bar, A.; Burgansky-Eliash, Z. Effects of dietary and lifestyle recommendations on patients with glaucoma: A randomized controlled pilot trial. *Eur. J. Integr. Med.* **2019**, *25*, 60–66. [CrossRef]
197. Jabbarpoor Bonyadi, M.H.; Yazdani, S.; Saadat, S. The ocular hypotensive effect of saffron extract in primary open angle glaucoma: A pilot study. *BMC Complement. Altern. Med.* **2014**, *14*, 399. [CrossRef]
198. Fernández-Albarral, J.A.; Ramírez, A.I.; de Hoz, R.; Lopez-Villarin, N.; Salobrar-Garcia, E.; Lopez-Cuenca, I.; Licastro, E.; Inarejos-Garcia, A.M.; Almodovar, P.; Pinazo-Duran, M.; et al. Neuroprotective and anti-inflammatory effects of a hydrophilic saffron extract in a model of glaucoma. *Int. J. Mol. Sci.* **2019**, *20*, 4110. [CrossRef]
199. Bucciantini, M.; Leri, M.; Nardiello, P.; Casamenti, F.; Stefani, M. Olive Polyphenols: Antioxidant and Anti-Inflammatory Properties. *Antioxidants* **2021**, *10*, 1044. [CrossRef]
200. Aggul, A.G.; Taslimi, P.; Kuzu, M.; Uzun, N.; Bilginer, S.; Gulcin, I. Gulcin, Oleuropein and verbascoside—Their inhibition effects on carbonic anhydrase and molecular docking studies. *J. Oleo Sci.* **2021**, *70*, ess21106. [CrossRef]
201. Scozzafava, A.; Supuran, C.T. Glaucoma and the applications of carbonic anhydrase inhibitors. *Subcell. Biochem.* **2014**, *75*, 349–359.
202. Patel, S.; Mathan, J.J.; Vaghefi, E.; Braakhuis, A.J. The effect of flavonoids on visual function in patients with glaucoma or ocular hypertension: A systematic review and meta-analysis. *Graefes Arch. Clin. Exp. Ophthalmol.* **2015**, *253*, 1841–1850. [CrossRef] [PubMed]

203. Chrysostomou, V.; van Wijngaarden, P.; Steinberg, G.R.; Crowston, J.G. A short term high-fat high-sucrose diet in mice impairs optic nerve recovery after injury and this is not reversed by exercise. *Exp. Eye Res.* **2017**, *162*, 104–109. [CrossRef] [PubMed]
204. Kong, Y.X.; van Bergen, N.; Bui, B.V.; Chrysostomou, V.; Vingrys, A.J.; Trounce, I.A.; Crowston, J.G. Impact of aging and diet restriction on retinal function during and after acute intraocular pressure injury. *Neurobiol. Aging* **2012**, *33*, e15–e25. [CrossRef]
205. Guo, X.; Kimura, A.; Azuchi, Y.; Akiyama, G.; Noro, T.; Harada, C.; Namekata, K.; Harada, T. Caloric restriction promotes cell survival in a mouse model of normal tension glaucoma. *Sci. Rep.* **2016**, *6*, 33950. [CrossRef]
206. Li, M.; Gao, Z.L.; Zhang, Q.P.; Luo, A.-X.; Xu, W.-Y.; Duan, T.-Q.; Wen, X.-P.; Zhang, R.-Q.; Zeng, R.; Huang, J.-F. Autophagy in glaucoma pathogenesis: Therapeutic potential and future perspectives. *Front. Cell Dev. Biol.* **2022**, *10*, 1068213. [CrossRef] [PubMed]
207. Russo, R.; Varano, G.P.; Adornetto, A.; Nazio, F.; Tettamanti, G.; Girardello, R.; Cianfanelli, V.; Cavaliere, F.; Morrone, L.A.; Corasaniti, M.T.; et al. Rapamycin and fasting sustain autophagy response activated by ischemia/reperfusion injury and promote retinal ganglion cell survival. *Cell Death Dis.* **2018**, *9*, 981. [CrossRef]
208. Alkozi, H.; Sánchez-Naves, J.; de Lara, M.J.; Carracedo, G.; Fonseca, B.; Martinez-Auguila, A.; Pintor, J. Elevated intraocular pressure increases melatonin levels in the aqueous humour. *Acta Ophthalmol.* **2017**, *95*, e185–e189. [CrossRef]
209. Hu, C.; Feng, Y.; Huang, G.; Cui, K.; Fan, M.; Xiang, W.; Shi, Y.; Ye, D.; Ye, H.; Bai, X.; et al. Melatonin prevents EAAC1 deletion-induced retinal ganglion cell degeneration by inhibiting apoptosis and senescence. *J. Pineal Res.* **2024**, *76*, e12916. [CrossRef]
210. Agoraatos, A.; Huber, C.G. The role of melatonin in glaucoma: Implications concerning pathophysiological relevance and therapeutic potential. *J. Pineal Res.* **2011**, *50*, 1–7. [CrossRef]
211. Romeo, A.; Kazsoki, A.; Musumeci, T.; Zelkó, R. A Clinical, Pharmacological, and Formulation Evaluation of Melatonin in the Treatment of Ocular Disorders-A Systematic Review. *Int. J. Mol. Sci.* **2024**, *25*, 3999. [CrossRef]
212. Aydin, E.; Sahin, S. Increased melatonin levels in aqueous humor of patients with proliferative retinopathy in type 2 diabetes mellitus. *Int. J. Ophthalmol.* **2016**, *9*, 721–724. [PubMed]
213. Garcia-Medina, J.J.; Rubio-Velazquez, E.; Lopez-Bernal, M.D.; Cobo-Martinez, A.; Zanon-Moreno, V.; Pinazo-Duran, M.D.; Del-Rio-Vellosillo, M. Glaucoma and Antioxidants: Review and Update. *Antioxidants* **2020**, *9*, 1031. [CrossRef] [PubMed]

nutrients

MDPI

Review

Diet and Nutrients in Rare Neurological Disorders: Biological, Biochemical, and Pathophysiological Evidence

Marilena Briglia [1,†], Fabio Allia [1,†], Rosanna Avola [1,†], Cinzia Signorini [2], Venera Cardile [3], Giovanni Luca Romano [1], Giovanni Giurdanella [1,*], Roberta Malaguarnera [1], Maria Bellomo [1] and Adriana Carol Eleonora Graziano [1,*]

[1] Department of Medicine and Surgery, "Kore" University of Enna, 94100 Enna, Italy; marilena.briglia@unikore.it (M.B.); fabio.allia@unikore.it (F.A.); rosanna.avola@unikore.it (R.A.); giovanniluca.romano@unikore.it (G.L.R.); roberta.malaguarnera@unikore.it (R.M.); maria.bellomo@unikore.it (M.B.)

[2] Department of Molecular and Developmental Medicine, University of Siena, 53100 Siena, Italy; cinzia.signorini@unisi.it

[3] Department of Biomedical and Biotechnological Sciences, University of Catania, 95123 Catania, Italy; cardile@unict.it

[*] Correspondence: giovanni.giurdanella@unikore.it (G.G.); adriana.graziano@unikore.it (A.C.E.G.)

[†] These authors equally contributed to this work.

Abstract: Background/Objectives: Rare diseases are a wide and heterogeneous group of multisystem life-threatening or chronically debilitating clinical conditions with reduced life expectancy and a relevant mortality rate in childhood. Some of these disorders have typical neurological symptoms, presenting from birth to adulthood. Dietary patterns and nutritional compounds play key roles in the onset and progression of neurological disorders, and the impact of alimentary needs must be enlightened especially in rare neurological diseases. This work aims to collect the *in vitro*, *in vivo*, and clinical evidence on the effects of diet and of nutrient intake on some rare neurological disorders, including some genetic diseases, and rare brain tumors. Herein, those aspects are critically linked to the genetic, biological, biochemical, and pathophysiological hallmarks typical of each disorder. **Methods**: By searching the major web-based databases (PubMed, Web of Science Core Collection, DynaMed, and Clinicaltrials.gov), we try to sum up and improve our understanding of the emerging role of nutrition as both first-line therapy and risk factors in rare neurological diseases. **Results**: In line with the increasing number of consensus opinions suggesting that nutrients should receive the same attention as pharmacological treatments, the results of this work pointed out that a standard dietary recommendation in a specific rare disease is often limited by the heterogeneity of occurrent genetic mutations and by the variability of pathophysiological manifestation. **Conclusions**: In conclusion, we hope that the knowledge gaps identified here may inspire further research for a better evaluation of molecular mechanisms and long-term effects.

Keywords: rare neurological disorders; nutritional compounds; dietary pattern; rare leukodystrophy; rare tumors; orphan disease; neurodegenerative disease

1. Introduction

According to the World Health Organization (WHO), neurological disorders (NDs) are actually the leading cause of illness and disability worldwide [1,2]. A recent study shows that in the last decades, the number of people suffering from nervous system (NS) disorders (43% of the world's population—3.4 billion people—affected in 2021) has sharply increased [3] and, even if the statistics report that tension-type headache and migraine represented the main disability-causing diseases, among the fatal conditions directly attributable to nervous system health loss, there were neonatal encephalopathy, Alzheimer's disease and other dementias, meningitis, epilepsy, and nervous system cancer. The nervous system (NS) is vulnerable to various disorders and can be damaged by inflammatory processes or by immunological-mediated mechanisms, genetic disorders, traumatic injuries, and/or cancers. Moreover, neurological health loss is often a consequence of various circumstances, such as neonatal (premature birth, jaundice, and sepsis) and congenital conditions (birth defects and chromosomal abnormalities), infectious diseases (COVID-19, echinococcosis, cystic, syphilis, malaria, and Zika virus disease), or comorbidity with metabolic disorders (diabetic neuropathy) [4]. Although the etiology is heterogeneous, the NDs show systemic and long-term organic disabilities with a high incidence of death, as supported by meta-analysis studies [5]. In this context, one of the most ambitious purposes of global health systems is to improve the quality of life for people with NDs by reducing the influence of NDs, as well as their associated mortality, morbidity, and disability [6], as established by the World Health Assembly with the Global Intersectoral Action Plan on Epilepsy and Other Neurological Disorders 2022–2031 (IGAP).

Although there is a high incidence of neurological disorders, unfortunately, to date, there are few effective cures. Despite the many advances made with the implementation of therapeutic solutions, there are problems and limitations that do not guarantee full control of neurological symptoms [7], suggesting that dietary nutrition may represent a useful tool. The dietary pattern and nutritional compounds play a key role in ensuring general physical wellness and specific neurological function. Daily dietary intake provides nutrients and molecules to support life and health maintenance through its general supply of substrates of homeostasis. Dysphagia, movement disorders, cognitive impairment, and depression associated with neurological disorders can directly or indirectly affect the nutritional status of patients. For instance, malnutrition causes delays in rehabilitation and induces an increase in both mortality and morbidity. Proper nutritional education could be a crucial intervention in ND treatment to minimize the risks of malnutrition with likely repercussions on the worsening of symptoms related to neurological diseases [8].

In recent years, epidemiological data have revealed the importance of a correlation between both healthy food intake and lifestyle and a clear reduction in the risk of central nervous system (CNS) diseases. These findings make diet and lifestyle interesting points on which we can focus the intervention research field [9,10]. Particularly, it has been scientifically proven that the intake of certain foods and nutrients guarantees clinical benefits for CNS disease. Currently, research on brain health and the effects of nutraceuticals receives high interest due to the potential neuroprotective effects.

In this work, we focus our attention on the effects of diet and/or nutrient intake on some rare diseases with neurological symptoms. The term "Rare diseases" (RDs) refers to a wide and heterogeneous group of multisystem life-threatening or chronically debilitating clinical conditions. They often correlate with high medical costs, poor quality of life, reduced life expectancy, and a significant mortality rate in childhood. Examples include genetic diseases, rare cancers, and infectious tropical diseases.

The only finding that RDs have in common is their low prevalence and low frequency in the population. According to the definition by the WHO, an RD is an illness or condition that occurs from 0.65 to 1 case per 1000 population, with a prevalence from 6.5 to 10 cases per 10.000 residents. Thus, their classification criterion is usually purely epidemiological. Thus, even if many pathologies reach just a prevalence of 0.001%, (1 case per 100,000), taken together, RDs affect approximately 6–8% of the world's population, and often, the rarity of a particular disease limits the process of drug development for economic factors [11].

In this background, we hypothesized that basic research, clinical practice, and evidence-based medicine could furnish us with the data to investigate new therapeutic perspectives by establishing a direct link between diet and neurological outcomes in a specific "rare" molecular context. Therefore, in this narrative review, we first introduce readers to the topic of rare diseases by focusing on rare neurological disorders. Then, we analyzed the concept of "nutritional approaches" in neurological illness, trying to summarize the main methodologies applied for the study of nutritional interventions in rare neurological disorders with a critical evaluation of the limits and strengths of *in vitro* and *in vivo* models and of clinical studies.

These preliminary indications supported the last part of the work in which we analyzed the scientific literature by reviewing the *in vitro*, *in vivo*, and clinical evidence on the effects of diet and of nutrient intake on some rare neurological disorders, including rare developmental disorders (Angelman Syndrome and Rett Syndrome), rare leukodystrophies (Krabbe disease and Pelizaeus–Merzbacher disease), rare genetic epilepsy, rare forms of ataxia, and rare brain tumors. Specifically, we review the evidence available on the following web-based databases: PubMed, Web of Science Core Collection, and DynaMed. Moreover, for clinical trials, we searched on clinicaltrials.gov. Globally, to focus our search, we applied the Boolean operator "AND" to combine the condition/disease with specific terms (diet, or nutrition, or nutrients). When a specific term turned results in line with the aims of this review, closer research was performed by the inclusion of the identified key work.

We try to sum up and improve our understanding of the emerging role of nutrition as both first-line therapy and risk factors in rare neurological diseases, in line with the increasing number of consensus opinions suggesting that nutrients should receive the same attention as pharmacological treatments. The limits, the challenges, and the knowledge gaps are identified in order to inspire further research for a better evaluation in the future.

2. Rare Disorders: Lights and Shadows

The term "Rare diseases" (RDs) refers to a narrow subset of diseases affecting a small group in the general population with a wide spectrum of clinical conditions. The findings shared by all the RDs are the low prevalence and low frequency in the population; they are orphan diseases (neglected diseases) [12]. Unfortunately, only a small group of these diseases can be predicted in possible treatment, in terms of relieving the symptoms and improving the quality of life [13]. From a medical perspective, rare neurological disorders include many rather complex and heterogeneous conditions, for which knowledge of natural history, diagnosis, prevention, neurobiology, and progress in treatment/drug strategies is almost entirely limited [14]. The designation is linked to disease prevalence and severity and the existence of alternative therapeutic options, but it varies across jurisdictions. A disease is listed as rare when its prevalence conventionally does not exceed a certain threshold. This latter varies among countries that adopt slightly different parameters. In the European Union (EU), this threshold is fixed at 0.05% of the population, (1 in 2000 people or 5 per 10,000). In the United States (US), a disease is considered rare when it does not exceed the prevalence of a 0.08% threshold and consequently, it affects fewer than 200,000 patients in the country (6.4 in 10,000 people); whereas, in Japan, a disease is rare if it includes less than 50,000 cases (4 cases per 10,000). Despite the low prevalence, the number of rare diseases for which no treatment is available is estimated to be between 6000 and 8000 worldwide. It is worrying that approximately 6% of the world's population is affected by potentially harmful and/or lethal rare diseases; and it is expected that the value could grow with the progress of science and, in particular, with advances in genetic research. Especially in conditions with a genetic etiology, next-generation sequencing techniques and the sequencing of the whole exome could play a key role [15,16]. Data show that a high prevalence of rare diseases has a genetic origin and mainly involves children [17]. It would seem, at least in part, that environmental exposure during pregnancy or in the early stages of life may affect genetic vulnerability. The main cause lies in rare tumors, autoimmune diseases, toxic and infectious diseases, or congenital malformations [18].

RDs constitute a priority public health topic by the European Commission and the agencies for public health due to their high number and their complex management that involves not only medical but also social issues [19]. In fact, RDs share a number of common features related to social and health burdens. Among these, the poor awareness among healthcare professionals and often the inadequate levels of care knowledge are consequences of their low prevalence. Often, all these findings turn into delayed diagnosis, misdiagnosis, or even un-diagnosis and troubles in disease management. Scientific evidence indicates that the delay in diagnosis is due to the lack of knowledge about the topic and to the gradual manifestation over time of clinical characteristics. In many cases, genetic testing would be necessary [20,21]. Moreover, the restricted market addressed to the single conditions reduces the attractivity of the pharmaceutical industry to invest in Research and Development (R&D) for new drugs and treatments that have huge development costs, resulting perhaps in the most expensive produced drugs [22]. For all these reasons, extraordinary support in providing high-quality information on rare diseases by increasing knowledge and ensuring improvement in healthcare and research is carried out by several international health committees. For example, the Orphanet provides many data and specialist services around the world through the ORPHA code that classifies rare diseases [23]. Moreover, the European Reference Network for Rare Neurological Diseases [24], the European Reference Network [25], and the ITHACA network (on rare congenital malformations and rare intellectual disabilities) [26] provide excellent support to the cooperation of researchers and health professionals in this field.

Rare Neurological Disorders

Rare neurological diseases could have an hereditary, post-infectious, iatrogenic, or unknown etiology. These pathological conditions cause damage to the brain, spinal cord, or peripheral nerves. Since the nervous system controls more functions of the body, depending on the pathology, the symptoms of a rare neurological disease can be many, and they range from mild tremors to severe motor and cognitive damage. Usually, it has been commonly described as the progressive cognitive decline toward dementia associated with personality changes and psychiatric disorders, the loss of coordination, mobility, muscle strength, and balance with tremors and dystonia, as well as vision and hearing problems and involuntary movements [27].

According to the Orphanet, the term "Rare neurological disorders" does not characterize a disease but a group of diseases with genetic causes (ORPHA:71859). The number of rare neurological disorders that are also orphan diseases is about 8000, a number which tends to increase monthly with advances in genetic science and research [28]. The identification of rare disorders as well as the rare expression of common disorders associated with these pathological conditions is particularly difficult, even if technological, financial, and social changes have taken place in recent decades [29]. To better characterize patient populations and delimit target populations, natural history studies are highly recommended. These latter are epidemiological studies focusing on frequency description and the characteristics and evolution of disease by collecting real data from groups of patients affected by these diseases. These studies are often performed early in the clinical development process to support and guide the design of clinical trials and drug development [29–31].

Therapy is often supportive. In recent years, pre-clinical and/or clinical studies have addressed more attention toward personalized medicine and global health with specific attention dedicated to nutritional interventions as both first-line therapy and risk factors, suggesting that nutrients and alimentary wellness should receive the same attention as pharmacological treatments [32–34]. Figure 1 reports the main characteristics of rare neurological diseases and possible investigations for symptomatic treatments.

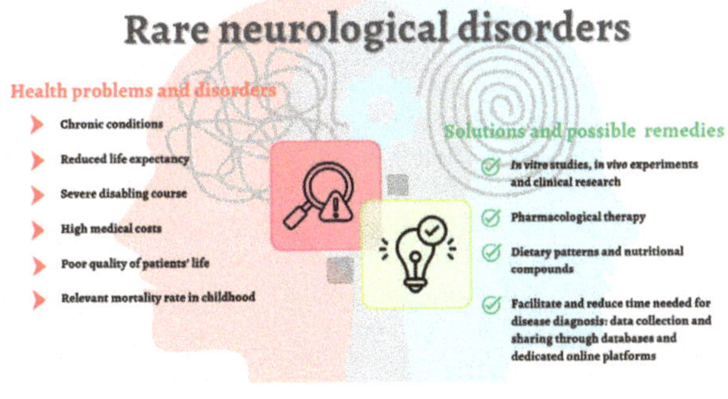

Figure 1. Main characteristics of rare neurological diseases and possible investigations for symptomatic treatments (created with BioRender.com, accessed on 3 September 2024).

3. The Alimentary Wellness for a Global Health

A varied and balanced diet is the basis of a healthy life. On the other hand, an unbalanced diet affects psycho-physical human health and represents one of the main risk factors for the onset of chronic diseases. In the last two decades, lifestyle and wrong eating habits have heavily affected the increased prevalence of metabolic diseases such as diabetes, obesity, cardiovascular disease, and fatty liver [35–38]. To address this growing public problem, health organizations have provided recommendations on proper nutrition intake [39].

Recent advances in research have improved the understanding of metabolism and highlighted the active involvement of nutrients and their metabolites in the regulation of gene expression and cell functions.

The assimilation and transformation of food including the action of nutrients and non-nutritive components are closely related to extrinsic factors (e.g., food, xenobiotics, and environment), intrinsic factors (e.g., gender, age, and gene changes), and the host/microbiota interaction (Figure 2) [33,40,41]. The aforementioned factors influence nutrient metabolism as well as the risk of developing various metabolic diseases. Extrinsic factors play an important role in the metabolic activity of nutrients and in ensuring a healthy condition.

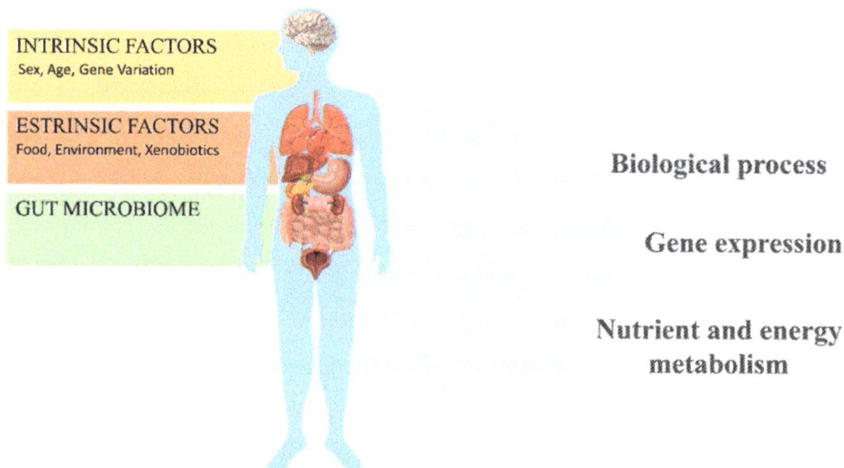

Figure 2. Factors that influence alimentary wellness by modulator effects on the main homeostatic functions (created with BioRender.com, accessed on 27 June 2024).

For instance, physical factors such as photoperiod and temperature are involved in the regulation of metabolic activity. This latter is influenced by circadian endogenous rhythms [42]. Exposure to harmful environmental conditions induces the release of stress hormones, which can impair the body's ability to perceive and respond to metabolic changes [43–45]. Moreover, extrinsic factors can stimulate the alterations of the epigenome with consequent lasting effects on the energy and nutrient metabolism, determining the development of metabolic disorders such as in the heart (coronary heart disease) and in the brain (for example, Alzheimer's disease) [46–48]. The intrinsic factors, including gene changes, gender, and age, influence the proper functioning of metabolic pathways [49].

It is clear that the modulation of central nervous system homeostasis by nutrients could be both a risk factor for the development of neurological disease [50] as well as an opportunity to improve the health status by nutritional interventions.

3.1. Nutritional Interventions: What Are They?

Nutritional interventions were defined as "purposefully planned actions intended to positively change a nutrition-related behavior, environmental condition, or aspect of health status" [51,52]. Any specific strategy designed to improve the health status of a patient by modifying its dietary habits is a nutritional intervention. The suggestion of specific nutrient intake or the consumption of a particular food might bring some benefits regarding CNS health. In clinical terms, nutritional approaches can be categorized into two major types: nutrient supplementation and dietary modifications.

Some clinical benefits on metabolic health, neural function, and longevity seem to be related to nutritional interventions in cases of neurological conditions with high healthcare and social costs.

In this section, a brief overview of nutritional approaches for the most common neurological diseases was conducted in order to better compare this scenario with that of rare neurological disorders.

3.1.1. The Nutrient Supplementation in Neurological Illness

In the case of nutrient supplementation, the intake of specific nutrients by dietary supplement is conducted from different sources (plant extracts, prebiotics, vitamins, amino acids, fibers, metals, fatty acids, etc.) as complementary and integrative treatments. Considering that the pathophysiology of neurological disorders is related to oxidative stress, neuroinflammation, and mitochondrial dysfunction [53], it has been shown that the action of some dietary factors on mitochondrial dysfunction, epigenetic modification, and neuroinflammation are mechanisms that would seem to underlie the action of nutrients on brain health [54].

Micronutrients (vitamins and trace elements) are essential components for metabolic processes. Micronutrients play an important role through catalytic action in enzyme systems like cofactors and components of metalloenzymes. They are also involved in antioxidant activity, modulation of cellular immunity, and wound healing [55]. Acute and chronic changes in micronutrient levels can cause complications in neurological diseases [56]. Actually, micronutrient dysregulation can cause damage to peripheral nerves (demyelination or axonal damage), impairment to the central nervous system, and a typical category of myeloneuropathy (damage to the peripheral and central nervous system) [32]. Globally, vitamins and inorganic ions seem to be useful—as complementary nutrients—for the prevention and management of neurological diseases [56].

Several pieces of scientific evidence show encouraging results on the synthesis of neurotrophic factors and neurotransmitters, neuroplasticity, myelination, and microglial activity as nutritional effects related to the intake of vitamins and minerals [57]. Specifically, as for vitamins, it has been reported that the vitamins of the B group (B6 pyridoxine, B9 folic acid, and B12 cobalamin) may slow the progression of cognitive decline in patients with mild cognitive impairment [58,59] by a reduction in homocysteine levels. A slower progression of Alzheimer's disease was reported in patients at disease onset when supplemented with vitamin E, which, as an antioxidant, protects neurons from oxidative damage [60]. Supplementation with vitamin D has been reported to reduce the frequency of relapses and to slow disease progression in subjects affected by Multiple Sclerosis [61,62].

Among the inorganic elements of bio-importance to health, magnesium and iron have been largely evaluated in neurological dysfunctions [63], and their supplementation seemed to support neurological function [64–66].

Dietary intake of fish and omega-3 fatty acids has been associated with a lower risk of Alzheimer's disease [67,68].

A wide variety of natural plant substances are known as "neuro-nutraceutical" substances [34]. *In vitro* studies showed neuroprotection effects given by seed extracts as a result of their antioxidant, anti-inflammatory, and anti-apoptotic properties [69]. Potential beneficial effects are related to several cannabinoid compounds extracted from *Cannabis sativa*. For instance, cannabisin F in a model of inflammation and oxidative stress induced by lipopolysaccharides in BV2 microglial cells showed a significant reduction in both inflammatory responses and production of reactive oxygen species associated with the expression pathway of the enzyme sirtuin 1/nuclear factor kappa B and the nuclear factor

2 related to erythroid-2. Moreover, the antioxidant activity of *Cannabis sativa* seed extract plays a role in the reduction of reactive oxygen species and in the expression of nuclear factor 2 related to erythroid-2 and oxygenase-1 of heme (HO-1) [70,71]. Bhuiyan et al. demonstrated the neuroprotective effect of anthocyanins extracted from black soybean seed coat using a model of ischemia induced by oxygen–glucose deprivation and cell death induced by glutamate in primary cortical neurons of rats [72].

Globally, we want to highlight that the major limit seemed to be that often, in the analyzed data, the baseline of nutrient levels was not fully reported at a specific stage of disease progression. So, the supplementation can have an individual effectiveness according to the severity of the specific neurological condition. Moreover, the epidemiological analysis of the possible correlation between nutrient consumption and cognitive decline is complex and laborious. It is unlikely that a single component, alone, plays an important role considering that many factors throughout life affect brain function. Therefore, multi-domain interventions could be more promising in the attempt to prevent cognitive decline. Nowadays, designing this type of trial is challenging for researchers [32].

3.1.2. The Dietary Modification in Neurological Illness

Dietary modification involves the change in the types or quantities of consumed food. Often, it is a diet restriction (DR) of particular nutrients (carbohydrates, amino acids, etc.) or a time-limited diet, such as intermittent fasting (IF) or a fasting-mimicking diet [7].

According to the literature analysis, in order to manage the most common neurological illnesses and potentially delay their progression, the principal dietary modifications are the Mediterranean Diet, the Ketogenic Diet, the Paleo diet, and the Low-Fat or the Low-Protein Diets [7,73,74].

Globally, the major reported limit is the adherence to a specific diet. Moreover, dietary preferences can be conditioned by single nucleotide polymorphisms in genes coding for taste receptors [75].

The Mediterranean Diet is characterized by a high intake of fruits, vegetables, monounsaturated fat, fish, whole grains, legumes, and nuts. Alcohol consumption is moderate, as well as the consumption of red meat, saturated fat, and refined grains. At the nutritional level, the contents of omega-3 unsaturated fatty acids, resveratrol, and micronutrients are high. It has been associated with a reduced risk of cognitive decline, Alzheimer's disease [76–78], and Parkinson's disease [79–81]. In the case of Multiple Sclerosis, the Mediterranean diet has been reported to be effective for prevention and for a reduction in comorbid disease severity, but the reduction in symptoms has been related to different diets such as the low saturated fat (Swank), low fat vegan (McDougall), modified Paleolithic (Wahls), and gluten-free diets as well as intermittent fasting, calorie restriction, and intermittent calorie restriction (fasting mimicking diet) [82].

The ketogenic diet is a high-fat low-carbohydrate diet that induces ketosis. It seems to improve cognitive function in Parkinson's disease [83,84]. It seems that the produced ketones may serve as an alternative energy source for the brain, potentially reducing the impact of glucose hypometabolism in Alzheimer's patients [85]. A lot of papers supported the role of the ketogenic diet in the treatment of epilepsy [86–89], probably associated with the stabilization of neuronal activity and reduction in seizure frequency by shifting the body's metabolism from glucose to ketone.

Early dietary interventions in children with Down syndrome have been reported as an opportunity for decreasing the risk or delaying some symptoms, ameliorating their quality of life [90].

4. Nutritional Interventions for Rare Neurological Disorders: What Is the Right Model?

Studying rare diseases is challenging, primarily because of their different etiology, the huge variety of symptoms, the variations in the time of onset, and the low prevalence in the total population.

This complexity necessitates innovative approaches and *ad hoc* models applicable both to better clarify the disease mechanisms and to screen and develop potential treatments or to evaluate supporting nutritional care.

In preclinical studies, *in vitro* models are useful tools for basic research and therapeutic development, while *in vivo* animal models are valuable resources to confirm *in vitro* speculations, especially when investigation in human subjects may be limited by ethical concerns or the limited availability of patients. In order to highlight the importance of tailored research methodologies, we analyzed the literature background to define how these diseases are studied/treated and to evaluate how preclinical research could mirror the genetic context of the patient with specific regard toward nutritional interventions.

4.1. In Vitro Models for Testing Nutritional Interventions in Rare Neurological Disorders

Concerning neurological diseases, the effects of nutrition and the mechanisms involved are still unclear. As it is still hard to understand the mechanism of rare neurological diseases on the basis of only clinical patients or animal models; cell models cultured *in vitro* play a key role, representing excellent tools in the research for disease-causing mechanisms and in the therapeutic development for screening and testing potential treatments.

The common and most long-lasting used *in vitro* model is the two-dimensional (2D) culture system. According to the examined disorder, it consists of the selection of the appropriate cell type (or types) that is cultured *in vitro* and exposed to the neurotoxic stimulus, a hallmark of each disease. In this context, the effects of nutrients could be investigated. This cell model allows the use of human cells, unraveling the doubts due to species differences. The advantages of using *in vitro* models are the reduced costs, the ability to study specific biological endpoints, and the medium and high throughput analysis of the experiments [91], but they fail in the reproduction of the microenvironment of human tissues. Major progress has been made in this field by the emerging three-dimensional (3D) culture [92] and induced pluripotent stem cell (iPSC) technologies.

Compared to 2D culture systems, in 3D culture systems, coherent cells within an extracellular matrix better reflect physiological behavior as in a real tissue environment, enabling the study of the interaction of different cell types [93]. In particular, iPSCs derived from the patient can be used to create patient-specific cell lines. These cell lines accurately reflect the disease phenotype at the cellular level, allowing a relatively accurate assessment of pharmacological responses and toxicity in the patient's genetic field [94,95]. This application strategy not only improves the relevance of preclinical results but may also lead to personalized therapeutic approaches.

In recent years, the major innovation for both 2D and 3D *in vitro* models is the application of High Throughput Screening (HTS) technology [96–99]. It consists of a miniaturized assay and large-scale data analysis to rapidly identify active compounds, antibodies, or genes modulating the molecular pathways involved.

At a glance, the major remarks are reported in Figure 3. Although *in vitro* studies ensure the possibility of prediction of clinical outcomes, they are limited as they cannot reflect the physiological processes of absorption, distribution, metabolism, and excretion.

4.2. In Vivo Models for Testing Nutritional Interventions in Rare Neurological Disorders

Several model organisms have been chosen to better understand the characteristics of rare neurological disorders and identify causative genes. Moreover, with respect to *in vitro* models, the *in vivo* models admit the evaluation of nervous functions, such as motor function, learning and memory, and physiological development. In this context, the effects of nutrients, diet, and nutraceuticals should better mimic the clinical outcomes.

As reported in Figure 4, the classical model organisms for *in vivo* studies are mammalian, especially mice; however, studies on non-mammalian model organisms are becoming increasingly common. Fruit flies (*Drosophila melanogaster*), nematode worms (*Caenorhabditis elegans*), and zebrafish (*Danio rerio*) mainly provide advantages in terms of a rapid and economic evaluation of the effects of nutrition correlation with gene variants. Subsequently, data obtained can be validated in mammalian model organisms, such as mice, and in human cells [100].

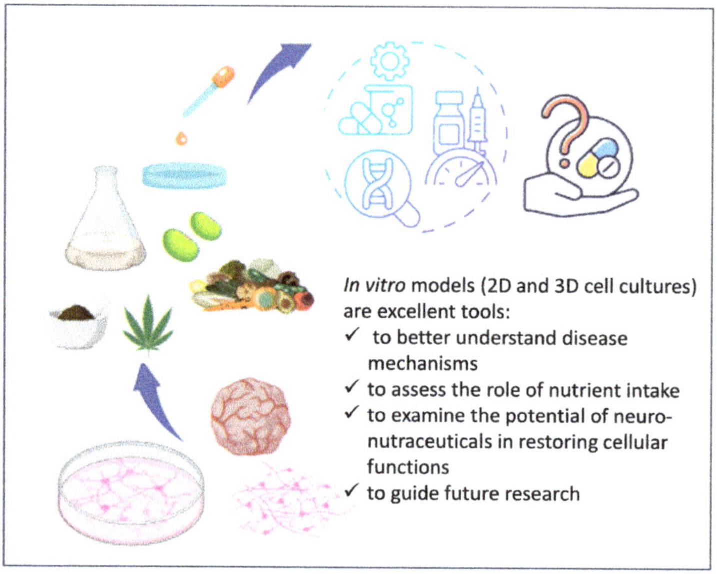

Figure 3. *In vitro* models as tools in rare neurological disease research and therapeutic development (partially created with BioRender.com, accessed on 7 August 2024).

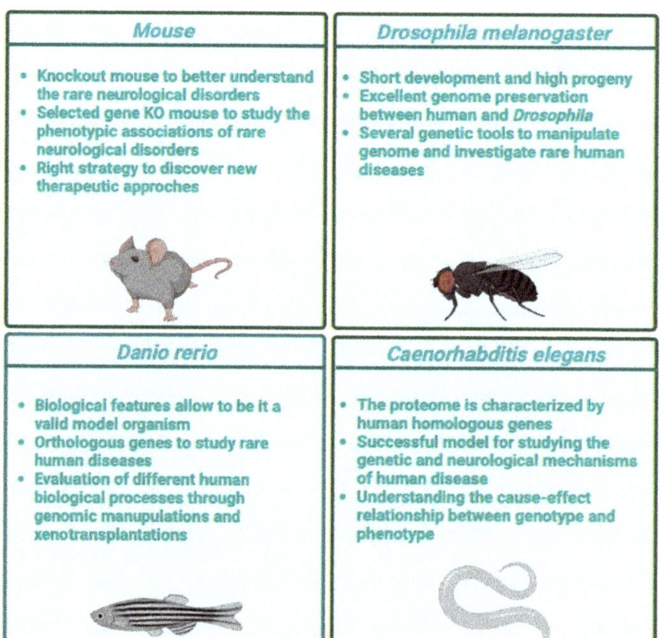

Figure 4. Animal models for rare neurological diseases (created with BioRender.com, accessed on 3 September 2024).

Functional studies in model organisms with informatics support can also be used to make diagnoses. The Model Organism Screening Center uses model organisms to provide evidence of the pathogenicity of genetic variants identified in Undiagnosed Diseases Network patients. Worms (*Caenorhabditis elegans*) and zebrafish (*Danio rerio*) are considered excellent models due to numerous advantages including high gene homology with humans, low maintenance costs, and rapid development. In addition, transparency—in certain stages—allows the detection of cellular and morphological defects [58].

Of course, when the disease-causing agent is a genetic mutation, gene knockout/knockin, chimeric, or chemically-induced organism models have been developed in all taxa [101].

4.2.1. Nonhuman Primate Genetic Models

As recently reviewed by Vallender and co-workers, naturally occurring nonhuman primate models (macaques) of rare human neurological diseases are being discovered and developed [102]. This is the case for late infantile neuronal ceroid lipofuscinosis, Krabbe disease, Leukoencephalopathy with Ataxia, and Pelizaeus–Merzbacher disease. Even if the macaques physiology is closely similar to the human one, the application of primate models—especially for nutritional intervention—is limited as a consequence of their cost and ethical affairs [103].

4.2.2. Rodents

Rodents (mice, rats, etc.) have been largely applied for studying rare neurological disorders. Their application is sustained by some advances: (i) small size, (ii) short generation times, (iii) short life cycle, (iv) easy maintaining and breeding, (v) similarities to humans in terms of anatomy and physiology, and (vi) possibility of genetic manipulation [104,105]. Meanwhile, some advantages could also represent limits, for example, the small size is linked with difficulties in experimental procedures and differences in metabolism [106]. A naturally occurring mouse model has been reported for some rare disorders like Krabbe disease [107], but knockout mouse models seemed to be the major resource for the study of rare diseases [108].

A system of synergistic data-sharing has been developed to limit the number of animals and to ensure the opportunity to discover still unknown therapeutic approaches (for example, dietary intervention or drug therapy) useful for patients [109,110].

4.2.3. *Caenorhabditis elegans* (*C. elegans*)

The microscopic nematode worm *Caenorhabditis elegans* is a model organism suitable for rare neurological disease modeling [111,112]. The whole genome of this organism has been sequenced and it is the only organism to have completed its connectome (the "wiring scheme" of neurons). It has also been shown that 83% of the proteome is characterized by human homologous genes [113]. These data show that *C. elegans* is suitable for functional human gene research, especially for studies on neural mechanisms and molecular learning, memory, coupling behavior, chemotaxis, thermotaxis, and mechanical transduction [114]. Recent advances in genetic manipulation have facilitated the precise manipulation of genes at a single nucleotide level using the CRISPR-Cas9 gene editing technology [115,116]. In addition, investigative studies on *C. elegans* are supported by well-established knowledge, which are valuable publicly accessible resources (e.g., WormBase, https://wormbase.org//#012-34-5; Caenorhabditis Genetics Center https:/cgc.umn.edu) to ensure the integration of new findings on disease pathologies.

To date, the *C. elegans* model aided the characterization of a novel gain-of-function mutation in Sodium Leak Channel Non-selective (NALCN) identified in a girl with intellectual disability, episodic and persistent ataxia, and arthrogryposis [117]. Moreover, genetic defects associated with rare ciliopathy disorders (Joubert syndrome, Meckel syndrome, and nephronophthisis) have been modeled in *C. elegans*, showing severely disrupted ciliary function and structure [118,119].

A nutritional intervention was tested in a *C. elegans* model for maple syrup urine disease, a rare genetic disorder of aminoacidic metabolism characterized by a deficiency of an enzyme complex (branched-chain alpha-keto acid dehydrogenase) that is required to break down the branched-chain amino acids (BCAAs) [120].

However, it should be noted that the application of *C. elegans* in an *in vivo* nutrigenomic model is limited by the high-protein low-fat low-carbohydrate diet, as it naturally feeds on living bacteria.

However, it should be noted that the application of *C. elegans* as an *in vivo* nutrigenomic model is limited by the high-protein low-fat low-carbohydrate diet, as it naturally feeds on live bacteria.

4.2.4. *Drosophila melanogaster*

The fruit fly *Drosophila melanogaster* is a very attractive *in vivo* model for basic research and large-scale screening experiments [121], as a consequence of the high progeny production, and the short time generation for each reproduction cycle. The genome has been fully sequenced [122] and it shows excellent preservation between human and fly genomes (65% for coding genes and 80% for human genes associated with disease) [123,124]. In addition, highly sophisticated genetic tools are available to manipulate the genes of interest in a controlled manner, making investigational studies on rare human variants with unknown pathogenicity in an appropriate cellular or molecular enviroment [125–127].

Interestingly, *Drosophila* has been reported as a diet discovery tool for treating rare disorders of amino acid metabolism disorders, offering the opportunity to generate and test the disease-relevant phenotypes and to perform high-throughput targeted diet screening [128,129].

4.2.5. Zebrafish (*Danio rerio*)

The zebrafish (*Danio Rerio*) is a small tropical freshwater fish easy to maintain in an animal facility at low costs. It is a valid model organism also for the transparency of embryos, the rapid development, and the opportunity to make real-time imaging of cells and internal structures under a microscope. In addition, the embryos develop externally and a pair of fish is able to lay several hundred embryos in a single brood [130]. It has been demonstrated that 82% of known human genes related to the disease and 76% of human genes involved in genomic association studies have orthologues in zebrafish [131]. With the impactful developments of CRISPR and next-generation sequencing technology, zebrafish models have gained increasing success and high approval in biomedical research [132], especially for rare neurological disorders [133,134].

As for the evaluation of dietary intervention in the zebrafish model of rare disease, the assessment of supplementation with cobalamin in the rare combined methylmalonic aciduria and homocystinuria has been performed by adding supplements in water [135]. Globally, the major obstacle is the solubility of some nutrients that could leach into the tank water.

4.3. Clinical Research for Testing Nutritional Interventions in Rare Neurological Disorders

The power of clinical studies is often limited by the small participant pool that is often restricted by rigid inclusion and exclusion criteria. In addition, the severity of symptoms often varies over time and test results only reflect the patient's condition at a specific time [136]. These pieces of evidence suggest that it is necessary (i) to conduct global multi-center recruitment; (ii) to improve the effectiveness of study projects with the cooperation of doctors, pharmaceutical companies, government agencies, clinical research methodologies, biologists, patients, and their families [137]; and (iii) to accelerate the development and application of shared and objective outcome measurements by biomarkers identification. To overcome these limitations, the National Institutes of Health (NIH) supported the Clinical Research Network on Rare Diseases (RDCRN), which was established to promote the diagnosis, management, and treatment of rare diseases and to ensure highly effective and efficient collaboration multi-site, patient-centered, pro-translational, and clinical research. It also works in close contact with patient advocacy groups. Up to now, the RDCRN has published several articles on topics ranging from natural history study results and case reports to practical guidelines and clinical trials. In addition, the RDCRN has been involved in work, which, as a result, has produced 10 treatments approved by the Food and Drug Administration (FDA) including the approval of trofinetide as the first treatment for Rett syndrome [138–142]. An increasing number of technologies are now being used for remote health monitoring by reducing cohort sizes and endpoint response times for clinical trials and by having a clear picture of the patient's status [143]. Digital measures have mainly been developed for monitoring neurological disorders that are not classified as rare [144–146], but these measures are increasingly being used for evaluations in patients with rare diseases. The use of digital measures of results is expected to increase with the development of innovative technologies and artificial intelligence [147].

As recently reported by Liu et al. [148], the research on diet therapy in patients with rare diseases focuses on diet therapy methods, diet therapy management, guidelines for diet therapy, and the impact of diet therapy on patients. They are mostly related to rare inborn errors of metabolism [149].

5. Diet and Nutrient Intake in Rare Neurological Disorders

5.1. Rare Pervasive Developmental Disorder

5.1.1. Angelman Syndrome

The clinical manifestations of Angelman syndrome include mental retardation, movement or balance disorder, typical abnormal behaviors, and severe limitations in speech and language. The etiology resides primarily in *de novo* maternal deletions involving chromosome 15q11.2-q13; but it is also associated with paternal uniparental disomy for the same chromosome, imprinting defects, or mutations in the gene encoding the ubiquitin-protein ligase E3A [150]. As for diet and nutrient intake in Angelman Syndrome, we found 2 clinical trials and 19 articles. Among the latter, only 13 dealt with nutritional interventions. Considering that Angelman syndrome is often associated with refractory epilepsy [151], the ketogenic diet [152,153] and the low glycemic index diet [154,155] have been suggested in clinical practice.

In mouse models of disease, (i) the ketone ester (R,S-1,3-butanediol acetoacetate diester) supplementation attenuated seizure activity and enhanced synaptic plasticity, improving the motor coordination, learning, and memory [156] and (ii) the supplementation with linoleic acid ameliorated the mechano-sensory deficits by acting on increasing the reduced activity of the PIEZO2 ion channel [157].

These results strongly seemed to support a clinical trial (ClinicalTrials.gov ID NCT03644693) that proved the safety and tolerability of a nutritional formulation containing exogenous ketones [158,159]. Moreover, dietary supplementation with methylation-promoting agents (betaine, metafolin, creatine, and vitamin B 12) was evaluated in a clinical trial (ClinicalTrials.gov ID NCT00348933) but resulted in being ineffective in decreasing the severity of the disease [160,161].

5.1.2. Rett Syndrome

Rett syndrome is a neurodevelopmental disorder first observed by the Austrian pediatrician Andreas Rett in two girls having typical hand-wringing stereotypes. It affects mainly, but not exclusively [162], female subjects [163].

The progression of pathology has been fully characterized [164–167]. Briefly, normal growth up to the first 6 months of life is followed by failure to reach the physiological developmental stages between 6 and 18 months. At 12–30 months, a period of regression appears with gait dysfunction, loss of acquired hand skills, and spoken language, and the onset of repetitive hand stereotypies. From approximately 5 years of age through adulthood, the pseudo-stationary stage appears with no continued skill regression, with the exception of some loss of ambulation in the teen years. At last, the late motor deterioration stage can extend for years or decades. Common features include scoliosis, decreased mobility, muscle weakness, spasticity, or stiffness. Sometimes walking stops. The reduction in communication skills also triggers the appearance of autistic traits. Other severe clinical defects have been described such as apnea, hyperventilation, scoliosis, weight loss, and cardiac abnormalities in affected girls [168].

Rett syndrome (OMIM #312750) is primarily caused by loss-of-function mutations in the methyl-CpG-binding protein 2 (*MECP2*) gene located on the X chromosome (Xq28) [169]. Over 500 different MECP2 mutations have been identified. As reviewed by Good et al., there are missense, nonsense, frameshift, splice site, and start codon mutations as well as larger deletions [170]. A great effort has been made to evaluate the genotype–phenotype relationship [171] and genetic therapies [172]. Despite a wide phenotypic variability, RTT is commonly associated with epilepsy, sleep disturbances, and gastrointestinal dysfunction.

MECP2 encodes for the methyl-CpG-binding protein 2 (MeCP2), a DNA-binding protein with acts as a transcriptional modulator and epigenetic regulator of gene expression [169] (Figure 5). The protein is strongly expressed in the CNS on a time course that correlates with neuronal maturation and synaptogenesis [173]. Thus, it plays a critical role in brain development, particularly in regulating the expression of other genes important for synaptic function [174,175], as well as in chromatin organization, alternative splicing, and miRNA processing [176]. Moreover, mutations in other genes can also result in Rett-like syndromes, such as the X-linked Cyclin-Dependent Kinase-Like 5 (CDKL5; OMIM #300203) [177] or the Forkhead box G1 (FOXG1; OMIM #164874) [178] genes.

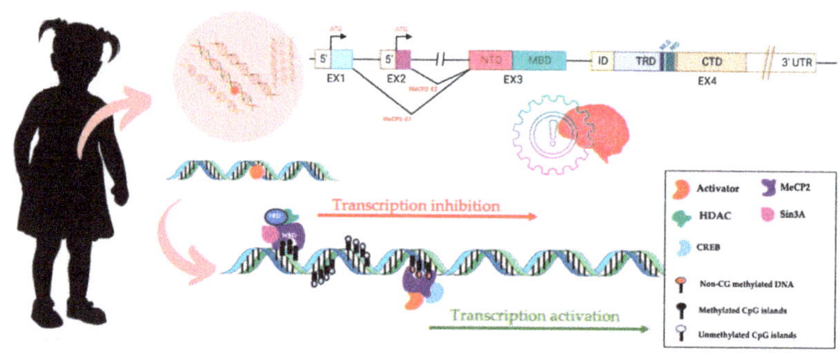

Figure 5. MeCP2 gene structure and its activity on target genes. The meCP2 gene has N-terminal (NTD); methyl binding (MBD); intervening (ID); transcription repression (TRD); and C-terminal (CTD) domains. MeCP2 recruits a transcriptional corepressor complex containing Sin3A and histone deacetylase (HDAC) to methylated CpG islands and induces transcription inhibition in the target gene (TRD, transcriptional repression domain; MBD, methyl-CpG-binding domain). MeCP2 can activate gene transcription by recruiting CREB and other transcriptional factors to non-methylated CG DNA regions (partially created with BioRender.com, accessed on 10 August 2024).

As for the cures, Trofinetide [179] has been recently approved for Rett syndrome [180]; and there are four recruiting clinical trials for gene therapy (ClinicalTrials.gov IDs: NCT05740761, NCT05898620, NCT05606614, and NCT06152237). Ongoing treatments are meant to alleviate disease symptoms, such as antiepileptic drugs, occupational and physical therapy, and scoliosis equipment.

Nutritional interventions (Table 1) for Rett disease have been suggested. Patients are characterized by weight deficiencies and poor growth, and are at risk of food shortages [181,182], vitamin D deficiency [183], osteopenia [184], and low bone mineral mass [185]. Moreover, changes in gut microbiota have been reported both in humans [186] and in MeCP2-deficient rats [187]. Recently, it has been reported that probiotic supplementation ameliorates neurological outcomes in Rett syndrome [188].

Motor and cognitive impairments may be related to cholinergic hypofunction abnormalities, and choline supplementation was found to alleviate the synaptic defects in iPSC-derived neurons [189] and to improve locomotor function in mouse models [190,191].

By searching for clinical trials with dietary intervention in Rett syndrome, we found two completed trials without published results: the NCT05352373 on the role of dietary calcium and the NCT 05420805 on pre- and post-biotics strategies. The creatine monohydrate supplementation was tested on Rett patients as a source of labile methyl groups for different methylation reactions (NCT01147575), but even if the DNA methylation increased, no significant differences were found in clinical parameters with respect to placebo [192].

From a biological and biochemical point of view, in zebrafish and in mutant mouse models as well as in humans, the major protein expression changes point out defects in energy metabolism, mitochondrial function, redox status imbalance, and muscle function [193,194], with annexed inflammation and oxidative stress [195,196]. In this context, the alteration in cholesterol and lipid metabolism [197] with a significant increase in lipid peroxidation [198,199] has been reported, and the diet supplementation with the ω-3 polyunsaturated fatty acids was found to improve the patient's subclinical inflammatory status, partially restoring membrane fatty acids and correcting the redox status [200,201]. Moreover, an anaplerotic triheptanoin diet—concerning the stimulation of mitochondrial function—revealed a significantly increased longevity and improved motor function and social interaction in KO mice [202].

As reviewed by Mouro et al. [203], most studies focused on the control of epilepsy by supplementation with derivatives of the cannabis plant or by diet modification. Concerning cannabinoids, Cannabidivarin seemed to be useful to rescue cognitive deficits and to delay neurological and motor defects in MeCP2-mutant mice [204,205]. Due to the pandemic period and the recruitment challenges, a long-term safety study of cannabidiol oral solution in patients with Rett syndrome was terminated (NCT0425286), and the posted results report a reduction in seizure frequency in 43% of patients, while 5% became seizure-free. However, patients suffered from serious adverse side effects such as diarrhea, vomiting, fatigue, pyrexia, and somnolence.

As for diet modification, the ketogenic diet has shown good results on seizure frequency and behavior, as well as in the case of refractory epilepsy [206–208]. The antiepileptic actions seem due to the GABAergic signaling mechanism [209], or to an increase in adenosine and BDNF signaling [210].

Table 1. Nutritional interventions for Rett syndrome.

	Nutritional Interventions for Rett Syndrome	
	Nutrients or Neuro-Nutraceutical Substance	**Effects**
Nutrient supplementation	Vitamin D [1]	Reduction in vitamin D deficiency [183]
	ω-3 polyunsaturated fatty acids [1]	Improvement in inflammatory status [200,201]
	Triheptanoin [2,3]	Amelioration of mitochondrial function, motor function and social interaction [202]
	Choline [2]	Modulation of neuronal plasticity, possibly leading to behavioral changes.
	Probiotic [1]	Amelioration of neurological outcomes [188]
	Cannabidivarin [2]	Rescue cognitive deficits and delay of neurological and motor defects [204,205]
	Cannabidiol [1]	Antiepileptic actions
	Creatine monohydrate [1]	Increase in DNA methylation [192]
	Type of diet	**Effects**
Dietary modification	Ketogenic diet [1,2]	Antiepileptic actions [206–208]

[1] Tested in patients. [2] Tested in animal models. [3] Tested in *in vitro* model.

5.2. Rare Leukodystrophies

Although there are different causes of leukodystrophies, the patients share the manifestations of neurological symptoms. Among these are muscular spasticity, ataxia, seizures, cognitive developmental delay, dystonia, and dyskinesias. Moreover, swallowing dysfunction and pulmonary problems result in feeding limitations.

5.2.1. Krabbe Disease

Krabbe disease (OMIM #245200) is one of the classic genetic lysosomal storage diseases with autosomal recessive inheritance. It is characterized by demyelination in the white matter of the central and peripheral nervous system. Clinical manifestations report infantile and late-onset forms. This pathology is due to mutations in the *galc* gene (Ch. 14q31), which encodes for galactocerebrosidase, the lysosomal enzyme that catalyzes the hydrolysis of galactose from galactocerebrosides and galactosyl-sphingosine (psychosine). The enzyme loses its function and, according to the "psychosine hypothesis" [211,212], the accumulation psychosine causes the death of the myelinating cells by triggering cell signaling pathways that induce oxidative stress, mitochondrial dysfunction, apoptosis, inflammation, endothelial/vascular dysfunctions, and neuronal and axonal damage [213–217].

To date, there is no cure. The only disease-modifying treatment currently available is hematopoietic stem cell transplantation, which is effective only when performed before symptoms appear, while other treatment options are symptomatic [218].

Gene therapy-based clinical trials (NCT04693598 and NCT05739643) are active; no clinical trials for the evaluation of dietary intervention exist. Through searches about diet and nutrient intake in Krabbe disease, the biomedical databases retrieved up to 18 results, but only 4 results really fitted within the aim of this work. Preclinical studies performed in the twitcher mouse (the naturally occurring animal model of the disease) suggested (i) a galactose-free diet enriched in soy isoflavones and antioxidants (coenzyme Q10, glutathione, and isoflavonoids) [219] and supplementation with vitamin D3 [220] for delaying the onset of symptoms. Recently, dietary supplementation with n-3 polyunsaturated fatty acids led to a slowing of the phenotypic presentation of the disease and restoration of lipid mediator production [221], as oxidative stress induces an increase in isoprostanoids levels in mouse brains [222]. In cellular systems, it has been reported that the 3′,4′,7-trihydroxyisoflavone holds a GALC-addressed chaperoning activity [223,224], useful to increase residual enzymatic activity in fibroblasts from Krabbe patients.

Although there are little data in the literature, such scientific evidence conducted on *in vivo* models gives a possible guarantee that a correct supplement of foods containing antioxidant nutrients can be a potential treatment to improve the quality of life in Krabbe syndrome (Table 2).

Table 2. Nutritional interventions for Krabbe disease.

	Nutritional Interventions Krabbe Disease	
	Nutrients or Neuro-Nutraceutical Substance	**Effects**
Nutrient supplementation	Vitamin D3 [1]	Delay symptoms onset [220]
	n-3 polyunsaturated fatty acids [1]	Slow the phenotypic presentation [221]
	Soy isoflavones and antioxidants (coenzyme Q10, glutathione and isoflavonoids [1]	Delay symptoms onset [219]
	3′,4′,7-Trihydroxyisoflavone [2]	Pharmacological chaperone [223,224]
	Type of diet	**Effects**
Dietary modification	Galactose-free diet [1]	Delay symptoms onset [219]

[1] Tested in animal models. [2] Tested in *in vitro* model.

5.2.2. Pelizaeus–Merzbacher Disease

Pelizaeus–Merzbacher disease (OMIM # 312080) is an X-linked leukodystrophy characterized by developmental delay, nystagmus, hypotonia, spasticity, and variable intellectual deficit [225]. The disorder is due to mutations or dosage alterations of the proteolipid protein 1 (PLP1) gene, located at chromosome Xq22.2 [226,227].

Treatments are symptomatic, including drugs for seizures, and spasticity.

Non-clinical trials matched the query as nutritional intervention. Biomedical databases returned to our search six results, but only three of these really fitted within the aim of this work. It has been shown in a mouse model of disease that cholesterol supplementation can enhance myelination [228], but dietary cholesterol was ineffective in myelination of patients [229]. The authors explain this phenomenon by suggesting that mice have a disturbance of blood–brain barrier (BBB) integrity that allows access of cholesterol from the circulation into the brain. Moreover, the ketogenic diet seemed to improve myelination and axonal damage in a mouse model with preserved BBB integrity [229], while the supplementation of curcumin has been reported as an antioxidant therapy in a mouse model of disease-carrying additional copies of PLP1 [230].

5.2.3. Miscellaneous

As reported above, hypomyelinating conditions have a general phenotypic description and genetic heterogeneity in common. In this context, it is not surprising that the same nutritional intervention has been proven as effective for multiple conditions.

It is the case of the ketogenic diet that was reported as beneficial for seizure control in patients with hypomyelinating leukodystrophy-14 (OMIM # 617899) and drug-resistant seizures [231] and for the alleviation of psychomotor regression in KARS-related mitochondrial dysfunction and progressive leukodystrophy (OMIM # 619147) [232].

5.3. Rare Genetic Epilepsy

Developmental and epileptic encephalopathies represent a clinically and genetically heterogeneous group of age-dependent neurologic disorders characterized by the onset of refractory seizures in infancy or early childhood. Several disorders fitting into this group, such as the West syndrome (OMIM # 308350) with a common genetic mutation in the ARX gene (Ch. Xp21.3) [233], the Ohtahara syndrome [234], the Dravet syndrome (OMIM # 607208) with a common genetic mutation in the *SCN1A* gene (Ch. 2q24.3) [235], and the Lennox–Gastaut syndrome [236], have an often unclear etiology. Recently, the term "Infantile Spasm Syndrome" has been applied to include both West syndrome as well as conditions characterized by epileptic spasms in children even in the absence of all inclusion diagnostic criteria for West syndrome [237]. As recently reported by Ramantani et al. [238], the main proposed nutritional intervention for these conditions—that are often refractory to the antiepileptic drugs [234]—is the ketogenic diet [239,240].

Up to now, we found four clinical studies based on diet for West syndrome/infantile spasm (NCT01006811; NCT05279118; NCT01549288; and NCT00968136). Among these (i) the trial about the modified Atkins diet in infantile spasms refractory to hormonal therapy was withdrawn (NCT01549288); (ii) the trial NCT05279118 is active, but not recruiting, and it aims to compare the effect of the ketogenic diet to those of Adrenocorticotropic hormone (selected as active comparator); and (iii) two trials have been completed (NCT00968136, and NCT01006811) without published results on clinicaltrial.org. However, searching the clinical trials' ID on the public database, the results of NCT01006811 were found and reported a 40% rate of spasm resolution after three months of modified Atkins diet (a less restrictive ketogenic diet) in children [241].

At the molecular level, it seems that the efficacy of the ketogenic diet could be strongly related to genetic variants, and consequently to rare disorders. Ko et al. [242] reported that patients identified as responders to the ketogenic diet were 77.8% in the case of Dravet syndrome; 35.8% in the case of West syndrome; 29.1% for the Lennox–Gastaut syndrome; and 64.7% in the case of Ohtahara syndrome. It must be pointed out that the nutritional intervention seemed to be effective in patients with SCN2A, STXBP1, KCNQ2, and SCN1A mutations that reported a responder rate of 100, 100, 83.3, and 77.8%, respectively. However, it was not effective in patients with CDKL5 mutations (responder rate = 0.0%) 3 months after implementation [86,242].

The reported results strongly suggest that the application of technologies, such as next-generation sequencing [243], could improve the understanding of the pathophysiology of each genetic mutation, enhancing the development of precision medicine to identify in which patients the ketogenic diet can ensure an efficient outcome.

The effectiveness of the ketogenic diet in the control of patients' seizures has been also supported by preclinical studies in animal models, as reviewed by Griffin et al. [244] and demonstrated in mouse models for Dravet syndrome [245–247], and for West syndrome [248].

Recently, a trial has been approved on melatonin supplementation in the treatment of infantile spasms (https://trialsearch.who.int/Trial2.aspx?TrialID=ChiCTR2000036208 accessed on 6 September 2024).

Among the neuro-nutraceutical substances, cannabinoids or derivatives of *Cannabis sativa* have been largely investigated to treat rare epilepsy. The diet supplementation with cannabidiol-enriched extract has been reported to reduce seizures both in humans [249,250] and in animal models [251]. Specifically, cannabidiol oral solution was efficacious for the treatment of patients with drop seizures associated with Lennox–Gastaut syndrome (NCT02224690) [252], or with Dravet syndrome (NCT02091375) [253]. According to these results, cannabidiol (Epidyolex) received the orphan drug designation by the U.S. Food and Drug Administration and by the European Medicines Agency.

5.4. Rare Forms of Ataxia

Ataxia is a disease characterized by impaired muscle control during voluntary movements such as walking or grasping objects. It is usually the consequence of damage to the cerebellum, the brain structure responsible for muscle coordination. The etiology varies from sporadic, hereditary, and acquired forms.

Among the rare forms of ataxia, Friedreich ataxia, Stiff Person Syndrome, and Gluten ataxia benefit from nutritional intervention.

Friedreich ataxia is the most common of the inherited ataxias, caused mainly by homozygous GAA triplet repeat expansion in intron 1 of the FXN gene encoding for frataxin [254], a mitochondrial protein involved in iron metabolism. The supplementation with antioxidant agents, like vitamin E plus coenzyme Q10, and vitamin B1 decreases oxidative stress and enhances mitochondrial function [255]. Moreover, the antioxidant resveratrol increases FXN gene expression in cell models [256], and it seems useful in patients [257]. Recently, the clinical trial "Micronised resveratrol as a treatment for Friedreich Ataxia" (NCT03933163) has been completed, but no result was shared. Actually, there is a recruiting trial that will investigate the effect of combined physical exercise and dietary supplementation with Nicotinamide riboside (as NAD+ precursor) on skeletal muscle mitochondrial oxidative phosphorylation capacity, muscle mass, aerobic capacity, and glucose homeostasis (NCT04192136).

Stiff person syndrome is a rare autoimmune disease associated with cerebellar ataxia. Most patients have high levels of glutamic acid decarboxylase (GAD) antibodies. The etiology remains unclear. Gluten ataxia is often associated with Spinocerebellar ataxia type 35 with mutation in the Transglutaminases (TGM6) gene or autoantibodies against TGM6. In both diseases, a gluten-free diet has demonstrated good clinical effectiveness [258,259] (NCT00006492).

For the symptomatic treatment, some edible mushrooms have been suggested for their neuroprotective properties. *Pleurotus giganteus*, *Ganoderma lucidium*, and *Hericium erinaceus* contain bioactive compounds such as terpenoids, polysaccharides, alkaloids, and antibiotics that induce both the activation of the master regulator of the antioxidant defenses and the synthesis of Nerve Growth Factor, determining a slowing down of neuronal senescence [260].

Still under discussion today are the endocannabinoids that have been extensively studied in diseases associated with nerve cell damage, demonstrating that endocannabinoids show a broad spectrum of neuroprotective activities (antioxidant, anti-inflammatory, and pro-neurogenic) [261]. The endocannabinoid system is under investigation in spinocerebellar Ataxia Type-3 and other autosomal-dominant cerebellar ataxias [262,263]. As summarized by Gómez-Ruiz et al. [262], elevated levels of cannabinoid type-2 and cannabinoid type-1 receptors were found in spinocerebellar ataxias animal models and patients' brain tissue. Specifically, in *post mortem* tissue of patients, the expression of cannabinoid type-2 receptors was found to be elevated in surviving neuronal cerebellar subpopulations (e.g., Purkinje cells, neurons of the dentate nucleus) [264] in which these receptors usually have a low physiological expression. The cannabinoid type-1 receptors were elevated in several cerebellar cells both in mouse models [265] and in *post mortem* patient tissue [264]. Globally, the collected evidence suggests that the alterations in endocannabinoid receptors may be related to the pathogenesis of ataxia and may be therapeutic targets. Results from *in vivo* studies suggested the potential role of cannabinoid type-1 receptor modulation for the management of ataxia symptoms in a rat model, but results were not fully satisfying, probably because the receptor may not act solely and other receptors should be considered [266].

5.5. Rare Brain Tumors

Based on the guidelines of the World Health Organization (WHO), rare brain tumors are classified according to several criteria, first and foremost, the histotype. To support the classification, several parameters are added such as morphological characteristics, growth characteristics, and ultimately, the molecular pattern involved in the genesis and development of the tumor [267,268]. As reported by Louis et al. [267], new tumor types and subtypes have been classified thanks to novel diagnostic technologies such as DNA methylome profiling.

Among the rare brain tumors, the most common malignant tumors that affect children are embryonal tumors (Medulloblastoma) [269], Pineal Region Tumors (Pineoblastoma) [270], Choroid Plexus Carcinoma [271], and Glial Tumors [272].

The nutritional factors on glioma incidence have been recently reviewed by Bielecka and Zukowsaka [273], suggesting a balanced diet containing fruits, vegetables, antioxidant-rich foods, omega-3 fatty acids, and adequate protein. As for the effect of dietary antioxidant vitamin intake on glioma risk in humans, published data seem to be controversial: some reports sustained their protective role [274,275], while other researchers described an apparent positive or null influence [276,277].

A recent systematic review and meta-analysis synthesized the associations between dietary antioxidant vitamin intake and risk of glioma by reviewing the available data up to March 2024 [278]. Even if the reading of this work is recommended [278], the major findings were that a (i) high intake of vitamin C was significantly associated with a lower risk of glioma and that (ii) high intake of vitamin A and vitamin E were not associated with the risk of glioma. In line with these points in a meta-analysis of 15 studies (including 13 control cases) assessing the impact of vitamin C intake on the risk of glioma, a higher intake of vitamin C than lower consumption was significantly related to a lower risk [279], while in contrast, previous studies reported that dietary vitamin A intake could reduce the glioma risk [280].

Thus, it seems clear that further clinical studies with detailed doses and large-scale prospective studies will be necessary, considering the encouraging results from preclinical analyses. To date, it was shown that (i) vitamin A (all-trans retinoic acid) stimulated apoptosis and had an inhibitory effect on the migration, invasion, and proliferation of glioma cell lines (U87 and SHG44) [281] and of primary cultures established from biopsies [282] and that (ii) the lipophilic derivative of vitamin C (ascorbile stearate) had a proapoptotic and antiproliferative effect in human glioma cells [283].

The ketogenic diet has been reported as useful for gliomas (including glioblastomas) by pieces of evidence from *in vitro, in vivo*, and clinical observations [284,285]. The rationale behind using a ketogenic diet for gliomas started from the observation that cancer cells, including glioma cells, often rely heavily on glucose for energy through the process of glycolysis (the "Warburg effect") [286]. By restricting glucose availability, the ketogenic diet may impair the energy supply to tumor cells while providing an alternative energy source (ketones) to normal brain cells. Preclinical studies reported that the ketogenic diet not only enhanced survival and slowed tumor growth, but it also potentiated the effect of radiation by extending the survival of a mouse model of malignant glioma [287]. From a mechanistic point of view, the improved survivability seemed possibly due to a reduction in reactive oxygen species levels, and a modulation—toward the physiological levels—of gene expression involved in oxidative stress and antioxidant defense pathways [288]. In recent years, numerous clinical studies have been conducted as reported in Table 3. The data from a recent metanalysis pointed out a global positive effect of the ketogenic diet on patient survival as an adjuvant therapy of malignant gliomas [289].

Table 3. Clinical trials for the ketogenic diet in gliomas.

Clinical Trials for Ketogenic Diet in Gliomas		
Status	**Clinical Trial ID/Name**	**Results**
Withdrawn	NCT05373381 The KetoGlioma (Ketogenic Glioma) Study	Study is on hold indefinitely due to funding issues
Unknown status	NCT03278249 Feasibility Study of Modified Atkins Ketogenic Diet in the Treatment of Newly Diagnosed Malignant Glioma	
	NCT02939378 Ketogenic Diet Adjunctive to Salvage Chemotherapy for Recurrent Glioblastoma: a Pilot Study	
Terminated	NCT02046187 Ketogenic Diet with Radiation and Chemotherapy for Newly Diagnosed Glioblastoma	Excessive protocol deviations due to strict nature of diet requirements
Completed	NCT02286167 Glioma Modified Atkins-based Diet in Patients with Glioblastoma	Production of ketonuria and significant systemic and cerebral metabolic changes in participants [290]
	NCT03075514 Ketogenic Diets as an Adjuvant Therapy in Glioblastoma	No posted results, only a linked paper with trial description [291]
	NCT02302235 Ketogenic Diet Treatment Adjunctive to Radiation and Chemotherapy in Glioblastoma Multiforme: a Pilot Study NCT01865162 Ketogenic Diet as Adjunctive Treatment in Refractory/End-stage Glioblastoma Multiforme: a Pilot Study (KGDinGBM)	Diet was well tolerated. The small sample size limits efficacy conclusions [292].
	NCT00575146 Ketogenic Diet for Recurrent Glioblastoma	Diet was feasible and safe but probably has no significant clinical activity when used as single agent in recurrent glioma [293].
	NCT01754350 Calorie-restricted, Ketogenic Diet and Transient Fasting During Reirradiation for Patients With Recurrent Glioblastoma	Diet was feasible and effective in inducing ketosis in heavily pretreated patients with recurrent glioma, but failed to increase the efficacy of reirradiation [294].
Recruiting	NCT05708352 A Phase 2 Study of the Ketogenic Diet vs. Standard Anti-cancer Diet Guidance for Patients With Glioblastoma in Combination With Standard-of-care Treatment	
	NCT04691960 A Pilot Study of Ketogenic Diet and Metformin in Glioblastoma: Feasibility and Metabolic Imaging	
	NCT05564949 A Ketogenic Diet as a Complementary Treatment on Patients With High-grade Gliomas and Brain Metastases	
	NCT05183204 Paxalisib With a High Fat, Low Carb Diet and Metformin for Glioblastoma	
	NCT04730869 Metabolic Therapy Program In Conjunction With Standard Treatment For Glioblastoma	

Table 3. *Cont.*

Clinical Trials for Ketogenic Diet in Gliomas		
Status	**Clinical Trial ID/Name**	**Results**
Active, not recruiting	NCT03451799 Ketogenic Diet in Combination With Standard-of-care Radiation and Temozolomide for Patients With Glioblastoma	
	NCT01535911 Pilot Study of a Metabolic Nutritional Therapy for the Management of Primary Brain Tumors	

The Li–Fraumeni syndrome is a rare genetic condition, also acting as a predisposition to cancer development in the brain. It is caused by point mutations in the p53 gene (TP53 on chromosome 17p13), which phenotypically cause uncontrolled cell proliferation [295]. Dietary supplementation with Nicotinamide riboside—a vitamin B3 dietary supplement—is under investigation (ClinicalTrials.gov ID NCT03789175).

Ongoing pieces of research are investigating the direct link between maternal diet during pregnancy and the development of rare brain tumors in offspring [296–298], suggesting that (i) foods generally associated with increased risk were cured meats, eggs/dairy, and oil products; foods generally associated with decreased risk were yellow-orange vegetables, fresh fish, and grains; (ii) maternal passive smoking and consumption of caffeinated beverages during pregnancy should be considered as a risk for glioma; (iii) oil products increased the risk of medulloblastoma; and (iv) diets higher in fruit and lower in fried foods and cured meats during pregnancy may reduce the risk of unilateral retinoblastoma.

6. Conclusions

The relationship between nutritional intervention and rare neurological diseases is an emerging area of research that may help to manage symptoms, slow disease progression, or even impact the underlying mechanisms of certain rare neurological disorders. The complex interactions between diet and neurological health require closer interdisciplinary collaboration between neurologists, dietitians, and researchers. It is crucial to conduct clinical trials to establish evidence-based dietary guidelines tailored to these unique conditions.

The major limits of this study reside in (i) the limited availability of data as accessed by the reduced number of published papers and completed clinical trials; (ii) the wide variety of the disease's pathophysiology; and (iii) the underestimated influence of individual variability due to genetic differences. Specifically, as reported above, the number of published articles about the role of diet/nutrients on rare neurological disorders is very small. Clinical data often derive from case reports or small cohort studies, and large-scale clinical trials are scarce due to the rarity of these disorders and the huge heterogeneity of clinical manifestations. Moreover, even if preclinical studies give us more detailed information about the impact of a specific nutrient on cell function or on disease progression in animal models, they could not reflect the patients' individual genetic differences and their global health status (e.g., integrity of anatomic structure and physiological functions as blood circulation, pressure, constipation, and deglutition). These latter aspects may influence how patients assume, metabolize, and distribute nutrients as has been discussed above in the case of cholesterol supplementation for Pelizaeus–Merzbacher disease.

Moreover, when searching for dietary recommendations for a specific rare disease, a great limit is the heterogeneity of occurring genetic mutations that turn into different phenotypes and clinical manifestations. Briefly, what could work for one patient may not be effective for another one, as enlightened here by reviewing the activity of trihydroxy-isoflavone as a GALC-addressed chaperone.

However, despite these limitations, our analysis has, as a strength, the identification of a specific dietary intervention for the management of some symptoms. It is the case of ketogenic diet efficacy in different rare neurological diseases that share epilepsy in clinical manifestation. This finding may drive new trials for the evaluation of the ketogenic diet in epileptic patients with different rare disorders. It should be useful in order to identify if a putative leitmotif of treatment is possible and if the diet interacts with the specific pharmacological medications already in use. Moreover, the evidence reviewed here may also open the suggestion of a more detailed nutritional plan during pregnancy.

Globally, our work suggests that the field of rare diseases is a "societal laboratory" that is predicting future trends in patient-centered human healthcare, and as such, it is a model for personalized medicine in some aspects. Personalized nutrition, which considers the individual's genetics, the microbiome, and the metabolic profile, is likely to play a pivotal role in future therapies. With advancements in technology, such as metabolomics and neuroimaging, it may be possible to identify biomarkers that predict dietary responses. According to these observations, the main challenge for the future may be to tailor nutrition and diet based on the individual's specific genetic and biochemical profile. It may be difficult, especially in the case of rare neurological diseases, because the approach may be expensive, difficult for the caregivers, and not available in all healthcare settings due to the required involvement of specialized healthcare professionals and sophisticated tools.

This work critically describes the actual scenario, hoping to inspire future investigations for a deeper understanding of these diseases and to address the actual knowledge gaps. As described, in many cases, the mechanisms by which specific nutrients or diets impact neurological health are poorly understood. The lack of mechanistic understanding hampers the development of more targeted nutritional therapies, as in the case of the ketogenic diet and cannabinoid supplementation. Moreover, we found that the majority of studies focus on short-term outcomes (seizure reduction, myelinization, or symptom improvements), but the long-term effects are often unknown and unexplored.

Taken together, in some rare neurological diseases, diet may become an integral part of holistic care strategies, and it needs to receive the same attention as pharmacological treatments, offering patients not only improved quality of life but also potentially new avenues for disease prevention, management, and treatment.

Author Contributions: Conceptualization, A.C.E.G., R.M., G.G., G.L.R., C.S., and V.C.; methodology, A.C.E.G., R.M., M.B. (Marilena Briglia), F.A., R.A., and M.B. (Maria Bellomo); software, M.B. (Marilena Briglia), F.A., and R.A.; validation, M.B. (Marilena Briglia), F.A., R.A., and M.B. (Maria Bellomo); formal analysis, all; investigation, M.B. (Marilena Briglia), F.A., and R.A.; data curation, M.B., F.A., R.A., R.M., and A.C.E.G.; writing—original draft preparation, M.B. (Marilena Briglia), F.A., R.A., and A.C.E.G.; writing—review and editing, A.C.E.G., R.M., M.B. (Maria Bellomo), G.G., G.L.R., C.S., and V.C.; supervision, A.C.E.G., R.M., and G.G.; project administration, A.C.E.G.; funding acquisition, A.C.E.G. All authors have read and agreed to the published version of the manuscript.

Funding: This research was partially funded by MUR—PRIN 2022 PNRR (Next Generation EU, componente M4C2, investimento 1.1.), grant number P202272FJC.

Data Availability Statement: Data sharing is not applicable.

Acknowledgments: The authors want to thank the staff of the University of Enna, especially the "Area della Didattica e della Ricerca" and its director Giovanni Lavalle, for the administrative support. The authors also thank the Italian Association for Krabbe Leukodystrophy "Progetto Grazia" for the growing attention toward rare neurological disease.

Conflicts of Interest: The authors declare no conflicts of interest.

References

1. Vigo, D.; Thornicroft, G.; Atun, R. Estimating the True Global Burden of Mental Illness. *Lancet Psychiatry* **2016**, *3*, 171–178. [CrossRef] [PubMed]
2. WHO Over 1 in 3 People Affected by Neurological Conditions, the Leading Cause of Illness and Disability Worldwide. Available online: https://www.who.int/news/item/14-03-2024-over-1-in-3-people-affected-by-neurological-conditions--the-leading-cause-of-illness-and-disability-worldwide (accessed on 30 May 2024).
3. Steinmetz, J.D.; Seeher, K.M.; Schiess, N.; Nichols, E.; Cao, B.; Servili, C.; Cavallera, V.; Cousin, E.; Hagins, H.; Moberg, M.E.; et al. Global, Regional, and National Burden of Disorders Affecting the Nervous System, 1990–2021: A Systematic Analysis for the Global Burden of Disease Study 2021. *Lancet Neurol.* **2024**, *23*, 344–381. [CrossRef]

4. Kouzani, A.Z.; Sillitoe, R.V. Medical Devices in Neurology. In *Encyclopedia of Biomedical Engineering*; Narayan, R., Ed.; Elsevier: Oxford, UK, 2019; pp. 409–413; ISBN 978-0-12-805144-3.
5. Chan, J.K.N.; Correll, C.U.; Wong, C.S.M.; Chu, R.S.T.; Fung, V.S.C.; Wong, G.H.S.; Lei, J.H.C.; Chang, W.C. Life Expectancy and Years of Potential Life Lost in People with Mental Disorders: A Systematic Review and Meta-Analysis. *EClinicalMedicine* **2023**, *65*, 102294. [CrossRef] [PubMed]
6. *Intersectoral Global Action Plan on Epilepsy and Other Neurological Disorders*; WHO: Geneva, Switzerland, 2023.
7. Mao, X.-Y.; Yin, X.-X.; Guan, Q.-W.; Xia, Q.-X.; Yang, N.; Zhou, H.-H.; Liu, Z.-Q.; Jin, W.-L. Dietary Nutrition for Neurological Disease Therapy: Current Status and Future Directions. *Pharmacol. Ther.* **2021**, *226*, 107861. [CrossRef] [PubMed]
8. Lee, H. The Importance of Nutrition in Neurological Disorders and Nutrition Assessment Methods. *Brain Neurorehabilit.* **2022**, *15*, e1. [CrossRef]
9. Kivipelto, M.; Mangialasche, F.; Ngandu, T. Lifestyle Interventions to Prevent Cognitive Impairment, Dementia and Alzheimer Disease. *Nat. Rev. Neurol.* **2018**, *14*, 653–666. [CrossRef]
10. Provensi, G.; Schmidt, S.D.; Boehme, M.; Bastiaanssen, T.F.S.; Rani, B.; Costa, A.; Busca, K.; Fouhy, F.; Strain, C.; Stanton, C.; et al. Preventing Adolescent Stress-Induced Cognitive and Microbiome Changes by Diet. *Proc. Natl. Acad. Sci. USA* **2019**, *116*, 9644–9651. [CrossRef]
11. Franco, P. Orphan Drugs: The Regulatory Environment. *Drug Discov. Today* **2013**, *18*, 163–172. [CrossRef] [PubMed]
12. Aronson, J. Rare Diseases, Orphan Drugs, and Orphan Diseases. *BMJ* **2006**, *333*, 127. [CrossRef]
13. Jinnah, H.A.; Albanese, A.; Bhatia, K.P.; Cardoso, F.; Da Prat, G.; De Koning, T.J.; Espay, A.J.; Fung, V.; Garcia-Ruiz, P.J.; Gershanik, O.; et al. Treatable Inherited Rare Movement Disorders. *Mov. Disord.* **2018**, *33*, 21–35. [CrossRef]
14. Federico, A. Rare Neurological Diseases: A Pandora's Box for Neurology (an European and Italian Perspective). *Rev. Neurol.* **2013**, *169*, S12–S17. [CrossRef] [PubMed]
15. Maroilley, T.; Tarailo-Graovac, M. Uncovering Missing Heritability in Rare Diseases. *Genes* **2019**, *10*, 275. [CrossRef] [PubMed]
16. García, J.-C.; Bustos, R.-H. The Genetic Diagnosis of Neurodegenerative Diseases and Therapeutic Perspectives. *Brain Sci.* **2018**, *8*, 222. [CrossRef]
17. Schaaf, J.; Sedlmayr, M.; Schaefer, J.; Storf, H. Diagnosis of Rare Diseases: A Scoping Review of Clinical Decision Support Systems. *Orphanet J. Rare Dis.* **2020**, *15*, 263. [CrossRef]
18. Matilla-Dueñas, A.; Corral-Juan, M.; Rodríguez-Palmero Seuma, A.; Vilas, D.; Ispierto, L.; Morais, S.; Sequeiros, J.; Alonso, I.; Volpini, V.; Serrano-Munuera, C.; et al. Rare Neurodegenerative Diseases: Clinical and Genetic Update. In *Rare Diseases Epidemiology: Update and Overview*; Posada De La Paz, M., Taruscio, D., Groft, S.C., Eds.; Advances in Experimental Medicine and Biology; Springer International Publishing: Cham, Switzerland, 2017; Volume 1031, pp. 443–496; ISBN 978-3-319-67142-0.
19. Schieppati, A.; Henter, J.I.; Daina, E.; Aperia, A. Why Rare Diseases Are an Important Medical and Social Issue. *Lancet* **2008**, *371*, 2039–2041. [CrossRef]
20. The Lancet Neurology. Rare Advances for Rare Diseases. *Lancet Neurol.* **2017**, *16*, 1. [CrossRef]
21. Sintila, S.-A.; Boziki, M.; Bakirtzis, C.; Stardeli, T.; Smyrni, N.; Nikolaidis, I.; Parissis, D.; Afrantou, T.; Karapanayiotides, T.; Koutroulou, I.; et al. The Experience of a Tertiary Reference Hospital in the Study of Rare Neurological Diseases. *Medicina* **2023**, *59*, 266. [CrossRef] [PubMed]
22. Meekings, K.N.; Williams, C.S.M.; Arrowsmith, J.E. Orphan Drug Development: An Economically Viable Strategy for Biopharma R&D. *Drug Discov. Today* **2012**, *17*, 660–664. [CrossRef]
23. Orphanet. Available online: https://www.orpha.net.
24. Painous, C.; Van Os, N.J.H.; Delamarre, A.; Michailoviene, I.; Marti, M.J.; Van De Warrenburg, B.P.; Meissner, W.G.; Utkus, A.; Reinhard, C.; Graessner, H.; et al. Management of Rare Movement Disorders in Europe: Outcome of Surveys of the European Reference Network for Rare Neurological Diseases. *Eur. J. Neurol.* **2020**, *27*, 1493–1500. [CrossRef]
25. European Reference Networks. Available online: http://www.Ern-Rnd.Eu.
26. ERN-ITHACA. European Reference Network for Rare Malformation Syndromes, Intellectual and Other Neurodevelopmental Disorders. Available online: https://Ern-Ithaca.Eu.
27. March, P.A. Diseases of the Brain and Cranial Nerves. In *Saunders Manual of Small Animal Practice*; Elsevier: Amsterdam, The Netherlands, 2006; pp. 1249–1276. ISBN 978-0-7216-0422-0.
28. Frederiksen, S.D.; Avramović, V.; Maroilley, T.; Lehman, A.; Arbour, L.; Tarailo-Graovac, M. Rare Disorders Have Many Faces: In Silico Characterization of Rare Disorder Spectrum. *Orphanet J. Rare Dis.* **2022**, *17*, 76. [CrossRef]
29. Krieg, S.I.; Krägeloh-Mann, I.; Groeschel, S.; Beck-Wödl, S.; Husain, R.A.; Schöls, L.; Kehrer, C. Natural History of Krabbe Disease–a Nationwide Study in Germany Using Clinical and MRI Data. *Orphanet J. Rare Dis.* **2020**, *15*, 243. [CrossRef] [PubMed]
30. Amelot, A.; Terrier, L.-M.; Cognacq, G.; Jecko, V.; Marlier, B.; Seizeur, R.; Emery, E.; Bauchet, L.; Roualdes, V.; Voirin, J.; et al. Natural History of Spinal Cord Metastasis from Brain Glioblastomas. *J. Neurooncol.* **2023**, *162*, 373–382. [CrossRef] [PubMed]
31. Lim, A.Z.; Ng, Y.S.; Blain, A.; Jiminez-Moreno, C.; Alston, C.L.; Nesbitt, V.; Simmons, L.; Santra, S.; Wassmer, E.; Blakely, E.L.; et al. Natural History of Leigh Syndrome: A Study of Disease Burden and Progression. *Ann. Neurol.* **2022**, *91*, 117–130. [CrossRef]
32. Gillette-Guyonnet, S.; Secher, M.; Vellas, B. Nutrition and Neurodegeneration: Epidemiological Evidence and Challenges for Future Research. *Br. J. Clin. Pharmacol.* **2013**, *75*, 738–755. [CrossRef]
33. Chen, Y.; Michalak, M.; Agellon, L.B. Importance of Nutrients and Nutrient Metabolism on Human Health. *Yale J. Biol. Med.* **2018**, *91*, 95–103.

34. Nasri, H.; Baradaran, A.; Shirzad, H.; Rafieian-Kopaei, M. New Concepts in Nutraceuticals as Alternative for Pharmaceuticals. *Int. J. Prev. Med.* **2014**, *5*, 1487–1499.
35. Kearney, P.M.; Whelton, M.; Reynolds, K.; Muntner, P.; Whelton, P.K.; He, J. Global Burden of Hypertension: Analysis of Worldwide Data. *Lancet* **2005**, *365*, 217–223. [CrossRef]
36. Kelly, T.; Yang, W.; Chen, C.-S.; Reynolds, K.; He, J. Global Burden of Obesity in 2005 and Projections to 2030. *Int. J. Obes.* **2008**, *32*, 1431–1437. [CrossRef] [PubMed]
37. Shaw, J.E.; Sicree, R.A.; Zimmet, P.Z. Global Estimates of the Prevalence of Diabetes for 2010 and 2030. *Diabetes Res. Clin. Pract.* **2010**, *87*, 4–14. [CrossRef]
38. Stanaway, J.D.; Afshin, A.; Gakidou, E.; Lim, S.S.; Abate, D.; Abate, K.H.; Abbafati, C.; Abbasi, N.; Abbastabar, H.; Abd-Allah, F.; et al. Global, Regional, and National Comparative Risk Assessment of 84 Behavioural, Environmental and Occupational, and Metabolic Risks or Clusters of Risks for 195 Countries and Territories, 1990–2017: A Systematic Analysis for the Global Burden of Disease Study 2017. *Lancet* **2018**, *392*, 1923–1994. [CrossRef]
39. World Health Organization. Healthy Diet 2015. Available online: https://www.who.int/news-room/fact-sheets/detail/healthy-diet/ (accessed on 2 August 2024).
40. Van Der Knaap, J.A.; Verrijzer, C.P. Undercover: Gene Control by Metabolites and Metabolic Enzymes. *Genes Dev.* **2016**, *30*, 2345–2369. [CrossRef] [PubMed]
41. Donald, B. Chapter5: Nutrients as Regulators of Gene Expression. In *Scientific Advances in Animal Nutrition: Promise for the New Century: Proceedings of a Symposium*; National Academies Press: Washington, DC, USA, 2001; p. 102997152024. Available online: https://nap.nationalacademies.org/read/10299/chapter/7 (accessed on 15 June 2024).
42. Johnston, J.D.; Ordovás, J.M.; Scheer, F.A.; Turek, F.W. Circadian Rhythms, Metabolism, and Chrononutrition in Rodents and Humans. *Adv. Nutr.* **2016**, *7*, 399–406. [CrossRef] [PubMed]
43. Dallman, M.F. Stress-Induced Obesity and the Emotional Nervous System. *Trends Endocrinol. Metab.* **2010**, *21*, 159–165. [CrossRef]
44. Feil, R.; Fraga, M.F. Epigenetics and the Environment: Emerging Patterns and Implications. *Nat. Rev. Genet.* **2012**, *13*, 97–109. [CrossRef]
45. Kyrou, I.; Tsigos, C. Stress Hormones: Physiological Stress and Regulation of Metabolism. *Curr. Opin. Pharmacol.* **2009**, *9*, 787–793. [CrossRef]
46. Jaenisch, R.; Bird, A. Epigenetic Regulation of Gene Expression: How the Genome Integrates Intrinsic and Environmental Signals. *Nat. Genet.* **2003**, *33*, 245–254. [CrossRef] [PubMed]
47. Nicolia, V.; Lucarelli, M.; Fuso, A. Environment, Epigenetics and Neurodegeneration: Focus on Nutrition in Alzheimer's Disease. *Exp. Gerontol.* **2015**, *68*, 8–12. [CrossRef]
48. Ordovás, J.M.; Smith, C.E. Epigenetics and Cardiovascular Disease. *Nat. Rev. Cardiol.* **2010**, *7*, 510–519. [CrossRef]
49. Teslovich, T.M.; Musunuru, K.; Smith, A.V.; Edmondson, A.C.; Stylianou, I.M.; Koseki, M.; Pirruccello, J.P.; Ripatti, S.; Chasman, D.I.; Willer, C.J.; et al. Biological, Clinical and Population Relevance of 95 Loci for Blood Lipids. *Nature* **2010**, *466*, 707–713. [CrossRef]
50. Estrada, J.A.; Contreras, I. Nutritional Modulation of Immune and Central Nervous System Homeostasis: The Role of Diet in Development of Neuroinflammation and Neurological Disease. *Nutrients* **2019**, *11*, 1076. [CrossRef]
51. Bhatara, V.S.; Bernstein, B.; Fazili, S. Complementary and Integrative Treatments of Aggressiveness/Emotion Dysregulation: Associated with Disruptive Disorders and Disruptive Mood Dysregulation Disorder. *Child Adolesc. Psychiatr. Clin. N. Am.* **2023**, *32*, 297–315. [CrossRef] [PubMed]
52. Swierz, M.J.; Storman, D.; Jasinska, K.W.; Storman, M.; Staskiewicz, W.; Gorecka, M.; Skuza, A.; Tobola, P.; Bala, M.M. Systematic Review and Meta-Analysis of Perioperative Behavioral Lifestyle and Nutritional Interventions in Bariatric Surgery: A Call for Better Research and Reporting. *Surg. Obes. Relat. Dis.* **2020**, *16*, 2088–2104. [CrossRef] [PubMed]
53. Cobley, J.N.; Fiorello, M.L.; Bailey, D.M. 13 Reasons Why the Brain Is Susceptible to Oxidative Stress. *Redox Biol.* **2018**, *15*, 490–503. [CrossRef]
54. Gómez-Pinilla, F. Brain Foods: The Effects of Nutrients on Brain Function. *Nat. Rev. Neurosci.* **2008**, *9*, 568–578. [CrossRef]
55. Holton, K.F. Micronutrients May Be a Unique Weapon Against the Neurotoxic Triad of Excitotoxicity, Oxidative Stress and Neuroinflammation: A Perspective. *Front. Neurosci.* **2021**, *15*, 726457. [CrossRef]
56. Lahoda Brodska, H.; Klempir, J.; Zavora, J.; Kohout, P. The Role of Micronutrients in Neurological Disorders. *Nutrients* **2023**, *15*, 4129. [CrossRef] [PubMed]
57. Poulose, S.M.; Miller, M.G.; Scott, T.; Shukitt-Hale, B. Nutritional Factors Affecting Adult Neurogenesis and Cognitive Function. *Adv. Nutr.* **2017**, *8*, 804–811. [CrossRef]
58. Smith, A.D.; Smith, S.M.; de Jager, C.A.; Whitbread, P.; Johnston, C.; Agacinski, G.; Oulhaj, A.; Bradley, K.M.; Jacoby, R.; Refsum, H. Homocysteine-Lowering by B Vitamins Slows the Rate of Accelerated Brain Atrophy in Mild Cognitive Impairment: A Randomized Controlled Trial. *PLoS ONE* **2010**, *5*, e12244. [CrossRef]
59. Zhang, L.; Chen, X.; Chen, Y.; Yan, J.; Huang, G.; Li, W. A Comparative Study Evaluating the Effectiveness of Folate-Based B Vitamin Intervention on Cognitive Function of Older Adults under Mandatory Folic Acid Fortification Policy: A Systematic Review and Meta-Analysis of Randomized Controlled Trials. *Nutrients* **2024**, *16*, 2199. [CrossRef]
60. Grundman, M. Vitamin E and Alzheimer Disease: The Basis for Additional Clinical Trials. *Am. J. Clin. Nutr.* **2000**, *71*, 630S–636S. [CrossRef]

61. Pierrot-Deseilligny, C.; Souberbielle, J.-C. Vitamin D and Multiple Sclerosis: An Update. *Mult. Scler. Relat. Disord.* **2017**, *14*, 35–45. [CrossRef] [PubMed]
62. Mavar, M.; Sorić, T.; Bagarić, E.; Sarić, A.; Matek Sarić, M. The Power of Vitamin D: Is the Future in Precision Nutrition through Personalized Supplementation Plans? *Nutrients* **2024**, *16*, 1176. [CrossRef]
63. Dales, J.-P.; Desplat-Jégo, S. Metal Imbalance in Neurodegenerative Diseases with a Specific Concern to the Brain of Multiple Sclerosis Patients. *Int. J. Mol. Sci.* **2020**, *21*, 9105. [CrossRef]
64. Barbagallo, M.; Dominguez, L.J. Magnesium and Aging. *Curr. Pharm. Des.* **2010**, *16*, 832–839. [CrossRef] [PubMed]
65. McCann, J.C.; Ames, B.N. An Overview of Evidence for a Causal Relation between Iron Deficiency during Development and Deficits in Cognitive or Behavioral Function. *Am. J. Clin. Nutr.* **2007**, *85*, 931–945. [CrossRef]
66. Parkin, P.C.; Borkhoff, C.M.; Macarthur, C.; Abdullah, K.; Birken, C.S.; Fehlings, D.; Koroshegyi, C.; Maguire, J.L.; Mamak, E.; Mamdani, M.; et al. Randomized Trial of Oral Iron and Diet Advice versus Diet Advice Alone in Young Children with Nonanemic Iron Deficiency. *J. Pediatr.* **2021**, *233*, 233–240.e1. [CrossRef] [PubMed]
67. Morris, M.C.; Evans, D.A.; Tangney, C.C.; Bienias, J.L.; Wilson, R.S. Fish Consumption and Cognitive Decline with Age in a Large Community Study. *Arch. Neurol.* **2005**, *62*, 1849–1853. [CrossRef]
68. Zhang, Y.; Chen, J.; Qiu, J.; Li, Y.; Wang, J.; Jiao, J. Intakes of Fish and Polyunsaturated Fatty Acids and Mild-to-Severe Cognitive Impairment Risks: A Dose-Response Meta-Analysis of 21 Cohort Studies. *Am. J. Clin. Nutr.* **2016**, *103*, 330–340. [CrossRef]
69. Duarte, G.M.; De Araújo, F.E.A.; Da Rocha, J.M.C.; Idalina Neta, F.; Do Rego, A.C.M.; Araújo Filho, I.; Pinheiro, F.I.; De Azevedo, E.P.; Cobucci, R.N.; Guzen, F.P. Neuroprotective Potential of Seed Extracts: Review of In Vitro and In Vivo Studies. *Nutrients* **2023**, *15*, 2502. [CrossRef]
70. Wang, S.; Luo, Q.; Zhou, Y.; Fan, P. CLG from Hemp Seed Inhibits LPS-Stimulated Neuroinflammation in BV2 Microglia by Regulating NF-κB and Nrf-2 Pathways. *ACS Omega* **2019**, *4*, 16517–16523. [CrossRef]
71. Wang, S.; Luo, Q.; Fan, P. Cannabisin F from Hemp (*Cannabis sativa*) Seed Suppresses Lipopolysaccharide-Induced Inflammatory Responses in BV2 Microglia as SIRT1 Modulator. *Int. J. Mol. Sci.* **2019**, *20*, 507. [CrossRef] [PubMed]
72. Bhuiyan, M.I.H.; Kim, J.Y.; Ha, T.J.; Kim, S.Y.; Cho, K.-O. Anthocyanins Extracted from Black Soybean Seed Coat Protect Primary Cortical Neurons against *in Vitro* Ischemia. *Biol. Pharm. Bull.* **2012**, *35*, 999–1008. [CrossRef]
73. Ramírez-Salazar, S.A.; Herren, C.; McCartney, J.; Ortiz García, J.G. Dietary Insights in Neurological Diseases. *Curr. Neurol. Neurosci. Rep.* **2021**, *21*, 55. [CrossRef] [PubMed]
74. Francis, H.M.; Stevenson, R.J. Potential for Diet to Prevent and Remediate Cognitive Deficits in Neurological Disorders. *Nutr. Rev.* **2018**, *76*, 204–217. [CrossRef]
75. Chamoun, E.; Mutch, D.M.; Allen-Vercoe, E.; Buchholz, A.C.; Duncan, A.M.; Spriet, L.L.; Haines, J.; Ma, D.W.L.; on behalf of the Guelph Family Health Study. A Review of the Associations between Single Nucleotide Polymorphisms in Taste Receptors, Eating Behaviors, and Health. *Crit. Rev. Food Sci. Nutr.* **2018**, *58*, 194–207. [CrossRef]
76. Gardener, H.; Caunca, M.R. Mediterranean Diet in Preventing Neurodegenerative Diseases. *Curr. Nutr. Rep.* **2018**, *7*, 10–20. [CrossRef]
77. Estruch, R. Anti-Inflammatory Effects of the Mediterranean Diet: The Experience of the PREDIMED Study. *Proc. Nutr. Soc.* **2010**, *69*, 333–340. [CrossRef]
78. Scarmeas, N.; Stern, Y.; Tang, M.-X.; Mayeux, R.; Luchsinger, J.A. Mediterranean Diet and Risk for Alzheimer's Disease. *Ann. Neurol.* **2006**, *59*, 912–921. [CrossRef] [PubMed]
79. Alcalay, R.N.; Gu, Y.; Mejia-Santana, H.; Cote, L.; Marder, K.S.; Scarmeas, N. The Association between Mediterranean Diet Adherence and Parkinson's Disease. *Mov. Disord.* **2012**, *27*, 771–774. [CrossRef]
80. Maraki, M.I.; Yannakoulia, M.; Stamelou, M.; Stefanis, L.; Xiromerisiou, G.; Kosmidis, M.H.; Dardiotis, E.; Hadjigeorgiou, G.M.; Sakka, P.; Anastasiou, C.A.; et al. Mediterranean Diet Adherence Is Related to Reduced Probability of Prodromal Parkinson's Disease. *Mov. Disord.* **2019**, *34*, 48–57. [CrossRef]
81. Metcalfe-Roach, A.; Yu, A.C.; Golz, E.; Cirstea, M.; Sundvick, K.; Kliger, D.; Foulger, L.H.; Mackenzie, M.; Finlay, B.B.; Appel-Cresswell, S. MIND and Mediterranean Diets Associated with Later Onset of Parkinson's Disease. *Mov. Disord.* **2021**, *36*, 977–984. [CrossRef] [PubMed]
82. Wahls, T.L. Dietary Approaches to Treating Multiple Sclerosis-Related Symptoms. *Phys. Med. Rehabil. Clin.* **2022**, *33*, 605–620. [CrossRef] [PubMed]
83. Krikorian, R.; Shidler, M.D.; Summer, S.S.; Sullivan, P.G.; Duker, A.P.; Isaacson, R.S.; Espay, A.J. Nutritional Ketosis for Mild Cognitive Impairment in Parkinson's Disease: A Controlled Pilot Trial. *Clin. Park. Relat. Disord.* **2019**, *1*, 41–47. [CrossRef] [PubMed]
84. Zhang, W.; Chen, S.; Huang, X.; Tong, H.; Niu, H.; Lu, L. Neuroprotective Effect of a Medium-Chain Triglyceride Ketogenic Diet on MPTP-Induced Parkinson's Disease Mice: A Combination of Transcriptomics and Metabolomics in the Substantia Nigra and Fecal Microbiome. *Cell Death Discov.* **2023**, *9*, 251. [CrossRef]
85. Krikorian, R.; Shidler, M.D.; Dangelo, K.; Couch, S.C.; Benoit, S.C.; Clegg, D.J. Dietary Ketosis Enhances Memory in Mild Cognitive Impairment. *Neurobiol. Aging* **2012**, *33*, 425.e19–425.e27. [CrossRef]
86. Ko, A.; Kwon, H.E.; Kim, H.D. Updates on the Ketogenic Diet Therapy for Pediatric Epilepsy. *Biomed. J.* **2022**, *45*, 19–26. [CrossRef]
87. Ruan, Y.; Chen, L.; She, D.; Chung, Y.; Ge, L.; Han, L. Ketogenic Diet for Epilepsy: An Overview of Systematic Review and Meta-Analysis. *Eur. J. Clin. Nutr.* **2022**, *76*, 1234–1244. [CrossRef]

88. Verrotti, A.; Iapadre, G.; Di Francesco, L.; Zagaroli, L.; Farello, G. Diet in the Treatment of Epilepsy: What We Know So Far. *Nutrients* **2020**, *12*, 2645. [CrossRef]
89. Wells, J.; Swaminathan, A.; Paseka, J.; Hanson, C. Efficacy and Safety of a Ketogenic Diet in Children and Adolescents with Refractory Epilepsy-A Review. *Nutrients* **2020**, *12*, 1809. [CrossRef]
90. Mazurek, D.; Wyka, J. Down Syndrome--Genetic and Nutritional Aspects of Accompanying Disorders. *Rocz. Panstw. Zakl. Hig.* **2015**, *66*, 189–194.
91. Duarte, A.C.; Costa, E.C.; Filipe, H.A.L.; Saraiva, S.M.; Jacinto, T.; Miguel, S.P.; Ribeiro, M.P.; Coutinho, P. Animal-Derived Products in Science and Current Alternatives. *Biomater. Adv.* **2023**, *151*, 213428. [CrossRef] [PubMed]
92. Jensen, C.; Teng, Y. Is It Time to Start Transitioning from 2D to 3D Cell Culture? *Front. Mol. Biosci.* **2020**, *7*, 33. [CrossRef]
93. Kim, J.; Koo, B.-K.; Knoblich, J.A. Human Organoids: Model Systems for Human Biology and Medicine. *Nat. Rev. Mol. Cell Biol.* **2020**, *21*, 571–584. [CrossRef]
94. Lee, H.-K.; Sanchez, C.V.; Chen, M.; Morin, P.J.; Wells, J.M.; Hanlon, E.B.; Xia, W. Three Dimensional Human Neuro-Spheroid Model of Alzheimer's Disease Based on Differentiated Induced Pluripotent Stem Cells. *PLoS ONE* **2016**, *11*, e0163072. [CrossRef] [PubMed]
95. Benchoua, A.; Lasbareilles, M.; Tournois, J. Contribution of Human Pluripotent Stem Cell-Based Models to Drug Discovery for Neurological Disorders. *Cells* **2021**, *10*, 3290. [CrossRef]
96. Ma, K.; Gauthier, L.O.; Cheung, F.; Huang, S.; Lek, M. High-Throughput Assays to Assess Variant Effects on Disease. *Dis. Model. Mech.* **2024**, *17*, dmm050573. [CrossRef]
97. Lee, D.D.H.; Cardinale, D.; Nigro, E.; Butler, C.R.; Rutman, A.; Fassad, M.R.; Hirst, R.A.; Moulding, D.; Agrotis, A.; Forsythe, E.; et al. Higher Throughput Drug Screening for Rare Respiratory Diseases: Readthrough Therapy in Primary Ciliary Dyskinesia. *Eur. Respir. J.* **2021**, *58*, 2000455. [CrossRef] [PubMed]
98. Bellomo, F.; Medina, D.L.; De Leo, E.; Panarella, A.; Emma, F. High-Content Drug Screening for Rare Diseases. *J. Inherit. Metab. Dis.* **2017**, *40*, 601–607. [CrossRef] [PubMed]
99. Lee, D.W.; Doh, I.; Nam, D.-H. Unified 2D and 3D Cell-Based High-Throughput Screening Platform Using a Micropillar/Microwell Chip. *Sens. Actuators Chem.* **2016**, *228*, 523–528. [CrossRef]
100. Yamamoto, S.; Kanca, O.; Wangler, M.F.; Bellen, H.J. Integrating Non-Mammalian Model Organisms in the Diagnosis of Rare Genetic Diseases in Humans. *Nat. Rev. Genet.* **2024**, *25*, 46–60. [CrossRef]
101. Cabasso, O.; Kuppuramalingam, A.; Lelieveld, L.; Van der Lienden, M.; Boot, R.; Aerts, J.M.; Horowitz, M. Animal Models for the Study of Gaucher Disease. *Int. J. Mol. Sci.* **2023**, *24*, 16035. [CrossRef] [PubMed]
102. Vallender, E.J.; Hotchkiss, C.E.; Lewis, A.D.; Rogers, J.; Stern, J.A.; Peterson, S.M.; Ferguson, B.; Sayers, K. Nonhuman Primate Genetic Models for the Study of Rare Diseases. *Orphanet J. Rare Dis.* **2023**, *18*, 20. [CrossRef] [PubMed]
103. Phillips, K.A.; Bales, K.L.; Capitanio, J.P.; Conley, A.; Czoty, P.W.; 't Hart, B.A.; Hopkins, W.D.; Hu, S.-L.; Miller, L.A.; Nader, M.A.; et al. Why Primate Models Matter. *Am. J. Primatol.* **2014**, *76*, 801–827. [CrossRef] [PubMed]
104. Bryda, E.C. The Mighty Mouse: The Impact of Rodents on Advances in Biomedical Research. *Mo. Med.* **2013**, *110*, 207–211.
105. Doyle, A.; McGarry, M.P.; Lee, N.A.; Lee, J.J. The Construction of Transgenic and Gene Knockout/Knockin Mouse Models of Human Disease. *Transgenic Res.* **2012**, *21*, 327–349. [CrossRef]
106. Bergen, W.G.; Mersmann, H.J. Comparative Aspects of Lipid Metabolism: Impact on Contemporary Research and Use of Animal Models1. *J. Nutr.* **2005**, *135*, 2499–2502. [CrossRef]
107. Duchen, L.W.; Eicher, E.M.; Jacobs, J.M.; Scaravilli, F.; Teixeira, F. Hereditary Leucodystrophy in the Mouse: The New Mutant Twitcher. *Brain* **1980**, *103*, 695–710. [CrossRef]
108. Da Silva-Buttkus, P.; Spielmann, N.; Klein-Rodewald, T.; Schütt, C.; Aguilar-Pimentel, A.; Amarie, O.V.; Becker, L.; Calzada-Wack, J.; Garrett, L.; Gerlini, R.; et al. Knockout Mouse Models as a Resource for the Study of Rare Diseases. *Mamm. Genome* **2023**, *34*, 244–261. [CrossRef]
109. Julkowska, D.; Austin, C.P.; Cutillo, C.M.; Gancberg, D.; Hager, C.; Halftermeyer, J.; Jonker, A.H.; Lau, L.P.L.; Norstedt, I.; Rath, A.; et al. The Importance of International Collaboration for Rare Diseases Research: A European Perspective. *Gene Ther.* **2017**, *24*, 562–571. [CrossRef]
110. Monaco, L.; Zanello, G.; Baynam, G.; Jonker, A.H.; Julkowska, D.; Hartman, A.L.; O'Connor, D.; Wang, C.M.; Wong-Rieger, D.; Pearce, D.A. Research on Rare Diseases: Ten Years of Progress and Challenges at IRDiRC. *Nat. Rev. Drug Discov.* **2022**, *21*, 319–320. [CrossRef]
111. Kropp, P.A.; Bauer, R.; Zafra, I.; Graham, C.; Golden, A. *Caenorhabditis elegans* for Rare Disease Modeling and Drug Discovery: Strategies and Strengths. *Dis. Model. Mech.* **2021**, *14*, dmm049010. [CrossRef] [PubMed]
112. O'Kane, C.J. Modelling Human Diseases in *Drosophila* and *Caenorhabditis*. *Semin. Cell Dev. Biol.* **2003**, *14*, 3–10. [CrossRef]
113. Lai, C.-H.; Chou, C.-Y.; Ch'ang, L.-Y.; Liu, C.-S.; Lin, W. Identification of Novel Human Genes Evolutionarily Conserved in *Caenorhabditis elegans* by Comparative Proteomics. *Genome Res.* **2000**, *10*, 703–713. [CrossRef] [PubMed]
114. Schafer, W.R. Deciphering the Neural and Molecular Mechanisms of *C. elegans* Behavior. *Curr. Biol.* **2005**, *15*, R723–R729. [CrossRef]
115. Paix, A.; Folkmann, A.; Rasoloson, D.; Seydoux, G. High Efficiency, Homology-Directed Genome Editing in *Caenorhabditis elegans* Using CRISPR-Cas9 Ribonucleoprotein Complexes. *Genetics* **2015**, *201*, 47–54. [CrossRef]

116. Paix, A.; Schmidt, H.; Seydoux, G. Cas9-Assisted Recombineering in *C. elegans*: Genome Editing Using in vivo Assembly of Linear DNAs. *Nucleic Acids Res.* **2016**, *44*, e128. [CrossRef]
117. Aoyagi, K.; Rossignol, E.; Hamdan, F.F.; Mulcahy, B.; Xie, L.; Nagamatsu, S.; Rouleau, G.A.; Zhen, M.; Michaud, J.L. A Gain-of-Function Mutation in *NALCN* in a Child with Intellectual Disability, Ataxia, and Arthrogryposis. *Hum. Mutat.* **2015**, *36*, 753–757. [CrossRef] [PubMed]
118. Lange, K.I.; Tsiropoulou, S.; Kucharska, K.; Blacque, O.E. Interpreting the Pathogenicity of Joubert Syndrome Missense Variants in *Caenorhabditis elegans*. *Dis. Model. Mech.* **2021**, *14*, dmm046631. [CrossRef]
119. Bachmann-Gagescu, R.; Dempsey, J.C.; Phelps, I.G.; O'Roak, B.J.; Knutzen, D.M.; Rue, T.C.; Ishak, G.E.; Isabella, C.R.; Gorden, N.; Adkins, J.; et al. Joubert Syndrome: A Model for Untangling Recessive Disorders with Extreme Genetic Heterogeneity. *J. Med. Genet.* **2015**, *52*, 514–522. [CrossRef]
120. Jia, F.; Cui, M.; Than, M.T.; Han, M. Developmental Defects of *Caenorhabditis elegans* Lacking Branched-Chain α-Ketoacid Dehydrogenase Are Mainly Caused by Monomethyl Branched-Chain Fatty Acid Deficiency. *J. Biol. Chem.* **2016**, *291*, 2967–2973. [CrossRef]
121. Verheyen, E.M. The Power of *Drosophila* in Modeling Human Disease Mechanisms. *Dis. Model. Mech.* **2022**, *15*, dmm049549. [CrossRef] [PubMed]
122. Adams, M.D.; Celniker, S.E.; Holt, R.A.; Evans, C.A.; Gocayne, J.D.; Amanatides, P.G.; Scherer, S.E.; Li, P.W.; Hoskins, R.A.; Galle, R.F.; et al. The Genome Sequence of *Drosophila melanogaster*. *Science* **2000**, *287*, 2185–2195. [CrossRef] [PubMed]
123. Hu, Y.; Flockhart, I.; Vinayagam, A.; Bergwitz, C.; Berger, B.; Perrimon, N.; Mohr, S.E. An Integrative Approach to Ortholog Prediction for Disease-Focused and Other Functional Studies. *BMC Bioinform.* **2011**, *12*, 357. [CrossRef]
124. Reiter, L.T.; Potocki, L.; Chien, S.; Gribskov, M.; Bier, E. A Systematic Analysis of Human Disease-Associated Gene Sequences in *Drosophila melanogaster*. *Genome Res.* **2001**, *11*, 1114–1125. [CrossRef]
125. Bellen, H.J.; Wangler, M.F.; Yamamoto, S. The Fruit Fly at the Interface of Diagnosis and Pathogenic Mechanisms of Rare and Common Human Diseases. *Hum. Mol. Genet.* **2019**, *28*, R207–R214. [CrossRef]
126. Harnish, J.M.; Deal, S.L.; Chao, H.-T.; Wangler, M.F.; Yamamoto, S. In Vivo Functional Study of Disease-Associated Rare Human Variants Using *Drosophila*. *J. Vis. Exp.* **2019**, *150*. [CrossRef]
127. Kanca, O.; Bellen, H.J.; Schnorrer, F. Gene Tagging Strategies to Assess Protein Expression, Localization, and Function in *Drosophila*. *Genetics* **2017**, *207*, 389–412. [CrossRef]
128. Mele, S.; Martelli, F.; Lin, J.; Kanca, O.; Christodoulou, J.; Bellen, H.J.; Piper, M.D.W.; Johnson, T.K. *Drosophila* as a Diet Discovery Tool for Treating Amino Acid Disorders. *Trends Endocrinol. Metab.* **2023**, *34*, 85–105. [CrossRef] [PubMed]
129. Martelli, F.; Quig, A.; Mele, S.; Lin, J.; Fulton, T.L.; Wansbrough, M.; Barlow, C.K.; Schittenhelm, R.B.; Johnson, T.K.; Piper, M.D.W. A Defined Diet for Pre-Adult *Drosophila melanogaster*. *Sci. Rep.* **2024**, *14*, 6974. [CrossRef]
130. Baxendale, S.; Van Eeden, F.; Wilkinson, R. The Power of Zebrafish in Personalised Medicine. In *Personalised Medicine*; El-Khamisy, S., Ed.; Advances in Experimental Medicine and Biology; Springer International Publishing: Cham, Switzerland, 2017; Volume 1007, pp. 179–197. ISBN 978-3-319-60731-3.
131. Howe, K.; Clark, M.D.; Torroja, C.F.; Torrance, J.; Berthelot, C.; Muffato, M.; Collins, J.E.; Humphray, S.; McLaren, K.; Matthews, L.; et al. The Zebrafish Reference Genome Sequence and Its Relationship to the Human Genome. *Nature* **2013**, *496*, 498–503. [CrossRef]
132. Choi, T.-Y.; Choi, T.-I.; Lee, Y.-R.; Choe, S.-K.; Kim, C.-H. Zebrafish as an Animal Model for Biomedical Research. *Exp. Mol. Med.* **2021**, *53*, 310–317. [CrossRef] [PubMed]
133. Son, M.; Kim, D.Y.; Kim, C.-H. Disease Modeling of Rare Neurological Disorders in Zebrafish. *Int. J. Mol. Sci.* **2022**, *23*, 3946. [CrossRef]
134. Crouzier, L.; Richard, E.M.; Sourbron, J.; Lagae, L.; Maurice, T.; Delprat, B. Use of Zebrafish Models to Boost Research in Rare Genetic Diseases. *Int. J. Mol. Sci.* **2021**, *22*, 13356. [CrossRef] [PubMed]
135. Sloan, J.L.; Achilly, N.P.; Arnold, M.L.; Catlett, J.L.; Blake, T.; Bishop, K.; Jones, M.; Harper, U.; English, M.A.; Anderson, S.; et al. The Vitamin B12 Processing Enzyme, Mmachc, Is Essential for Zebrafish Survival, Growth and Retinal Morphology. *Hum. Mol. Genet.* **2020**, *29*, 2109–2123. [CrossRef] [PubMed]
136. Rath, A.; Salamon, V.; Peixoto, S.; Hivert, V.; Laville, M.; Segrestin, B.; Neugebauer, E.A.M.; Eikermann, M.; Bertele, V.; Garattini, S.; et al. A Systematic Literature Review of Evidence-Based Clinical Practice for Rare Diseases: What Are the Perceived and Real Barriers for Improving the Evidence and How Can They Be Overcome? *Trials* **2017**, *18*, 556. [CrossRef]
137. Chang, X.; Zhang, J.; Jiang, Y.; Shang, M.; Wu, Y. A Survey of Registered Pharmacological Clinical Trials on Rare Neurological Diseases in Children in 2010–2020. *Front. Pediatr.* **2022**, *10*, 963601. [CrossRef]
138. Lumsden, J.M.; Urv, T.K. The Rare Diseases Clinical Research Network: A Model for Clinical Trial Readiness. *Ther. Adv. Rare Dis.* **2023**, *4*, 26330040231219272. [CrossRef]
139. U.S. Food & Drug Administration. Rare Diseases: Natural History Studies for Drug Development 2019. Available online: https://www.Fda.Gov/Regulatory-Information/Search-Fda-Guidance-Documents/Rare-Diseases-Natural-History-Studies-Drug-Development (accessed on 28 June 2024).
140. U.S. Food & Drug Administration. FDA Approves Givosiran for Acute Hepatic Porphyria: U.S. Food and Drug Administration. Available online: https://www.Fda.Gov/Drugs/Resources-Information-Approved-Drugs/Fda-Approves-Givosiran-Acute-Hepatic-Porphyria (accessed on 18 December 2023).

141. Al-Horani, R.A.; Chiles, R. First Therapeutic Approval for Eosinophilic Esophagitis. *Gastroenterol. Insights* **2022**, *13*, 238–244. [CrossRef]
142. U.S. Food & Drug Administration. FDA Approves First Treatment for Rett Syndrome. Available online: https://www.fda.gov/drugs/news-events-human-drugs/fda-approves-first-treatment-rett-syndrome (accessed on 18 June 2024).
143. Piwek, L.; Ellis, D.A.; Andrews, S.; Joinson, A. The Rise of Consumer Health Wearables: Promises and Barriers. *PLoS Med.* **2016**, *13*, e1001953. [CrossRef]
144. Youn, B.-Y.; Ko, Y.; Moon, S.; Lee, J.; Ko, S.-G.; Kim, J.-Y. Digital Biomarkers for Neuromuscular Disorders: A Systematic Scoping Review. *Diagnostics* **2021**, *11*, 1275. [CrossRef]
145. Sica, M.; Tedesco, S.; Crowe, C.; Kenny, L.; Moore, K.; Timmons, S.; Barton, J.; O'Flynn, B.; Komaris, D.-S. Continuous Home Monitoring of Parkinson's Disease Using Inertial Sensors: A Systematic Review. *PLoS ONE* **2021**, *16*, e0246528. [CrossRef] [PubMed]
146. Vienne-Jumeau, A.; Quijoux, F.; Vidal, P.-P.; Ricard, D. Wearable Inertial Sensors Provide Reliable Biomarkers of Disease Severity in Multiple Sclerosis: A Systematic Review and Meta-Analysis. *Ann. Phys. Rehabil. Med.* **2020**, *63*, 138–147. [CrossRef] [PubMed]
147. Ricotti, V.; Kadirvelu, B.; Selby, V.; Festenstein, R.; Mercuri, E.; Voit, T.; Faisal, A.A. Wearable Full-Body Motion Tracking of Activities of Daily Living Predicts Disease Trajectory in Duchenne Muscular Dystrophy. *Nat. Med.* **2023**, *29*, 95–103. [CrossRef]
148. Liu, X.; Chen, H.; Zhong, Y.; Lee, T.-Y.; Han, W.; Yu, D.; Liu, H.; Ji, J. Diet Therapy in Patients with Rare Diseases: A Scoping Review. *J. Hum. Nutr. Diet.* **2023**, *36*, 742–753. [CrossRef] [PubMed]
149. Camp, K.M.; Lloyd-Puryear, M.A.; Yao, L.; Groft, S.C.; Parisi, M.A.; Mulberg, A.; Gopal-Srivastava, R.; Cederbaum, S.; Enns, G.M.; Ershow, A.G.; et al. Expanding Research to Provide an Evidence Base for Nutritional Interventions for the Management of Inborn Errors of Metabolism. *Mol. Genet. Metab.* **2013**, *109*, 319–328. [CrossRef]
150. Kishino, T.; Lalande, M.; Wagstaff, J. UBE3A/E6-AP Mutations Cause Angelman Syndrome. *Nat. Genet.* **1997**, *15*, 70–73. [CrossRef] [PubMed]
151. Samanta, D. Epilepsy in Angelman Syndrome: A Scoping Review. *Brain Dev.* **2021**, *43*, 32–44. [CrossRef]
152. Stein, D.; Chetty, M.; Rho, J.M. A "Happy" Toddler Presenting with Sudden, Life-Threatening Seizures. *Semin. Pediatr. Neurol.* **2010**, *17*, 35–38. [CrossRef]
153. Evangeliou, A.; Doulioglou, V.; Haidopoulou, K.; Aptouramani, M.; Spilioti, M.; Varlamis, G. Ketogenic Diet in a Patient with Angelman Syndrome. *Pediatr. Int.* **2010**, *52*, 831–834. [CrossRef]
154. Grocott, O.R.; Herrington, K.S.; Pfeifer, H.H.; Thiele, E.A.; Thibert, R.L. Low Glycemic Index Treatment for Seizure Control in Angelman Syndrome: A Case Series from the Center for Dietary Therapy of Epilepsy at the Massachusetts General Hospital. *Epilepsy Behav. EB* **2017**, *68*, 45–50. [CrossRef]
155. Thibert, R.L.; Pfeifer, H.H.; Larson, A.M.; Raby, A.R.; Reynolds, A.A.; Morgan, A.K.; Thiele, E.A. Low Glycemic Index Treatment for Seizures in Angelman Syndrome. *Epilepsia* **2012**, *53*, 1498–1502. [CrossRef] [PubMed]
156. Ciarlone, S.L.; Grieco, J.C.; D'Agostino, D.P.; Weeber, E.J. Ketone Ester Supplementation Attenuates Seizure Activity, and Improves Behavior and Hippocampal Synaptic Plasticity in an Angelman Syndrome Mouse Model. *Neurobiol. Dis.* **2016**, *96*, 38–46. [CrossRef] [PubMed]
157. Romero, L.O.; Caires, R.; Kaitlyn Victor, A.; Ramirez, J.; Sierra-Valdez, F.J.; Walsh, P.; Truong, V.; Lee, J.; Mayor, U.; Reiter, L.T.; et al. Linoleic Acid Improves PIEZO2 Dysfunction in a Mouse Model of Angelman Syndrome. *Nat. Commun.* **2023**, *14*, 1167. [CrossRef] [PubMed]
158. Herber, D.L.; Weeber, E.J.; D'Agostino, D.P.; Duis, J. Evaluation of the Safety and Tolerability of a Nutritional Formulation in Patients with ANgelman Syndrome (FANS): Study Protocol for a Randomized Controlled Trial. *Trials* **2020**, *21*, 60. [CrossRef] [PubMed]
159. Carson, R.P.; Herber, D.L.; Pan, Z.; Phibbs, F.; Key, A.P.; Gouelle, A.; Ergish, P.; Armour, E.A.; Patel, S.; Duis, J. Nutritional Formulation for Patients with Angelman Syndrome: A Randomized, Double-Blind, Placebo-Controlled Study of Exogenous Ketones. *J. Nutr.* **2021**, *151*, 3628–3636. [CrossRef]
160. Bird, L.M.; Tan, W.-H.; Bacino, C.A.; Peters, S.U.; Skinner, S.A.; Anselm, I.; Barbieri-Welge, R.; Bauer-Carlin, A.; Gentile, J.K.; Glaze, D.G.; et al. A Therapeutic Trial of Pro-Methylation Dietary Supplements in Angelman Syndrome. *Am. J. Med. Genet. Part A* **2011**, *155*, 2956–2963. [CrossRef]
161. Han, J.; Bichell, T.J.; Golden, S.; Anselm, I.; Waisbren, S.; Bacino, C.A.; Peters, S.U.; Bird, L.M.; Kimonis, V. A Placebo-Controlled Trial of Folic Acid and Betaine in Identical Twins with Angelman Syndrome. *Orphanet J. Rare Dis.* **2019**, *14*, 232. [CrossRef]
162. Neul, J.L.; Benke, T.A.; Marsh, E.D.; Skinner, S.A.; Merritt, J.; Lieberman, D.N.; Standridge, S.; Feyma, T.; Heydemann, P.; Peters, S.; et al. The Array of Clinical Phenotypes of Males with Mutations in Methyl-CpG Binding Protein 2. *Am. J. Med. Genet. B Neuropsychiatr. Genet.* **2019**, *180*, 55–67. [CrossRef]
163. Hagberg, B. Rett's Syndrome: Prevalence and Impact on Progressive Severe Mental Retardation in Girls. *Acta Paediatr. Scand.* **1985**, *74*, 405–408. [CrossRef]
164. Neul, J.L.; Kaufmann, W.E.; Glaze, D.G.; Christodoulou, J.; Clarke, A.J.; Bahi-Buisson, N.; Leonard, H.; Bailey, M.E.S.; Schanen, N.C.; Zappella, M.; et al. Rett Syndrome: Revised Diagnostic Criteria and Nomenclature. *Ann. Neurol.* **2010**, *68*, 944–950. [CrossRef]

165. Neul, J.L.; Lane, J.B.; Lee, H.-S.; Geerts, S.; Barrish, J.O.; Annese, F.; Baggett, L.M.; Barnes, K.; Skinner, S.A.; Motil, K.J.; et al. Developmental Delay in Rett Syndrome: Data from the Natural History Study. *J. Neurodev. Disord.* **2014**, *6*, 20. [CrossRef] [PubMed]
166. Hagberg, B. Clinical Manifestations and Stages of Rett Syndrome. *Ment. Retard. Dev. Disabil. Res. Rev.* **2002**, *8*, 61–65. [CrossRef] [PubMed]
167. Percy, A.K.; Neul, J.L.; Glaze, D.G.; Motil, K.J.; Skinner, S.A.; Khwaja, O.; Lee, H.-S.; Lane, J.B.; Barrish, J.O.; Annese, F.; et al. Rett Syndrome Diagnostic Criteria: Lessons from the Natural History Study. *Ann. Neurol.* **2010**, *68*, 951–955. [CrossRef] [PubMed]
168. Chahrour, M.; Zoghbi, H.Y. The Story of Rett Syndrome: From Clinic to Neurobiology. *Neuron* **2007**, *56*, 422–437. [CrossRef]
169. Amir, R.E.; Van den Veyver, I.B.; Wan, M.; Tran, C.Q.; Francke, U.; Zoghbi, H.Y. Rett Syndrome Is Caused by Mutations in X-Linked MECP2, Encoding Methyl-CpG-Binding Protein 2. *Nat. Genet.* **1999**, *23*, 185–188. [CrossRef]
170. Good, K.V.; Vincent, J.B.; Ausió, J. MeCP2: The Genetic Driver of Rett Syndrome Epigenetics. *Front. Genet.* **2021**, *12*, 620859. [CrossRef]
171. Frullanti, E.; Papa, F.T.; Grillo, E.; Clarke, A.; Ben-Zeev, B.; Pineda, M.; Bahi-Buisson, N.; Bienvenu, T.; Armstrong, J.; Roche Martinez, A.; et al. Analysis of the Phenotypes in the Rett Networked Database. *Int. J. Genom.* **2019**, *2019*, 6956934. [CrossRef]
172. Palmieri, M.; Pozzer, D.; Landsberger, N. Advanced Genetic Therapies for the Treatment of Rett Syndrome: State of the Art and Future Perspectives. *Front. Neurosci.* **2023**, *17*, 1172805. [CrossRef]
173. Cohen, D.R.S.; Matarazzo, V.; Palmer, A.M.; Tu, Y.; Jeon, O.-H.; Pevsner, J.; Ronnett, G.V. Expression of MeCP2 in Olfactory Receptor Neurons Is Developmentally Regulated and Occurs before Synaptogenesis. *Mol. Cell. Neurosci.* **2003**, *22*, 417–429. [CrossRef]
174. Belichenko, P.V.; Wright, E.E.; Belichenko, N.P.; Masliah, E.; Li, H.H.; Mobley, W.C.; Francke, U. Widespread Changes in Dendritic and Axonal Morphology in Mecp2-Mutant Mouse Models of Rett Syndrome: Evidence for Disruption of Neuronal Networks. *J. Comp. Neurol.* **2009**, *514*, 240–258. [CrossRef]
175. Kishi, N.; Macklis, J.D. Dissecting MECP2 Function in the Central Nervous System. *J. Child Neurol.* **2005**, *20*, 753–759. [CrossRef] [PubMed]
176. Fasolino, M.; Zhou, Z. The Crucial Role of DNA Methylation and MeCP2 in Neuronal Function. *Genes* **2017**, *8*, 141. [CrossRef] [PubMed]
177. Evans, J.C.; Archer, H.L.; Colley, J.P.; Ravn, K.; Nielsen, J.B.; Kerr, A.; Williams, E.; Christodoulou, J.; Gécz, J.; Jardine, P.E.; et al. Early Onset Seizures and Rett-like Features Associated with Mutations in CDKL5. *Eur. J. Hum. Genet. EJHG* **2005**, *13*, 1113–1120. [CrossRef]
178. Philippe, C.; Amsallem, D.; Francannet, C.; Lambert, L.; Saunier, A.; Verneau, F.; Jonveaux, P. Phenotypic Variability in Rett Syndrome Associated with FOXG1 Mutations in Females. *J. Med. Genet.* **2010**, *47*, 59–65. [CrossRef] [PubMed]
179. Neul, J.L.; Percy, A.K.; Benke, T.A.; Berry-Kravis, E.M.; Glaze, D.G.; Marsh, E.D.; Lin, T.; Stankovic, S.; Bishop, K.M.; Youakim, J.M. Trofinetide for the Treatment of Rett Syndrome: A Randomized Phase 3 Study. *Nat. Med.* **2023**, *29*, 1468–1475. [CrossRef]
180. Singh, A.; Balasundaram, M.K.; Gupta, D. Trofinetide in Rett Syndrome: A Brief Review of Safety and Efficacy. *Intractable Rare Dis. Res.* **2023**, *12*, 262–266. [CrossRef]
181. Czerwonogrodzka-Senczyna, A.; Milewska, M.; Kwiecień, P.; Szczałuba, K. Diet and Nutritional Status of Polish Girls with Rett Syndrome-A Case-Control Study. *Nutrients* **2023**, *15*, 3334. [CrossRef]
182. Lotan, M.; Zysman, L. The Digestive System and Nutritional Considerations for Individuals with Rett Syndrome. *Sci. World J.* **2006**, *6*, 1737–1749. [CrossRef]
183. Motil, K.J.; Barrish, J.O.; Lane, J.; Geerts, S.P.; Annese, F.; McNair, L.; Percy, A.K.; Skinner, S.A.; Neul, J.L.; Glaze, D.G. Vitamin D Deficiency Is Prevalent in Girls and Women with Rett Syndrome. *J. Pediatr. Gastroenterol. Nutr.* **2011**, *53*, 569–574. [CrossRef]
184. Haas, R.H.; Dixon, S.D.; Sartoris, D.J.; Hennessy, M.J. Osteopenia in Rett Syndrome. *J. Pediatr.* **1997**, *131*, 771–774. [CrossRef]
185. Motil, K.J.; Barrish, J.O.; Neul, J.L.; Glaze, D.G. Low Bone Mineral Mass Is Associated with Decreased Bone Formation and Diet in Females with Rett Syndrome. *J. Pediatr. Gastroenterol. Nutr.* **2014**, *59*, 386. [CrossRef]
186. Borghi, E.; Borgo, F.; Severgnini, M.; Savini, M.N.; Casiraghi, M.C.; Vignoli, A. Rett Syndrome: A Focus on Gut Microbiota. *Int. J. Mol. Sci.* **2017**, *18*, 344. [CrossRef] [PubMed]
187. Gallucci, A.; Patterson, K.; Weit, A.; Van Der Pol, W.; Dubois, L.; Percy, A.; Morrow, C.; Campbell, S.; Olsen, M. Microbial Community Changes in a Female Rat Model of Rett Syndrome. *Prog. Neuropsychopharmacol. Biol. Psychiatry* **2021**, *109*, 110259. [CrossRef]
188. Wong, L.C.; Hsu, C.J.; Wu, Y.T.; Chu, H.F.; Lin, J.H.; Wang, H.P.; Hu, S.C.; Tsai, Y.C.; Tsai, W.C.; Lee, W.T. Investigating the impact of probiotic on neurological outcomes in Rett syndrome: A randomized, double-blind, and placebo-controlled pilot study. *Autism* **2024**, *28*, 2267–2281. [CrossRef] [PubMed]
189. Chin, E.W.M.; Marcy, G.; Yoon, S.-I.; Ma, D.; Rosales, F.J.; Augustine, G.J.; Goh, E.L.K. Choline Ameliorates Disease Phenotypes in Human iPSC Models of Rett Syndrome. *Neuromolecular Med.* **2016**, *18*, 364–377. [CrossRef] [PubMed]
190. Nag, N.; Berger-Sweeney, J.E. Postnatal Dietary Choline Supplementation Alters Behavior in a Mouse Model of Rett Syndrome. *Neurobiol. Dis.* **2007**, *26*, 473–480. [CrossRef]
191. Nag, N.; Ward, B.; Berger-Sweeney, J.E. Nutritional Factors in a Mouse Model of Rett Syndrome. *Neurosci. Biobehav. Rev.* **2009**, *33*, 586–592. [CrossRef]

192. Freilinger, M.; Dunkler, D.; Lanator, I.; Item, C.B.; Mühl, A.; Fowler, B.; Bodamer, O.A. Effects of Creatine Supplementation in Rett Syndrome: A Randomized, Placebo-Controlled Trial. *J. Dev. Behav. Pediatr. JDBP* **2011**, *32*, 454–460. [CrossRef]
193. Cortelazzo, A.; Pietri, T.; De Felice, C.; Leoncini, S.; Guerranti, R.; Signorini, C.; Timperio, A.M.; Zolla, L.; Ciccoli, L.; Hayek, J. Proteomic Analysis of the Rett Syndrome Experimental Model mecp2Q63X Mutant Zebrafish. *J. Proteom.* **2017**, *154*, 128–133. [CrossRef]
194. Gold, W.A.; Williamson, S.L.; Kaur, S.; Hargreaves, I.P.; Land, J.M.; Pelka, G.J.; Tam, P.P.L.; Christodoulou, J. Mitochondrial Dysfunction in the Skeletal Muscle of a Mouse Model of Rett Syndrome (RTT): Implications for the Disease Phenotype. *Mitochondrion* **2014**, *15*, 10–17. [CrossRef]
195. De Felice, C.; Ciccoli, L.; Leoncini, S.; Signorini, C.; Rossi, M.; Vannuccini, L.; Guazzi, G.; Latini, G.; Comporti, M.; Valacchi, G.; et al. Systemic Oxidative Stress in Classic Rett Syndrome. *Free Radic. Biol. Med.* **2009**, *47*, 440–448. [CrossRef] [PubMed]
196. De Felice, C.; Della Ragione, F.; Signorini, C.; Leoncini, S.; Pecorelli, A.; Ciccoli, L.; Scalabrì, F.; Marracino, F.; Madonna, M.; Belmonte, G.; et al. Oxidative Brain Damage in Mecp2-Mutant Murine Models of Rett Syndrome. *Neurobiol. Dis.* **2014**, *68*, 66–77. [CrossRef] [PubMed]
197. Nagy, G.; Ackerman, S. Cholesterol Metabolism and Rett Syndrome Pathogenesis. *Fac. Res. 2013* **2013**, *45*, 965–967. [CrossRef]
198. Signorini, C.; De Felice, C.; Leoncini, S.; Giardini, A.; D'Esposito, M.; Filosa, S.; Della Ragione, F.; Rossi, M.; Pecorelli, A.; Valacchi, G.; et al. F4-Neuroprostanes Mediate Neurological Severity in Rett Syndrome. *Clin. Chim. Acta* **2011**, *412*, 1399–1406. [CrossRef]
199. Leoncini, S.; De Felice, C.; Signorini, C.; Pecorelli, A.; Durand, T.; Valacchi, G.; Ciccoli, L.; Hayek, J. Oxidative Stress in Rett Syndrome: Natural History, Genotype, and Variants. *Redox Rep.* **2011**, *16*, 145–153. [CrossRef]
200. Signorini, C.; De Felice, C.; Leoncini, S.; Durand, T.; Galano, J.-M.; Cortelazzo, A.; Zollo, G.; Guerranti, R.; Gonnelli, S.; Caffarelli, C.; et al. Altered Erythrocyte Membrane Fatty Acid Profile in Typical Rett Syndrome: Effects of Omega-3 Polyunsaturated Fatty Acid Supplementation. *Prostaglandins Leukot. Essent. Fatty Acids* **2014**, *91*, 183–193. [CrossRef]
201. De Felice, C.; Signorini, C.; Durand, T.; Ciccoli, L.; Leoncini, S.; D'Esposito, M.; Filosa, S.; Oger, C.; Guy, A.; Bultel-Poncé, V.; et al. Partial Rescue of Rett Syndrome by ω-3 Polyunsaturated Fatty Acids (PUFAs) Oil. *Genes Nutr.* **2012**, *7*, 447–458. [CrossRef] [PubMed]
202. Park, M.J.; Aja, S.; Li, Q.; Degano, A.L.; Penati, J.; Zhuo, J.; Roe, C.R.; Ronnett, G.V. Anaplerotic Triheptanoin Diet Enhances Mitochondrial Substrate Use to Remodel the Metabolome and Improve Lifespan, Motor Function, and Sociability in MeCP2-Null Mice. *PLoS ONE* **2014**, *9*, e109527. [CrossRef]
203. Mouro, F.M.; Miranda-Lourenço, C.; Sebastião, A.M.; Diógenes, M.J. From Cannabinoids and Neurosteroids to Statins and the Ketogenic Diet: New Therapeutic Avenues in Rett Syndrome? *Front. Neurosci.* **2019**, *13*, 680. [CrossRef]
204. Zamberletti, E.; Gabaglio, M.; Piscitelli, F.; Brodie, J.S.; Woolley-Roberts, M.; Barbiero, I.; Tramarin, M.; Binelli, G.; Landsberger, N.; Kilstrup-Nielsen, C.; et al. Cannabidivarin Completely Rescues Cognitive Deficits and Delays Neurological and Motor Defects in Male *Mecp2* Mutant Mice. *J. Psychopharmacol.* **2019**, *33*, 894–907. [CrossRef]
205. Vigli, D.; Cosentino, L.; Raggi, C.; Laviola, G.; Woolley-Roberts, M.; De Filippis, B. Chronic Treatment with the Phytocannabinoid Cannabidivarin (CBDV) Rescues Behavioural Alterations and Brain Atrophy in a Mouse Model of Rett Syndrome. *Neuropharmacology* **2018**, *140*, 121–129. [CrossRef] [PubMed]
206. Haas, R.H.; Rice, M.A.; Trauner, D.A.; Merritt, T.A.; Opitz, J.M.; Reynolds, J.F. Therapeutic Effects of a Ketogenic Diet in Rett Syndrome. *Am. J. Med. Genet.* **1986**, *25*, 225–246. [CrossRef] [PubMed]
207. Liebhaber, G.M.; Riemann, E.; Matthias Baumeister, F.A. Ketogenic Diet in Rett Syndrome. *J. Child Neurol.* **2003**, *18*, 74–75. [CrossRef] [PubMed]
208. van der Louw, E.; van den Hurk, D.; Neal, E.; Leiendecker, B.; Fitzsimmon, G.; Dority, L.; Thompson, L.; Marchió, M.; Dudzińska, M.; Dressler, A.; et al. Ketogenic Diet Guidelines for Infants with Refractory Epilepsy. *Eur. J. Paediatr. Neurol.* **2016**, *20*, 798–809. [CrossRef] [PubMed]
209. Rogawski, M.A.; Löscher, W.; Rho, J.M. Mechanisms of Action of Antiseizure Drugs and the Ketogenic Diet. *Cold Spring Harb. Perspect. Med.* **2016**, *6*, a022780. [CrossRef] [PubMed]
210. Masino, S.A.; Li, T.; Theofilas, P.; Sandau, U.S.; Ruskin, D.N.; Fredholm, B.B.; Geiger, J.D.; Aronica, E.; Boison, D. A Ketogenic Diet Suppresses Seizures in Mice through Adenosine A1 Receptors. *J. Clin. Investig.* **2011**, *121*, 2679–2683. [CrossRef]
211. Miyatake, T.; Suzuki, K. Globoid Cell Leukodystrophy: Additional Deficiency of Psychosine Galactosidase. *Biochem. Biophys. Res. Commun.* **1972**, *48*, 539–543. [CrossRef]
212. Suzuki, K. Twenty Five Years of the "Psychosine Hypothesis": A Personal Perspective of Its History and Present Status. *Neurochem. Res.* **1998**, *23*, 251–259. [CrossRef]
213. Won, J.-S.; Singh, A.K.; Singh, I. Biochemical, Cell Biological, Pathological, and Therapeutic Aspects of Krabbe's Disease. *J. Neurosci. Res.* **2016**, *94*, 990–1006. [CrossRef]
214. Graziano, A.C.E.; Parenti, R.; Avola, R.; Cardile, V. Krabbe Disease: Involvement of Connexin43 in the Apoptotic Effects of Sphingolipid Psychosine on Mouse Oligodendrocyte Precursors. *Apoptosis Int. J. Program. Cell Death* **2016**, *21*, 25–35. [CrossRef]
215. Avola, R.; Graziano, A.C.E.; Pannuzzo, G.; Alvares, E.; Cardile, V. Krabbe's Leukodystrophy: Approaches and Models in Vitro. *J. Neurosci. Res.* **2016**, *94*, 1284–1292. [CrossRef] [PubMed]
216. Ida, H.; Eto, Y.; Maekawa, K. Biochemical Pathogenesis of Demyelination in Globoid Cell Leukodystrophy (Krabbe's Disease): The Effects of Psychosine upon Oligodendroglial Cell Culture. *Acta Paediatr. Jpn.* **1990**, *32*, 20–26. [CrossRef] [PubMed]

217. Gowrishankar, S.; Cologna, S.M.; Givogri, M.I.; Bongarzone, E.R. Deregulation of Signalling in Genetic Conditions Affecting the Lysosomal Metabolism of Cholesterol and Galactosyl-Sphingolipids. *Neurobiol. Dis.* **2020**, *146*, 105142. [CrossRef] [PubMed]
218. Escolar, M.L.; West, T.; Dallavecchia, A.; Poe, M.D.; LaPoint, K. Clinical Management of Krabbe Disease. *J. Neurosci. Res.* **2016**, *94*, 1118–1125. [CrossRef]
219. Pannuzzo, G.; Cardile, V.; Costantino-Ceccarini, E.; Alvares, E.; Mazzone, D.; Perciavalle, V. A Galactose-Free Diet Enriched in Soy Isoflavones and Antioxidants Results in Delayed Onset of Symptoms of Krabbe Disease in Twitcher Mice. *Mol. Genet. Metab.* **2010**, *100*, 234–240. [CrossRef]
220. Paintlia, M.K.; Singh, I.; Singh, A.K. Effect of Vitamin D3 Intake on the Onset of Disease in a Murine Model of Human Krabbe Disease. *J. Neurosci. Res.* **2015**, *93*, 28–42. [CrossRef] [PubMed]
221. Signorini, C.; Pannuzzo, G.; Graziano, A.C.E.; Moretti, E.; Collodel, G.; Cardile, V. Dietary Supplementation with N-3 Polyunsaturated Fatty Acids Delays the Phenotypic Manifestation of Krabbe Disease and Partially Restores Lipid Mediator Production in the Brain—Study in a Mouse Model of the Disease. *Int. J. Mol. Sci.* **2024**, *25*, 7149. [CrossRef]
222. Signorini, C.; Cardile, V.; Pannuzzo, G.; Graziano, A.C.E.; Durand, T.; Galano, J.-M.; Oger, C.; Leoncini, S.; Cortelazzo, A.; Lee, J.C.-Y.; et al. Increased Isoprostanoid Levels in Brain from Murine Model of Krabbe Disease–Relevance of Isoprostanes, Dihomo-Isoprostanes and Neuroprostanes to Disease Severity. *Free Radic. Biol. Med.* **2019**, *139*, 46–54. [CrossRef]
223. Berardi, A.S.; Pannuzzo, G.; Graziano, A.; Costantino-Ceccarini, E.; Piomboni, P.; Luddi, A. Pharmacological Chaperones Increase Residual β-Galactocerebrosidase Activity in Fibroblasts from Krabbe Patients. *Mol. Genet. Metab.* **2014**, *112*, 294–301. [CrossRef]
224. Graziano, A.C.E.; Pannuzzo, G.; Avola, R.; Cardile, V. Chaperones as Potential Therapeutics for Krabbe Disease. *J. Neurosci. Res.* **2016**, *94*, 1220–1230. [CrossRef]
225. Inoue, K. PLP1-Related Inherited Dysmyelinating Disorders: Pelizaeus-Merzbacher Disease and Spastic Paraplegia Type 2. *Neurogenetics* **2005**, *6*, 1–16. [CrossRef] [PubMed]
226. Bahrambeigi, V.; Song, X.; Sperle, K.; Beck, C.R.; Hijazi, H.; Grochowski, C.M.; Gu, S.; Seeman, P.; Woodward, K.J.; Carvalho, C.M.B.; et al. Distinct Patterns of Complex Rearrangements and a Mutational Signature of Microhomeology Are Frequently Observed in PLP1 Copy Number Gain Structural Variants. *Genome Med.* **2019**, *11*, 80. [CrossRef] [PubMed]
227. Cailloux, F.; Gauthier-Barichard, F.; Mimault, C.; Isabelle, V.; Courtois, V.; Giraud, G.; Dastugue, B.; Boespflug-Tanguy, O. Genotype-Phenotype Correlation in Inherited Brain Myelination Defects Due to Proteolipid Protein Gene Mutations. Clinical European Network on Brain Dysmyelinating Disease. *Eur. J. Hum. Genet. EJHG* **2000**, *8*, 837–845. [CrossRef] [PubMed]
228. Saher, G.; Rudolphi, F.; Corthals, K.; Ruhwedel, T.; Schmidt, K.-F.; Löwel, S.; Dibaj, P.; Barrette, B.; Möbius, W.; Nave, K.-A. Therapy of Pelizaeus-Merzbacher Disease in Mice by Feeding a Cholesterol-Enriched Diet. *Nat. Med.* **2012**, *18*, 1130–1135. [CrossRef]
229. Stumpf, S.K.; Berghoff, S.A.; Trevisiol, A.; Spieth, L.; Düking, T.; Schneider, L.V.; Schlaphoff, L.; Dreha-Kulaczewski, S.; Bley, A.; Burfeind, D.; et al. Ketogenic Diet Ameliorates Axonal Defects and Promotes Myelination in Pelizaeus–Merzbacher Disease. *Acta Neuropathol.* **2019**, *138*, 147–161. [CrossRef]
230. Epplen, D.B.; Prukop, T.; Nientiedt, T.; Albrecht, P.; Arlt, F.A.; Stassart, R.M.; Kassmann, C.M.; Methner, A.; Nave, K.-A.; Werner, H.B.; et al. Curcumin Therapy in a Plp1 Transgenic Mouse Model of Pelizaeus-Merzbacher Disease. *Ann. Clin. Transl. Neurol.* **2015**, *2*, 787–796. [CrossRef]
231. Ünalp, A.; Köse, M.; Karaoğlu, P.; Güzin, Y.; Yılmaz, Ü. A Rare Case of Hypomyelinating Leukodystrophy-14 Benefiting from Ketogenic Diet Therapy. *Turk. J. Pediatr.* **2022**, *64*, 747–753. [CrossRef]
232. Murofushi, Y.; Hayakawa, I.; Abe, Y.; Ohto, T.; Murayama, K.; Suzuki, H.; Takenouchi, T.; Kosaki, K.; Kubota, M. Ketogenic Diet for KARS-Related Mitochondrial Dysfunction and Progressive Leukodystrophy. *Neuropediatrics* **2021**, *53*, 65–68. [CrossRef] [PubMed]
233. Pavone, P.; Polizzi, A.; Marino, S.D.; Corsello, G.; Falsaperla, R.; Marino, S.; Ruggieri, M. West Syndrome: A Comprehensive Review. *Neurol. Sci.* **2020**, *41*, 3547–3562. [CrossRef]
234. Kim, H.J.; Yang, D.; Kim, S.H.; Kim, B.; Kim, H.D.; Lee, J.S.; Choi, J.R.; Lee, S.-T.; Kang, H.-C. The Phenotype and Treatment of SCN2A-Related Developmental and Epileptic Encephalopathy. *Epileptic Disord.* **2020**, *22*, 563–570. [CrossRef]
235. Anwar, A.; Saleem, S.; Patel, U.K.; Arumaithurai, K.; Malik, P. Dravet Syndrome: An Overview. *Cureus* **2019**, *11*, e5006. [CrossRef] [PubMed]
236. Strzelczyk, A.; Schubert-Bast, S. Expanding the Treatment Landscape for Lennox-Gastaut Syndrome: Current and Future Strategies. *CNS Drugs* **2021**, *35*, 61–83. [CrossRef] [PubMed]
237. Zuberi, S.M.; Wirrell, E.; Yozawitz, E.; Wilmshurst, J.M.; Specchio, N.; Riney, K.; Pressler, R.; Auvin, S.; Samia, P.; Hirsch, E.; et al. ILAE Classification and Definition of Epilepsy Syndromes with Onset in Neonates and Infants: Position Statement by the ILAE Task Force on Nosology and Definitions. *Epilepsia* **2022**, *63*, 1349–1397. [CrossRef] [PubMed]
238. Ramantani, G.; Bölsterli, B.K.; Alber, M.; Klepper, J.; Korinthenberg, R.; Kurlemann, G.; Tibussek, D.; Wolff, M.; Schmitt, B. Treatment of Infantile Spasm Syndrome: Update from the Interdisciplinary Guideline Committee Coordinated by the German-Speaking Society of Neuropediatrics. *Neuropediatrics* **2022**, *53*, 389–401. [CrossRef]
239. Neal, E.G.; Chaffe, H.; Schwartz, R.H.; Lawson, M.S.; Edwards, N.; Fitzsimmons, G.; Whitney, A.; Cross, J.H. The Ketogenic Diet for the Treatment of Childhood Epilepsy: A Randomised Controlled Trial. *Lancet Neurol.* **2008**, *7*, 500–506. [CrossRef]

240. Caraballo, R.; Vaccarezza, M.; Cersósimo, R.; Rios, V.; Soraru, A.; Arroyo, H.; Agosta, G.; Escobal, N.; Demartini, M.; Maxit, C.; et al. Long-Term Follow-up of the Ketogenic Diet for Refractory Epilepsy: Multicenter Argentinean Experience in 216 Pediatric Patients. *Seizure* **2011**, *20*, 640–645. [CrossRef]
241. Sharma, S.; Sankhyan, N.; Gulati, S.; Agarwala, A. Use of the Modified Atkins Diet in Infantile Spasms Refractory to First-Line Treatment. *Seizure* **2012**, *21*, 45–48. [CrossRef]
242. Ko, A.; Jung, D.E.; Kim, S.H.; Kang, H.-C.; Lee, J.S.; Lee, S.T.; Choi, J.R.; Kim, H.D. The Efficacy of Ketogenic Diet for Specific Genetic Mutation in Developmental and Epileptic Encephalopathy. *Front. Neurol.* **2018**, *9*, 530. [CrossRef] [PubMed]
243. Møller, R.S.; Dahl, H.A.; Helbig, I. The Contribution of next Generation Sequencing to Epilepsy Genetics. *Expert Rev. Mol. Diagn.* **2015**, *15*, 1531–1538. [CrossRef]
244. Griffin, A.; Hamling, K.R.; Hong, S.; Anvar, M.; Lee, L.P.; Baraban, S.C. Preclinical Animal Models for Dravet Syndrome: Seizure Phenotypes, Comorbidities and Drug Screening. *Front. Pharmacol.* **2018**, *9*, 573. [CrossRef]
245. Dutton, S.B.B.; Sawyer, N.T.; Kalume, F.; Jumbo-Lucioni, P.; Borges, K.; Catterall, W.A.; Escayg, A. Protective Effect of the Ketogenic Diet in Scn1a Mutant Mice. *Epilepsia* **2011**, *52*, 2050–2056. [CrossRef] [PubMed]
246. Jancovski, N.; Baldwin, T.; Orford, M.; Li, M.; Jones, G.D.; Burbano, L.E.; Rutherford, T.; Reid, C.; Heales, S.; Eaton, S.; et al. Protective Effects of Medium Chain Triglyceride Diet in a Mouse Model of Dravet Syndrome. *Epilepsia* **2021**, *62*, 3131–3142. [CrossRef] [PubMed]
247. Miljanovic, N.; Potschka, H. The Impact of *Scn1a* Deficiency and Ketogenic Diet on the Intestinal Microbiome: A Study in a Genetic Dravet Mouse Model. *Epilepsy Res.* **2021**, *178*, 106826. [CrossRef]
248. Mu, C.; Choudhary, A.; Mayengbam, S.; Barrett, K.T.; Rho, J.M.; Shearer, J.; Scantlebury, M.H. Seizure Modulation by the Gut Microbiota and Tryptophan-Kynurenine Metabolism in an Animal Model of Infantile Spasms. *EBioMedicine* **2022**, *76*, 103833. [CrossRef]
249. Maa, E.; Figi, P. The Case for Medical Marijuana in Epilepsy. *Epilepsia* **2014**, *55*, 783–786. [CrossRef]
250. Porter, B.E.; Jacobson, C. Report of a Parent Survey of Cannabidiol-Enriched Cannabis Use in Pediatric Treatment-Resistant Epilepsy. *Epilepsy Behav.* **2013**, *29*, 574–577. [CrossRef] [PubMed]
251. Kaplan, J.S.; Stella, N.; Catterall, W.A.; Westenbroek, R.E. Cannabidiol Attenuates Seizures and Social Deficits in a Mouse Model of Dravet Syndrome. *Proc. Natl. Acad. Sci. USA* **2017**, *114*, 11229–11234. [CrossRef]
252. Thiele, E.A.; Marsh, E.D.; French, J.A.; Mazurkiewicz-Beldzinska, M.; Benbadis, S.R.; Joshi, C.; Lyons, P.D.; Taylor, A.; Roberts, C.; Sommerville, K.; et al. Cannabidiol in Patients with Seizures Associated with Lennox-Gastaut Syndrome (GWPCARE4): A Randomised, Double-Blind, Placebo-Controlled Phase 3 Trial. *Lancet* **2018**, *391*, 1085–1096. [CrossRef]
253. Devinsky, O.; Cross, J.H.; Laux, L.; Marsh, E.; Miller, I.; Nabbout, R.; Scheffer, I.E.; Thiele, E.A.; Wright, S. Trial of Cannabidiol for Drug-Resistant Seizures in the Dravet Syndrome. *N. Engl. J. Med.* **2017**, *376*, 2011–2020. [CrossRef]
254. Filla, A.; De Michele, G.; Cavalcanti, F.; Pianese, L.; Monticelli, A.; Campanella, G.; Cocozza, S. The Relationship between Trinucleotide (GAA) Repeat Length and Clinical Features in Friedreich Ataxia. *Am. J. Hum. Genet.* **1996**, *59*, 554–560.
255. Tai, G.; Corben, L.A.; Yiu, E.M.; Milne, S.C.; Delatycki, M.B. Progress in the Treatment of Friedreich Ataxia. *Neurol. Neurochir. Pol.* **2018**, *52*, 129–139. [CrossRef] [PubMed]
256. Li, L.; Voullaire, L.; Sandi, C.; Pook, M.A.; Ioannou, P.A.; Delatycki, M.B.; Sarsero, J.P. Pharmacological Screening Using an FXN-EGFP Cellular Genomic Reporter Assay for the Therapy of Friedreich Ataxia. *PLoS ONE* **2013**, *8*, e55940. [CrossRef] [PubMed]
257. Yiu, E.M.; Tai, G.; Peverill, R.E.; Lee, K.J.; Croft, K.D.; Mori, T.A.; Scheiber-Mojdehkar, B.; Sturm, B.; Praschberger, M.; Vogel, A.P.; et al. An Open-Label Trial in Friedreich Ataxia Suggests Clinical Benefit with High-Dose Resveratrol, without Effect on Frataxin Levels. *J. Neurol.* **2015**, *262*, 1344–1353. [CrossRef] [PubMed]
258. Hadjivassiliou, M.; Zis, P.; Sanders, D.S.; Hoggard, N.; Sarrigiannis, P.G. Stiff Person Syndrome and Gluten Sensitivity. *Nutrients* **2021**, *13*, 1373. [CrossRef] [PubMed]
259. Hadjivassiliou, M. Gluten Ataxia in Perspective: Epidemiology, Genetic Susceptibility and Clinical Characteristics. *Brain* **2003**, *126*, 685–691. [CrossRef]
260. Yadav, S.K.; Ir, R.; Jeewon, R.; Doble, M.; Hyde, K.D.; Kaliappan, I.; Jeyaraman, R.; Reddi, R.N.; Krishnan, J.; Li, M.; et al. A Mechanistic Review on Medicinal Mushrooms-Derived Bioactive Compounds: Potential Mycotherapy Candidates for Alleviating Neurological Disorders. *Planta Med.* **2020**, *86*, 1161–1175. [CrossRef]
261. Kluger, B.; Triolo, P.; Jones, W.; Jankovic, J. The Therapeutic Potential of Cannabinoids for Movement Disorders. *Mov. Disord.* **2015**, *30*, 313–327. [CrossRef]
262. Gómez-Ruiz, M.; Rodríguez-Cueto, C.; Luna-Piñel, E.; Hernández-Gálvez, M.; Fernández-Ruiz, J. Endocannabinoid System in Spinocerebellar Ataxia Type-3 and Other Autosomal-Dominant Cerebellar Ataxias: Potential Role in Pathogenesis and Expected Relevance as Neuroprotective Targets. *Front. Mol. Neurosci.* **2019**, *12*, 94. [CrossRef]
263. Stephens, G.J. Does Modulation of the Endocannabinoid System Have Potential Therapeutic Utility in Cerebellar Ataxia? *J. Physiol.* **2016**, *594*, 4631–4641. [CrossRef]
264. Rodríguez-Cueto, C.; Benito, C.; Fernández-Ruiz, J.; Romero, J.; Hernández-Gálvez, M.; Gómez-Ruiz, M. Changes in CB(1) and CB(2) Receptors in the Post-Mortem Cerebellum of Humans Affected by Spinocerebellar Ataxias. *Br. J. Pharmacol.* **2014**, *171*, 1472–1489. [CrossRef]

265. Rodríguez-Cueto, C.; Hernández-Gálvez, M.; Hillard, C.J.; Maciel, P.; García-García, L.; Valdeolivas, S.; Pozo, M.A.; Ramos, J.A.; Gómez-Ruiz, M.; Fernández-Ruiz, J. Dysregulation of the Endocannabinoid Signaling System in the Cerebellum and Brainstem in a Transgenic Mouse Model of Spinocerebellar Ataxia Type-3. *Neuroscience* **2016**, *339*, 191–209. [CrossRef] [PubMed]
266. Ranjbar, H.; Soti, M.; Janahmadi, M.; Kohlmeier, K.A.; Sheibani, V.; Ahmadi-Zeidabadi, M.; Shabani, M. Modulation of the CB1 Cannabinoid Receptor Has Potential Therapeutic Utility in the 3-Acetylpyridine Cerebellar Ataxia Rat Model. *Exp. Brain Res.* **2022**, *240*, 2339–2348. [CrossRef] [PubMed]
267. Louis, D.N.; Perry, A.; Wesseling, P.; Brat, D.J.; Cree, I.A.; Figarella-Branger, D.; Hawkins, C.; Ng, H.K.; Pfister, S.M.; Reifenberger, G.; et al. The 2021 WHO Classification of Tumors of the Central Nervous System: A Summary. *Neuro-Oncology* **2021**, *23*, 1231–1251. [CrossRef] [PubMed]
268. Gu, J.; Liu, Y.; Kyritsis, A.P.; Bondy, M.L. Molecular Epidemiology of Primary Brain Tumors. *Neurotherapeutics* **2009**, *6*, 427–435. [CrossRef]
269. Archer, T.C.; Mahoney, E.L.; Pomeroy, S.L. Medulloblastoma: Molecular Classification-Based Personal Therapeutics. *Neurotherapeutics* **2017**, *14*, 265–273. [CrossRef]
270. Cavalheiro, S.; Valsechi, L.C.; Dastoli, P.A.; Nicácio, J.M.; Cappellano, A.M.; Saba da Silva, N.; Silva da Costa, M.D. Outcomes and surgical approaches for pineal region tumors in children: 30 years' experience. *J. Neurosurg. Pediatr.* **2023**, *32*, 184–193. [CrossRef]
271. Ogiwara, H.; Dipatri, A.J.; Alden, T.D.; Bowman, R.M.; Tomita, T. Choroid Plexus Tumors in Pediatric Patients. *Br. J. Neurosurg.* **2012**, *26*, 32–37. [CrossRef]
272. Viaene, A.N.; Santi, M.; Hawkins, C. Pediatric Glial Tumors. *Pediatr. Dev. Pathol.* **2022**, *25*, 46–58. [CrossRef]
273. Bielecka, J.; Markiewicz-Żukowska, R. The Influence of Nutritional and Lifestyle Factors on Glioma Incidence. *Nutrients* **2020**, *12*, 1812. [CrossRef]
274. Heydari, M.; Shayanfar, M.; Sharifi, G.; Saneei, P.; Sadeghi, O.; Esmaillzadeh, A. The Association between Dietary Total Antioxidant Capacity and Glioma in Adults. *Nutr. Cancer* **2021**, *73*, 1947–1956. [CrossRef]
275. Tedeschi-Blok, N.; Lee, M.; Sison, J.D.; Miike, R.; Wrensch, M. Inverse Association of Antioxidant and Phytoestrogen Nutrient Intake with Adult Glioma in the San Francisco Bay Area: A Case-Control Study. *BMC Cancer* **2006**, *6*, 148. [CrossRef] [PubMed]
276. Dubrow, R.; Darefsky, A.S.; Park, Y.; Mayne, S.T.; Moore, S.C.; Kilfoy, B.; Cross, A.J.; Sinha, R.; Hollenbeck, A.R.; Schatzkin, A.; et al. Dietary Components Related to N-Nitroso Compound Formation: A Prospective Study of Adult Glioma. *Cancer Epidemiol Biomarkers Prev.* **2010**, *19*, 1709–1722. [CrossRef] [PubMed]
277. Hu, J.; La Vecchia, C.; Negri, E.; Chatenoud, L.; Bosetti, C.; Jia, X.; Liu, R.; Huang, G.; Bi, D.; Wang, C. Diet and Brain Cancer in Adults: A Case-Control Study in Northeast China. *Int. J. Cancer* **1999**, *81*, 20–23. [CrossRef]
278. Ni, Y.-J.; Huang, Y.-Q.; Yu, L.; Zhang, X.-Y.; Zhu, Q.; Shu, L.; Zhang, L. Associations between Dietary Antioxidant Vitamins and Risk of Glioma: An Updated Systematic Review and Meta-Analysis of Observational Studies. *Front. Nutr.* **2024**, *11*, 1428528. [CrossRef]
279. Zhou, S.; Wang, X.; Tan, Y.; Qiu, L.; Fang, H.; Li, W. Association between Vitamin C Intake and Glioma Risk: Evidence from a Meta-Analysis. *Neuroepidemiology* **2015**, *44*, 39–44. [CrossRef]
280. Lv, W.; Zhong, X.; Xu, L.; Han, W. Association between Dietary Vitamin A Intake and the Risk of Glioma: Evidence from a Meta-Analysis. *Nutrients* **2015**, *7*, 8897–8904. [CrossRef]
281. LIANG, C.; YANG, L.; GUO, S. All-Trans Retinoic Acid Inhibits Migration, Invasion and Proliferation, and Promotes Apoptosis in Glioma Cells in Vitro. *Oncol. Lett.* **2015**, *9*, 2833–2838. [CrossRef]
282. Bouterfa, H.; Picht, T.; Keß, D.; Herbold, C.; Noll, E.; Black, P.M.; Roosen, K.; Tonn, J.C. Retinoids Inhibit Human Glioma Cell Proliferation and Migration in Primary Cell Cultures But Not in Established Cell Lines. *Neurosurgery* **2000**, *46*, 419. [CrossRef] [PubMed]
283. Naidu, K.A.; Tang, J.L.; Naidu, K.A.; Prockop, L.D.; Nicosia, S.V.; Coppola, D. Antiproliferative and Apoptotic Effect of Ascorbyl Stearate in Human Glioblastoma Multiforme Cells: Modulation of Insulin-like Growth Factor-I Receptor (IGF-IR) Expression. *J. Neurooncol.* **2001**, *54*, 15–22. [CrossRef]
284. Dal Bello, S.; Valdemarin, F.; Martinuzzi, D.; Filippi, F.; Gigli, G.L.; Valente, M. Ketogenic Diet in the Treatment of Gliomas and Glioblastomas. *Nutrients* **2022**, *14*, 3851. [CrossRef]
285. Clontz, A.D. Ketogenic Therapies for Glioblastoma: Understanding the Limitations in Transitioning from Mice to Patients. *Front. Nutr.* **2023**, *10*, 1110291. [CrossRef] [PubMed]
286. Poff, A.; Koutnik, A.P.; Egan, K.M.; Sahebjam, S.; D'Agostino, D.; Kumar, N.B. Targeting the Warburg Effect for Cancer Treatment: Ketogenic Diets for Management of Glioma. *Semin. Cancer Biol.* **2019**, *56*, 135–148. [CrossRef] [PubMed]
287. Abdelwahab, M.G.; Fenton, K.E.; Preul, M.C.; Rho, J.M.; Lynch, A.; Stafford, P.; Scheck, A.C. The Ketogenic Diet Is an Effective Adjuvant to Radiation Therapy for the Treatment of Malignant Glioma. *PLoS ONE* **2012**, *7*, e36197. [CrossRef]
288. Stafford, P.; Abdelwahab, M.G.; Kim, D.Y.; Preul, M.C.; Rho, J.M.; Scheck, A.C. The Ketogenic Diet Reverses Gene Expression Patterns and Reduces Reactive Oxygen Species Levels When Used as an Adjuvant Therapy for Glioma. *Nutr. Metab.* **2010**, *7*, 74. [CrossRef]
289. Sargaço, B.; Oliveira, P.A.; Antunes, M.L.; Moreira, A.C. Effects of the Ketogenic Diet in the Treatment of Gliomas: A Systematic Review. *Nutrients* **2022**, *14*, 1007. [CrossRef]

290. Schreck, K.C.; Hsu, F.-C.; Berrington, A.; Henry-Barron, B.; Vizthum, D.; Blair, L.; Kossoff, E.H.; Easter, L.; Whitlow, C.T.; Barker, P.B.; et al. Feasibility and Biological Activity of a Ketogenic/Intermittent-Fasting Diet in Patients with Glioma. *Neurology* **2021**, *97*, e953–e963. [CrossRef]
291. Martin-McGill, K.J.; Marson, A.G.; Tudur Smith, C.; Jenkinson, M.D. Ketogenic Diets as an Adjuvant Therapy in Glioblastoma (the KEATING Trial): Study Protocol for a Randomised Pilot Study. *Pilot Feasibility Stud.* **2017**, *3*, 67. [CrossRef] [PubMed]
292. Klein, P.; Tyrlikova, I.; Zuccoli, G.; Tyrlik, A.; Maroon, J.C. Treatment of Glioblastoma Multiforme with "Classic" 4:1 Ketogenic Diet Total Meal Replacement. *Cancer Metab.* **2020**, *8*, 24. [CrossRef]
293. Rieger, J.; Bähr, O.; Maurer, G.D.; Hattingen, E.; Franz, K.; Brucker, D.; Walenta, S.; Kämmerer, U.; Coy, J.F.; Weller, M.; et al. ERGO: A Pilot Study of Ketogenic Diet in Recurrent Glioblastoma. *Int. J. Oncol.* **2014**, *44*, 1843–1852. [CrossRef]
294. Voss, M.; Wagner, M.; von Mettenheim, N.; Harter, P.N.; Wenger, K.J.; Franz, K.; Bojunga, J.; Vetter, M.; Gerlach, R.; Glatzel, M.; et al. ERGO2: A Prospective, Randomized Trial of Calorie-Restricted Ketogenic Diet and Fasting in Addition to Reirradiation for Malignant Glioma. *Int. J. Radiat. Oncol. Biol. Phys.* **2020**, *108*, 987–995. [CrossRef]
295. Frebourg, T.; Barbier, N.; Yan, Y.X.; Garber, J.E.; Dreyfus, M.; Fraumeni, J.; Li, F.P.; Friend, S.H. Germ-Line P53 Mutations in 15 Families with Li-Fraumeni Syndrome. *Am. J. Hum. Genet.* **1995**, *56*, 608–615. [PubMed]
296. Lombardi, C.; Ganguly, A.; Bunin, G.R.; Azary, S.; Alfonso, V.; Ritz, B.; Heck, J.E. Maternal Diet during Pregnancy and Unilateral Retinoblastoma. *Cancer Causes Control* **2015**, *26*, 387–397. [CrossRef] [PubMed]
297. Pogoda, J.M.; Preston-Martin, S.; Howe, G.; Lubin, F.; Mueller, B.A.; Holly, E.A.; Filippini, G.; Peris-Bonet, R.; McCredie, M.R.E.; Cordier, S.; et al. An International Case-Control Study of Maternal Diet during Pregnancy and Childhood Brain Tumor Risk: A Histology-Specific Analysis by Food Group. *Ann. Epidemiol.* **2009**, *19*, 148–160. [CrossRef] [PubMed]
298. Hu, Z.; Ye, J.; Shi, S.; Luo, C.; Wang, T.; Liu, Y.; Ye, J.; Sun, X.; Ke, Y.; Hou, C. Maternal Smoking, Consumption of Alcohol, and Caffeinated Beverages during Pregnancy and the Risk of Childhood Brain Tumors: A Meta-Analysis of Observational Studies. *BMC Public Health* **2024**, *24*, 1238. [CrossRef]

nutrients

Article

Modulation of Gut Microbiome and Autism Symptoms of ASD Children Supplemented with Biological Response Modifier: A Randomized, Double-Blinded, Placebo-Controlled Pilot Study

Jela Hrnciarova [1,2], Klara Kubelkova [3,*], Vanda Bostik [3], Ivan Rychlik [4], Daniela Karasova [4], Vladimir Babak [4], Magdalena Datkova [1,2], Katerina Simackova [1,2] and Ales Macela [3]

[1] Faculty of Medicine, Charles University, 500 03 Hradec Kralove, Czech Republic; hrnciarovaj@lfhk.cuni.cz (J.H.); datkovam@lfhk.cuni.cz (M.D.); katerina.simackova@fnhk.cz (K.S.)
[2] Department of Psychiatry, University Hospital in Hradec Kralove, 500 03 Hradec Kralove, Czech Republic
[3] Military Faculty of Medicine, University of Defence, 500 03 Hradec Kralove, Czech Republic; vanda.bostikova@unob.cz (V.B.); amacela@seznam.cz (A.M.)
[4] Veterinary Research Institute, 621 00 Brno, Czech Republic; ivan.rychlik@vri.cz (I.R.); daniela.karasova@vri.cz (D.K.); vladimir.babak@vri.cz (V.B.)
* Correspondence: klara.kubelkova@unob.cz

Abstract: The etiology and mechanisms of autism and autism spectrum disorder (ASD) are not yet fully understood. There is currently no treatment for ASD for providing significant improvement in core symptoms. Recent studies suggest, however, that ASD is associated with gut dysbiosis, indicating that modulation of gut microbiota in children with ASD may thus reduce the manifestation of ASD symptoms. The aim of this pilot study (prospective randomized, double-blinded, placebo-controlled) was to evaluate efficacy of the biological response modifier Juvenil in modulating the microbiome of children with ASD and, in particular, whether Juvenil is able to alleviate the symptoms of ASD. In total, 20 children with ASD and 12 neurotypical children were included in our study. Supplementation of ASD children lasted for three months. To confirm Juvenil's impact on the gut microbiome, stool samples were collected from all children and the microbiome's composition was analyzed. This pilot study demonstrated that the gut microbiome of ASD children differed significantly from that of healthy controls and was converted by Juvenil supplementation toward a more neurotypical microbiome that positively modulated children's autism symptoms.

Keywords: autism; microbiome; biological response modifier; psychobiotics

1. Introduction

Autism spectrum disorder (ASD) is a behaviorally defined neurodevelopmental disorder. ASD lacks specific clinical biomarkers and has seen an evolving conceptualization through the decades since it was first described. Over the past four decades, there has been dramatic increase in the number of individuals diagnosed with ASD [1]. In general, ASD is diagnosed by 3 years of age in most of those children experiencing it, although roughly 40% of such children are not first evaluated until 4 years of age [2]. A psychiatric diagnosis of ASD, which has behavior as its basis of definition, relies heavily upon precise observation and clinical expertise because the condition lacks standardized biomarkers [3].

ASD is one of the most common and challenging neurodevelopmental disorders in children. Its prevalence rate worldwide now exceeds 1%. A small number of these children appear to develop normally in their first year and then go through a period of regression between 18 and 24 months of age. ASD is characterized by deficits in communication and social interaction, as well as a presence of repetitive and restrictive behaviors. Moreover, ASD often manifests with a wide range of comorbidities that include morphological, physiological, and psychiatric conditions. ASD's most commonly proposed causes are physiological and metabolic disorders involving immunity, oxidative stress, and mitochondrial dysfunction [4]. Co-occurrence of two or more disorders in the same individual has been observed, with comorbidities including anxiety, depression, attention deficit/hyperactivity disorder (AD/HD), epilepsy, gastrointestinal symptoms/problems, sleep disorders, learning disabilities, obsessive–compulsive disorder, intellectual disability, sensory problems, and immune disorders. The most prevalent comorbidity, at roughly 50%, is intellectual disability [5]. At least one comorbidity exists in about 70% of children with ASD, while 41% have two or more [6]. An estimated 20% of individuals diagnosed with ASD also have epilepsy [7].

Although ASD's etiology remains largely unexplained, a recent finding identifies specific gut microbiota composition in ASD patients. Post-mortem examination of ASD subjects' brain tissue and small intestines has revealed that the blood–brain barrier and gut barrier were disrupted, with significant neuroinflammation evidenced by increased expression of genes and markers associated with brain inflammation. It has further been inferred that the gut–brain axis disruption may be associated with non-self antigens that trigger a neuroinflammatory reaction by crossing the damaged gut barriers, thus leading to ASD in genetically susceptible subjects [8]. In a study involving 192 twins, however, genetic factors accounted for only 38% of ASD risk, whereas the remaining 68% was attributed to environmental factors [9]. A significant role was ascribed to the gut microbiota [10]. Microbiota is shaped by diet, lifestyle, and microbial exposure in the early developmental phase and infection, as well as by genetic makeup, metabolites, and immunological and hormonal aspects [11].

Analysis of gut microbiota is currently a growing area of research linked to neuropsychological disorders, including depression [12], metabolic disorders such as obesity [13], and gastrointestinal disorders, including inflammatory bowel disease or irritable bowel syndrome [14,15]. Many studies have identified that microbiota composition in ASD patients differs significantly from that in healthy controls [16–19]. Nutritional intervention, prebiotics, probiotics, and symbiotics, including fecal microbiota transplantation as a remedy to modulate the species composition of the intestinal microbiota of patients with ASD, already have been tested [20–24]. These studies generally have concluded, however, that their findings should be taken with caution because there still exist only limited data from studies examining different regimens of different remedy applications and there have not yet been double-blind studies demonstrating clinical significance of the effects of those remedies used. Here, therefore, we present the results of a double-blind, placebo-controlled study utilizing the biological response modifier Juvenil for influencing the gut microbiota of children suffering from ASD and for modulating their ASD symptoms.

2. Study Design, Materials, and Methods

2.1. Study Design

The pilot prospective double-blind randomized feasibility study enrolled 28 children (9 girls and 19 boys) of Czech nationality aged 3 to 7 years. Of these, 16 children under care of the Psychiatric Clinic of the Hradec Kralove University Hospital, Czech Republic met the criteria for a diagnosis of ASD and the remaining 12 children (control group) were neurotypical (NT), i.e., without any signs of ASD (Table 1). The ASD children were randomly selected for the study by their attending psychiatrists. The children forming the control (neurotypical) group were randomly included in the study on the basis of an agreement with parents living in the geographical area of this study. These children were never under the care of a psychiatrist.

Table 1. Demographic characteristics of the subjects involved in this study.

Characteristic	Value	
	ASD Group	Neurotypical Group
Average age at enrollment (years)	6 ± 3 *	5 ± 2 *
Age range (years)	3–9	3–9
Male/Female, (number)	13/3	8/4
Ethnicity/Location	White/East Bohemia, CZE	White/East Bohemia, CZE

Note: * mean ± SD.

The group of autistic children was randomly divided into two groups of 8 children each. The first group was administered Juvenil, while the second group of 8 autistic children was given a placebo throughout the study. Juvenil or placebo capsules were administered orally to the children by their parents at home once a day for 3 months. The inclusion of children with autism into the study did not affect their existing treatments, education, or rehabilitation. To evaluate the effect of Juvenil on the gut microbiome, stool samples were collected once from healthy children and from autistic children before and after providing Juvenil or placebo. Autistic children were evaluated using the Childhood Autism Rating Scale in its standard version (CARS2-ST) by a clinical psychiatrist and subjective information based on observations of the children's behavior was obtained by conducting interviews with the children's parents. The study was approved by the hospital's ethics committee and informed consent was signed by the parents of all study participants (see Figure 1).

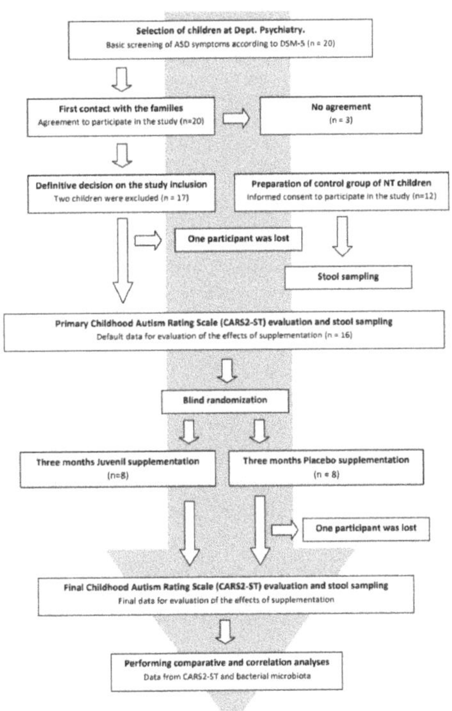

Figure 1. Flow diagram of the participant progress from recruitment to the end of the experiment. Include exclusions/dropouts.

2.2. *Juvenil and Placebo*

The Juvenil and placebo capsules were prepared by Uniregen, Ltd., Nachod, Czech Republic. Capsules of Juvenil contained 1.0 mg of Juvenil crude substance while placebo capsules contained redistilled water. Both capsule types looked identical and were stored at room temperature until use.

Juvenil is a nontoxic alcohol–ether extract of bovine tissue registered as a dietary supplement with modulatory activity on the immunity, dominantly on cells of the innate immune system, and on the organism's regeneration [25,26]. Juvenil is a complex mixture of peptides, nucleotides, free amino acids, and some other components of animal origin [27].

2.3. *Stool Samples Procedure*

The following recommendations were given to parents concerning fecal sample collection: (1) sample a stool from a clean container (e.g., potty) or from a piece of stool on toilet paper; (2) the stool must not be diluted with water or urine; (3) collect the stool sample using the scoop placed in the lid of the collection container; (4) a small amount of stool is sufficient for the examination (i.e., not larger than the size of a hazelnut or 1–2 mL in the case of a liquid stool sample); (5) return the sampling scoop to the container and screw on the cap; (6) place the collection container in a plastic bag, close the bag, and label it with the child's name and surname, not marking the sample container in any other way; (7) place the container with the stool sample in a closed plastic bag and keep it at 4 °C (in the refrigerator); (8) personally deliver the collection container with the sample to the doctor no later than 24 h after taking the sample (preferably the same day, but no later than the following day); and (9) stool samples should be collected one day before the start of supplementation and on the last day of supplementation (i.e., after 12 weeks from the start of supplementation).

The collection container and its contents were labeled by the attending physician and the samples were stored at −20 °C until subsequent examination.

2.4. *Microbiota Analysis*

Microbiota composition was determined as described previously [28]. The samples were homogenized in a MagNALyzer (Roche, Basel, Switzerland). Following homogenization, the DNA was extracted using a QIAamp DNA Stool Mini Kit (Qiagen, Hilden, Germany) according to the manufacturer's instructions and the DNA concentration was determined spectrophotometrically. DNA samples were diluted to 5 ng/mL and were used as template in polymerase chain reaction (PCR) with forward primer 5′-TCGTCGGCAGCGTCAGATGTGTATAAGAG ACAG-MID-GTCCTACGGGNGGC WGCAG-3′ and reverse primer 5′-GTCTCGTGGGCTC GGAGATGTGTATAAGAGACA G-MIDGTGACTACHVGGGTATCTAATCC-3′. MIDs shown above represent different sequences 5, 6, 7, or 9 base pairs in length that were used to identify individual samples within the sequencing groups. PCR amplification was performed using a HotStarTaq Plus Master Mix kit (Qiagen) and the resulting PCR products were purified using AMPure beads (Beckman Coulter, Prague, Czech Republic). In the next steps, the concentration of PCR products was determined spectrophotometrically, the DNA was diluted to 100 ng/μL, and groups of 14 PCR products with different MID sequences were indexed with the same indices using a Nextera XT Index Kit (Illumina, San Diego, CA, USA). Prior to sequencing, the concentrations of differently indexed samples were determined using a KAPA Library Quantification Complete kit (Kapa Biosystems, Wilmington, MA, USA), all indexed samples were diluted to 4 ng/μL, and 20 pM phiX DNA was added to final concentration of 5% (*v*/*v*). Sequencing was performed using a MiSeq Reagent Kit v3 and MiSeq apparatus (Illumina).

Sequencing data were analyzed using QIIME 2 [29]. Raw sequence data were demultiplexed and quality filtered; sequencing primers were then clipped using Je [30] and Fastp [31]. The resulting sequences were denoised with DADA2 [32]. Taxonomy was assigned to ASVs using the q2-feature-classifier [33] classify-sklearn naïve Bayes taxonomy classifier against the Silva 138 [34]. All the software tools were used with default settings.

2.5. Statistical Methods

Microbiota composition in different groups of patients was compared using permutational multivariate analysis of variance (PERMANOVA, R project, package vegan, function adonis2; Bray–Curtis dissimilarity, 9999 permutations). If PERMANOVA rejected the null hypothesis, then pairwise comparisons were made of all groups. Statistical significance was established at $p < 0.05$. LEfSe (linear discriminant analysis effect size) was used to determine taxa which most likely explained the differences between the compared groups [35]. To correlate the individual categories of ASD patient symptoms with bacterial taxa that may characterize the gut dysbiosis of children with ASD, the Covariance S tool in Microsoft Excel 2021 (v16.0) was used.

3. Results

3.1. Gut Microbiota Composition in Autistic and Neurotypical Children

Comparison of beta diversity using principal coordinate analysis (PCoA) confirmed differences in the clustering of samples from control and ASD children (Figure 2). All groups of ASD patients (NT, ASD Placebo, and ASD before and after Juvenil treatment) harbored microbiota different from those of the NT group (p = 0.001). At the bacterial phylum level, there were significant differences in the abundance of Actinobacteriota, Firmicutes, and Proteobacteria. While Actinobacteriota and Proteobacteria dominated in autistic children, Firmicutes were more abundant in neurotypical controls (Tables 2 and S1).

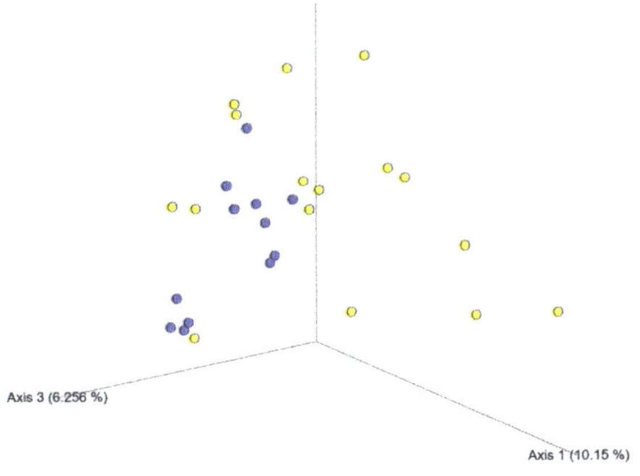

Figure 2. Fecal microbiota analysis of ASD and control children. Principal coordinate analysis (PCoA) using Bray–Curtis distance matrix separated samples of control and ASD children. Blue dots indicate samples from neurotypical control children; yellow dots indicate samples from ASD children.

LEfSe analysis (linear discriminant analysis effect size) identified bacterial taxa discriminating the ASD groups prior to supplementation from the NT group (Figure 3). *Bifidobacterium longum, Ruminococcus torque, Faecalibacterium, Flavonifractor, Pseudomonadas,* and *Clostridia* vadinBB60 were characteristic for the gut microbiota of autistic children while *Streptococcus parasanguinis, Monoglobus, Terrisporobacter,* or *Bacteroides cellulosilyticus* were more abundant in the microbiota of healthy controls.

Table 2. Relative abundance of gut microbiota between ASD group (without any treatment) and neurotypical control group (NT) at phylum levels.

Phylum	ASD	NT	*p*-Value
Actinobacteriota	3.04	1.18	0.03
Bacteroidota	49.02	49.48	0.60
Campylobacterota	0.03	0	0.35
Cyanobacteria	0.09	0.12	0.92
Desulfobacterota	0.29	0.17	0.39
Euryarchaeota	0.006	0.21	0.03
Firmicutes	41	46	0.004
Fusobacteriota	0.001	0.003	0.17
Patescibacteria	0.02	0.002	0.08
Proteobacteria	6	2	0.01
Synergistota	0.005	0	0.28
Verrucomicrobiota	0.84	0.69	0.80

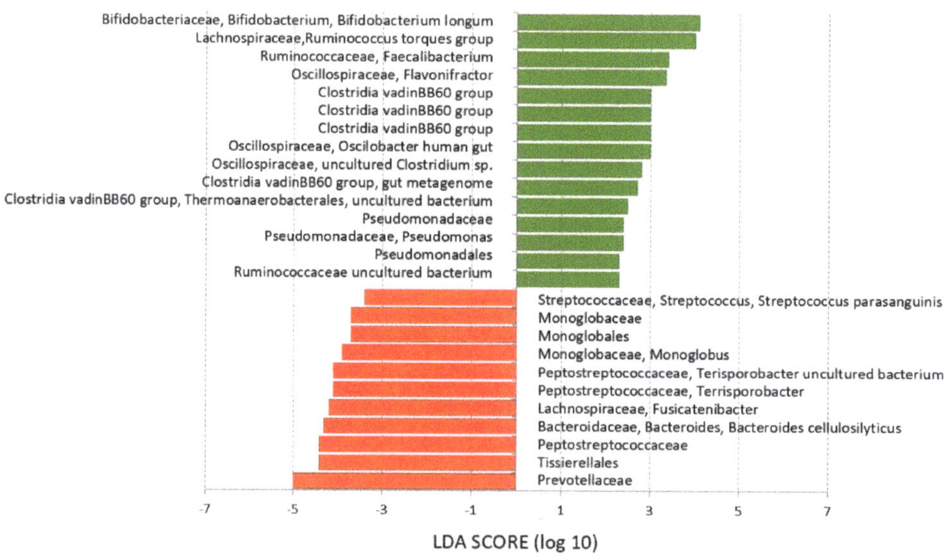

Figure 3. Taxa characteristic for ASD and control children. LEfSe analysis identified bacterial taxa in stool samples typical for ASD (green) and neurotypical control children (red).

3.2. *Gut Microbiome Modulation by Juvenil*

Pairwise comparison of NT, ASD patients before treatment, and ASD patients after Juvenil supplementation demonstrated there to be a significant difference in the microbiota composition between the NT and ASD patients before supplementation, but there was no significant difference in microbiota composition in a comparison of ASD patients after Juvenil either with NT healthy controls or with ASD patients before treatment. Juvenil administration thus shifted the profile of the gut microbiota composition of autistic children toward that of the neurotypical children, although it did not result in the restoration of a completely healthy type of microbiota. In addition, there were no significant differences in microbiota composition at the phylum level when comparing ASD patients either before and after placebo treatment or before and after Juvenil administration (Table 3).

Table 3. Comparison of microbiota composition (%) at the phylum level before and after Juvenil or placebo treatment.

Phylum	Juvenil			Placebo		
	Before	After	*p*-Value	Before	After	*p*-Value
Actinobacteriota	3.78	2.25	0.26	2.78	2.54	0.84
Bacteroidota	51.05	58.62	0.57	50.83	57.22	0.53
Campylobacterota	0.07	0.06	0.91	0.0006	0	0.32
Cyanobacteria	0.07	0.06	0.83	0.05	0.11	0.36
Desulfobacterota	0.29	0.13	0.48	0.33	0.36	0.79
Euryarcheota	0.005	0.02	0.37	0.01	0.02	0.54
Firmicutes	36.37	34.05	0.60	41.83	34.29	0.12
Fusobacteriota	0.0008	0	0.32	0.0009	0	0.32
Patescibacteria	0.006	0.006	0.95	0.02	0.01	0.44
Proteobacteria	6.97	4.11	0.22	3.48	4.38	0.59
Synergistota	0.005	0.002	0.56	0.007	0	0.32
Verrucomicrobiota	1.28	0.61	0.48	0.66	1.07	0.51

A moderate effect of Juvenil administration in comparison to placebo can be seen also from the comparison of operational taxonomic units (OTUs) completely lost or newly acquired during treatment (Table S2). The greatest losses or acquisition were recorded in OTUs belonging to families *Lachnospiraceae*, *Ruminococcaceae*, *Oscilospiraceae*, and *Christenellaceae*, all comprising spore-forming bacterial species. While children after placebo treatment lost 305 OTUs and newly acquired 132 OTUs, microbiota in Juvenil-treated ASD patients were more stable as there were only 200 OTUs lost and 152 newly appearing (Table S2).

3.3. Behavioral Status of Autistic Children and Juvenil

The behavioral status of autistic children of both groups was assessed using CARS2-ST at the time of entry into the study and after completion of supplementation. A comparison of the group supplemented with Juvenil and the group supplemented with placebo showed a positive shift in the values of the rating scale parameters during 3-month supplementation with Juvenil (Table 4 and Table S3). Nevertheless, the changes associated with Juvenil as well as placebo supplementation did not reach statistical significance (ASD placebo group $p = 0.62$, ASD Juvenil group $p = 0.19$). There were also no significant differences in the effect of supplementation in children with mild symptoms of ASD and severe symptoms of ASD in either group (Juvenil group $p = 0.95$, placebo group $p = 0.82$). Numerically, however, using the percentile parameter, there was a 12.4% reduction in autism symptoms associated with Juvenil supplementation, which was approximately double that associated with the placebo (6.6% reduction). In comparing the difference between Juvenil supplementation versus placebo using the T-score parameter, we can observe shifts of two or more points in favor of Juvenil for categories 4 (motor manifestations), 7 (visual reactions), 10 (fear and nervousness), 12 (nonverbal communication), and 13 (activity level). In addition, Juvenil showed a significant positive effect ($p = 0.009$) when comparing the index of individual CARS-2 ST categories modulation by Juvenil and placebo, respectively (Table 4).

Table 4. Comparison of individual CARS2-ST * categories in Juvenil- and placebo-supplemented groups. Data were collected before and after 3-months supplementation. Before and after numbers are sums of the CARS2-ST category values from all members of the given group (Juvenil, placebo).

	Childhood Autism Rating Scale (CARS2-ST) Category	Σ Juvenil Before	Σ Juvenil After	Σ Placebo Before	Σ Placebo After	Shift Juvenil/Placebo	Index Juvenil	Index Placebo
1	Relationship to people	26.5	24.5	21.5	19.5	2/2	0.916	0.921
2	Imitation	23.0	21.5	17.5	16.0	1.5/1.5	0.928	0.933
3	Emotional response	21.5	19.5	22.5	22.0	2/0.5	0.902	1
4	Body	21.0	19.0	20.0	20.0	2/0	0.900	1
5	Object use	20.5	19.0	15.5	15.0	1/0.5	0.921	1
6	Adaptation to change	19.5	18.5	21.5	19.5	1/2	0.944	0.916
7	Visual response	19.5	17.5	18.5	18.5	2/0	0.892	1
8	Listening response	22.0	21.0	17.5	17.5	1/0	1	1
9	Taste–smell–touch response and use	18.0	18.0	18.0	17.0	0/1	1	1
10	Fear and nervousness	20.0	17.0	20.0	19.5	3/0.5	0.823	0.972
11	Verbal communication	26.5	26.5	24.5	23.0	0/1.5	1	0.952
12	Nonverbal communication	24.0	22.5	16.5	18.0	1.5/−1.5	0.930	1
13	Activity level	24.0	22.0	21.5	21.5	2/0	0.904	1
14	Level and consistency of intellectual response	21.5	21.5	24.5	24.5	0/0	1	1
15	General impressions	25.0	24.5	24.0	24.0	0.5/0	0.978	1
	t-test						$p = 0.0095$	

Note: * Eric Schopler, Mary E. Van Bourgondien, Glenna Janette Wellman, Steven R. Love. (CARS™2) Childhood Autism Rating Scale™, Second Edition (https://www.wpspublish.com/cars-2-childhood-autism-rating-scale-second-edition.html, accessed on 12 May 2020).

3.4. Correlation between Abundance of Key Bacterial Genera and ASD Symptoms

To further explore whether specific microbiota composition can be associated with ASD symptoms, the abundance of 22 genera reported by other authors as associated with ASD symptoms were compared with results from CARS2-ST testing using a covariance S test (Figure 4). *Prevotella, Escherichia/Shigella, Veillonella, Streptococcus, Alistipes*, and *Bifidobacterium* had the highest positive correlation coefficients in relation to the total CARS2-ST score (i.e., the greater the abundance of these genera, the more severe were the autism symptoms). On the other hand, a negative correlation was observed for *Bacteroides, Faecalibacterium, Barnesiella*, and *Blautia* (Figure 4A), indicating that an increase in the abundance of these genera was associated with relief in autistic symptoms. Of the five tested CARS2-ST categories, the nonverbal communication category was characterized by a positive correlation with *Blautia* (Figure 4E). For all categories tested, including the total score, *Veillonella, Streptococcus*, and *Clostridium* repeatedly exhibited a positive correlation.

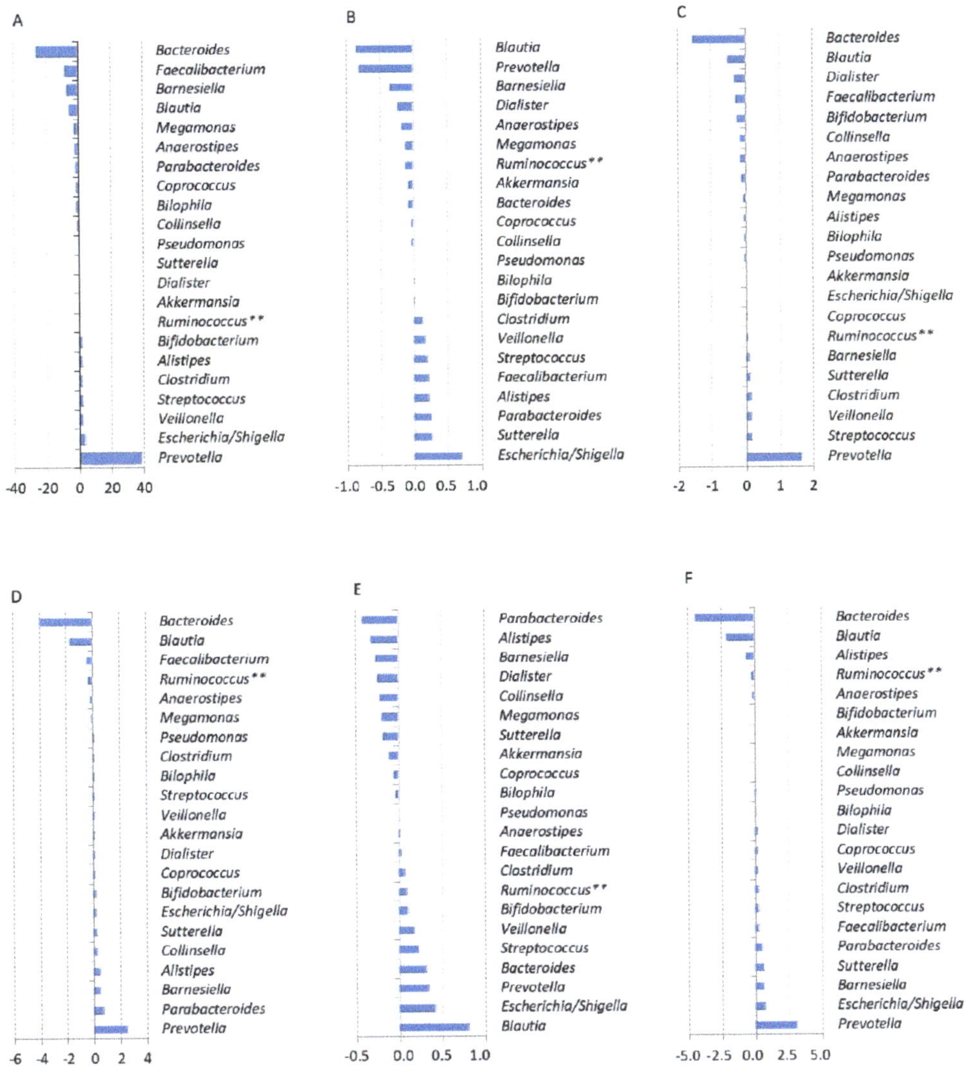

Figure 4. Comparison of microbiome data with total scores of CARS2-ST and its individual categories in which the total score shifted by at least two points after Juvenil supplementation. (**A**)—total score, (**B**)—body, (**C**)—visual response, (**D**)—fear and nervousness, (**E**)—nonverbal communication, (**F**)—activity level. ** only *Ruminococcus torques* group and/or *Ruminococcus gauvreauii*_group.

4. Discussion

The gut microbiota is a complex ecosystem that, through its metabolites or enteroendocrine cell products induced by those metabolites, affects the microbiota–gut–brain axis and thus homeostasis of the entire organism. Consistent with previously published studies [36–41], a different composition of gut microbiota was recorded for children with ASD compared to unrelated neurotypical controls. The significant changes in representation were demonstrated in the phyla *Actinobacteriota*, *Firmicutes*, and *Proteobacteria*. LEfSe analysis revealed an increased abundance of *Bifidobacterium longum*, *Ruminococcus torques* group, *Faecalibacterium*, *Flavonifractor*, and several taxa of *Clostiridiae* and *Pseudomonodaceae* as characteristic for the gut microbiota of autistic children. On the other hand, autistic children were characterized also by lower abundance of taxa from the *Prevotellaceae*, *Peptostreptococcaceae*, and *Monoglobaceae* families. *Faecalibacterium prausnitzii* (*F. prausnitzii*) is one of the main producers of butyrate in the intestine and, because butyrate is an inhibitor of NF-κB and IFN-γ [42], *F. prausnitzii* may interfere with the body's inflammatory responses. Moreover, *F. prausnitzii* is an of IL-10 inducer and may, therefore, be referred to as an anti-inflammatory gut bacterium [43]. Aside from *Faecalibacterium*, the remaining positively scored bacterial taxa (e.g., *Ruminococcus*, *Flavonifractor*, and *Bifidobacterium* species have been suggested as predictors of more adverse post-traumatic neuropsychiatric sequelae outcomes [44]. *Flavonifractor* prevalence has also been associated with another psychiatric diagnosis of affective disorder [45]. On the other hand, the bacterial taxa with decreased abundance in the gut microbiota of autistic children (e.g., *Prevotella*) have high genetic diversity and, therefore, it is difficult to predict their functional relationships to autism [46,47]. Whether *Prevotella* is or is not beneficial to health depends on many factors [48], so it cannot be used unambiguously as a predictive factor of gut dysbiosis in autism [49]. A lower abundance of the families *Monoglobaceae* and *Peptostreptococcaceae* in the gut microbiome has been associated with maternal prenatal stress or anxiety symptoms [50,51].

In this pilot study, the genuses *Bacteroides* and *Prevotella* were found to have the highest negative and positive correlation coefficients, respectively, in relation to total CAR2-ST score. *Bacteroides* and *Prevotella*, two quite closely related genera, were frequently associated with extreme opposite autism symptoms (Figure 4). Interestingly, *Prevotella* is usually enriched in African ethnics with a high proportion of plants in their food (enterotype 2) while *Bacteroides* enrichment is associated with a Western diet (enterotype 1) [52,53]. Another interesting observation for *Prevotella* and *Bacteroides* is that *Bacteroides* dominates the gut microbiota of piglets or humans under lactation and is replaced within a short time after weaning by the related *Prevotella* [54–56]. This may point to the importance of weaning, diet, and associated changes in the gut microbiota for the development of autism in children. In this respect, moreover, the mother's diet during pregnancy and lactation, especially from the viewpoint of consuming a Western diet with an unbalanced ratio of polyunsaturated fatty acids, may adversely affect brain development [57,58].

Microbiota transfer therapy or probiotic supplementation of ASD individuals has been tested as means to alleviate ASD symptoms by modulation of gut microbiota [59–66]. Microbiota transfer therapy treatment protocol, consisting of the application of vancomycin, the laxative MoviPrep, SHGM (Standardized Human Gut Microbiota), and Prilosec (Omeprazole), reduced the rates of core ASD symptoms [5,64]. The administration of probiotics, either alone or in combination with other biologically active substances such as colostrum or oxytocin, led to minor reductions in autism symptoms but without reaching statistical significance [62,63,67]. Only a study that was based on the application of four probiotic bacteria in combination with fructooligosaccharides provided a significant reduction in the severity of autism and gastrointestinal symptoms [65]. Although our results are in agreement with these studies, in that a biologically active substance may alleviate ASD symptoms, whether this is a direct immunostimulating effect or rather is caused by modification of the gut microbiota composition, remains uncertain.

Data from this pilot study presented here demonstrate significant changes in the microbiome, and in parallel, the effects on some categories of CARS2-ST. However, these results must be approached with caution, as children in home care were included in this study, and confounding variables such as diet in the family, passive smoking, the influence of complementary medicines used by children, or the physical and psychological family environment could have impacted microbiota profiles.

5. Conclusions

This pilot study confirms, albeit in a small number of children, that children with ASD have altered composition of the gut microbiota. A high abundance of *Bacteroides* was associated with weaker ASD symptoms while *Prevotella*, *Escherichia/Shigella*, *Veillonella*, *Streptococcus*, *Alistipes*, and *Bifidobacterium* were enriched in the gut microbiota of autistic children with strongest symptoms. An altered composition of ASD children's gut microbiota was shifted toward a neurotypical profile by Juvenil supplementation. Juvenil also positively modulated children's autism symptoms, namely in the categories of motor manifestations, visual reactions, fear and nervousness, nonverbal communication, and activity level. Juvenil supplementation of ASD children was safe, well-tolerated, and had no side effects.

Supplementary Materials: The following supporting information can be downloaded at: https://www.mdpi.com/article/10.3390/nu16131988/s1, Table S1: All basic data on the composition of gut microbiota of children with ASD and neurotypical children expressed in percentages; Table S2: Number of lost or acquired OTUs within individual bacterial families before and after supplementation of ASD children; Table S3: The results of behavioral status of autistic children of both groups assessed by the CARS2-ST at the time of entry into the study and after completion of supplementation. For the purposes of this study, Juvenil was designated as the moon and placebo as the sun.

Author Contributions: J.H.; participated in conceptualization, decisions on methodology, and writing of the original draft, K.K.; participated in conceptualization, decisions on methodology, investigation, and writing of the original draft, V.B. (Vanda Bostik); writing—review and editing, I.R.; participated in decisions on methodology, formal analysis, validation, and writing of the original draft, D.K.; participated in microbiome analysis, V.B. (Vladimir Babak); participated in microbiome analysis, M.D.; Methodology, Validation, K.S.; Methodology, Validation, A.M.; participated in conceptualization, decisions on methodology, investigation, data curation, supervision, and writing of the original draft. All authors have read and agreed to the published version of the manuscript.

Funding: This study was supported by the Ministry of Defence of the Czech Republic—long-term organization development plan Medical Aspects of Weapons of Mass Destruction of the Faculty of Military Health Sciences, University of Defence (DZRO-FVZ-ZHN-II).

Institutional Review Board Statement: The study was conducted according to the guidelines of the Declaration of Helsinki, and approved by the Institutional Review Board and Ethics Commission of the University Hospital Hradec Kralove, Czech Republic, No.: 202005 S09P, on 12 May 2020.

Informed Consent Statement: Informed consent was obtained from all subjects involved in the study.

Data Availability Statement: All data are available within the manuscript.

Acknowledgments: The authors wish to thank Juvenil Products, a.s. Prezletice, Czech Republic for providing extract from bovine tissue for our experiments.

Conflicts of Interest: The authors declare no conflicts of interest. There are, therefore, no financial or material benefits for author/co-authors. The purchase of extract were conducted by the standard way.

Abbreviations

ASD	autism spectrum disorder
ADHD	attention deficit/hyperactivity disorder
CARS2-ST	Childhood Autism Rating Scale in its standard version
DNA	deoxyribonucleic acid
F. prausnitzii	Faecalibacterium prausnitzii
IFN-γ	type II interferon
LEfSe	linear discriminant analysis effect size
NF-κB	nuclear factor Kappa-light-chain-enhancer of activated B-cells
NT	neurotypical
OTU	operational taxonomic units
PCoA	principal component analysis
IL-10	interleukin 10

References

1. Elsabbagh, M.; Divan, G.; Koh, Y.J.; Kim, Y.S.; Kauchali, S.; Marcin, C.; Montiel-Nava, C.; Patel, V.; Paula, C.S.; Wang, C.; et al. Global prevalence of autism and other pervasive developmental disorders. *Autism Res.* **2012**, *5*, 160–179. [CrossRef]
2. Baio, J.; Wiggins, L.; Christensen, D.L.; Maenner, M.J.; Daniels, J.; Warren, Z.; Kurzius-Spencer, M.; Zahorodny, W.; Rosenberg, C.R.; White, T. Prevalence of autism spectrum disorder among children aged 8 years-Autism and developmental disabilities monitoring network, 11 sites, United States, 2014. *MMWR Surveill. Summ.* **2018**, *67*, 1. [CrossRef] [PubMed]
3. Venigalla, H.; Mekala, H.M.; Hassan, M.; Ahmed, R.; Zain, H.; Dar, S.; Veliz, S. An update on biomarkers in psychiatric disorders–are we aware use in our clinical practice. *Ment. Health Fam. Med.* **2017**, *13*, 471–479.
4. Styles, M.; Alsharshani, D.; Samara, M.; Alsharshani, M.; Khattab, A.; Walid, M.; Al-Dewik, N.I. Risk factors, diagnosis, prognosis and treatment of autism. *Front. Biosci.* **2020**, *25*, 1682–1717.
5. Kang, D.W.; Adams, J.B.; Gregory, A.C.; Borody, T.; Chittick, L.; Fasano, A.; Khoruts, A.; Geis, E.; Maldonado, J.; McDonough-Means, S. Microbiota Transfer Therapy alters gut ecosystem and improves gastrointestinal and autism symptoms: An open-label study. *Microbiome* **2017**, *5*, 10. [CrossRef] [PubMed]
6. Mannion, A.; Leader, G. Comorbidity in autism spectrum disorder: A literature review. *Res. Autism Spectr. Disord.* **2013**, *7*, 1595–1616. [CrossRef]
7. Doshi-Velez, F.; Ge, Y.; Kohane, I. Comorbidity clusters in autism spectrum disorders: An electronic health record time-series analysis. *Pediatrics* **2014**, *133*, e54–e63. [CrossRef]
8. Fiorentino, M.; Sapone, A.; Senger, S.; Camhi, S.S.; Kadzielski, S.M.; Buie, T.M.; Kelly, D.L.; Cascella, N.; Fasano, A. Blood–brain barrier and intestinal epithelial barrier alterations in autism spectrum disorders. *Mol. Autism.* **2016**, *7*, 49. [CrossRef] [PubMed]
9. Hallmayer, J.; Cleveland, S.; Torres, A.; Phillips, J.; Cohen, B.; Torigoe, T.; Miller, J.; Fedele, A.; Collins, J.; Smith, K. Genetic heritability and shared environmental factors among twin pairs with autism. *Arch. Gen. Psychiatry* **2011**, *68*, 1095–1102. [CrossRef]
10. Ogunrinola, G.A.; Oyewale, J.O.; Oshamika, O.O.; Olasehinde, G.I. The Human Microbiome and Its Impacts on Health. *Int. J. Microbiol.* **2020**, *2020*, 8045646. [CrossRef]
11. Zhu, S.; Jiang, Y.; Xu, K.; Cui, M.; Ye, W.; Zhao, G.; Jin, L.; Chen, X. The progress of gut microbiome research related to brain disorders. *J. Neuroinflamm.* **2020**, *17*, 25. [CrossRef]
12. Liu, L.; Wang, H.; Chen, X.; Zhang, Y.; Zhang, H.; Xie, P. Gut microbiota and its metabolites in depression: From pathogenesis to treatment. *EBioMedicine* **2023**, *90*, 104527. [CrossRef]
13. Asadi, A.; Shadab, M.N.; Mohamadi, M.H.; Shokri, F.; Heidary, M.; Sadeghifard, N.; Khoshnood, S. Obesity and gut-microbiota-brain axis: A narrative review. *J. Clin. Lab. Anal.* **2022**, *36*, e24420. [CrossRef] [PubMed]
14. Quaglio, A.E.V.; Grillo, T.G.; De Oliveira, E.C.S.; Di Stasi, L.C.; Sassaki, L.Y. Gut microbiota, inflammatory bowel disease and colorectal cancer. *World J. Gastroenterol.* **2022**, *28*, 4053–4060. [CrossRef]
15. Canakis, A.; Haroon, M.; Weber, H.C. Irritable bowel syndrome and gut microbiota. *Curr. Opin. Endocrinol. Diabetes Obes.* **2020**, *27*, 28–35. [CrossRef]
16. Louis, P. Does the human gut microbiota contribute to the etiology of autism spectrum disorders? *Dig. Dis. Sci.* **2012**, *57*, 1987–1989. [CrossRef]
17. Sivamaruthi, B.S.; Suganthy, N.; Kesika, P.; Chaiyasut, C. The Role of Microbiome, Dietary Supplements, and Probiotics in Autism Spectrum Disorder. *Int. J. Environ. Res. Public Health* **2020**, *17*, 2647. [CrossRef] [PubMed]
18. Saurman, V.; Margolis, K.G.; Luna, R.A. Autism Spectrum Disorder as a Brain-Gut-Microbiome Axis Disorder. *Dig. Dis. Sci.* **2020**, *65*, 818–828. [CrossRef] [PubMed]
19. Alamoudi, M.U.; Hosie, S.; Shindler, A.E.; Wood, J.L.; Franks, A.E.; Hill-Yardin, E.L. Comparing the Gut Microbiome in Autism and Preclinical Models: A Systematic Review. *Front. Cell. Infect. Microbiol.* **2022**, *12*, 905841. [CrossRef]
20. Ristori, M.V.; Quagliariello, A.; Reddel, S.; Ianiro, G.; Vicari, S.; Gasbarrini, A.; Putignani, L. Autism, Gastrointestinal Symptoms and Modulation of Gut Microbiota by Nutritional Interventions. *Nutrients* **2019**, *11*, 2812. [CrossRef]
21. Martínez-González, A.E.; Andreo-Martínez, P. Prebiotics, probiotics and fecal microbiota transplantation in autism: A systematic review. *Rev. Psiquiatr. Salud Ment.* **2020**, *13*, 150–164. [CrossRef] [PubMed]
22. Ansari, F.; Pourjafar, H.; Tabrizi, A.; Homayouni, A. The Effects of Probiotics and Prebiotics on Mental Disorders: A Review on Depression, Anxiety, Alzheimer, and Autism Spectrum Disorders. *Curr. Pharm. Biotechnol.* **2020**, *21*, 555–565. [CrossRef] [PubMed]
23. Tan, Q.; Orsso, C.E.; Deehan, E.C.; Kung, J.Y.; Tun, H.M.; Wine, E.; Madsen, K.L.; Zwaigenbaum, L.; Haqq, A.M. Probiotics, prebiotics, synbiotics, and fecal microbiota transplantation in the treatment of behavioral symptoms of autism spectrum disorder: A systematic review. *Autism Res.* **2021**, *14*, 1820–1836. [CrossRef] [PubMed]
24. Taniya, M.A.; Chung, H.J.; Al Mamun, A.; Alam, S.; Aziz, M.A.; Emon, N.U.; Islam, M.M.; Hong, S.S.; Podder, B.R. Role of Gut Microbiome in Autism Spectrum Disorder and Its Therapeutic Regulation. *Front. Cell. Infect. Microbiol.* **2022**, *12*, 915701. [CrossRef] [PubMed]
25. Kubelkova, K.; Macela, A. A short history of Imuregne—An original tissue extract. *MMSL* **2019**, *88*, 115–120. [CrossRef]
26. Bostik, V.; Kubelková, K.; Macela, A. Juvenil, a natural immune booster affecting biological responses through modulation of gut microbiota composition. *MMSL*, 2023; *in press*. [CrossRef]
27. Kubelkova, K.; Hubalek, M.; Rehulka, P.; Rehulkova, H.; Friedecky, D.; Zakova, J.; Macela, A. Molecular characterization of alcohol/ether extract from bovine tissue. *MMSL* **2021**, *90*, 120–136. [CrossRef]

28. Kubelkova, K.; Rychlik, I.; Crhanova, M.; Karasova, D.; Slizova, D.; Zakova, J.; Luksikova, L.; Macela, A. Gut microbiota alterations by nutritional supplement Imuregen. *MMSL* **2020**, *89*, 114–125. [CrossRef]
29. Bolyen, E.; Rideout, J.R.; Dillon, M.R.; Bokulich, N.A.; Abnet, C.C.; Al-Ghalith, G.A.; Alexander, H.; Alm, E.J.; Arumugam, M.; Asnicar, F.; et al. Reproducible, interactive, scalable and extensible microbiome data science using QIIME 2. *Nat. Biotechnol.* **2019**, *37*, 852–857. [CrossRef]
30. Girardot, C.; Scholtalbers, J.; Sauer, S.; Su, S.Y.; Furlong, E.E. Je, a versatile suite to handle multiplexed NGS libraries with unique molecular identifiers. *BMC Bioinform.* **2016**, *17*, 419. [CrossRef]
31. Chen, S.; Zhou, Y.; Chen, Y.; Gu, J. An ultra-fast all-in-one FASTQ preprocessor. *Bioinformatics* **2018**, *34*, i884–i890. [CrossRef]
32. Callahan, B.J.; McMurdie, P.J.; Rosen, M.J.; Han, A.W.; Johnson, A.J.; Holmes, S.P. DADA2: High-resolution sample inference from Illumina amplicon data. *Nat. Methods* **2016**, *13*, 581–583. [CrossRef] [PubMed]
33. Bokulich, N.A.; Kaehler, B.D.; Rideout, J.R.; Dillon, M.; Bolyen, E.; Knight, R.; Huttley, G.A.; Gregory Caporaso, J. Optimizing taxonomic classification of marker-gene amplicon sequences with QIIME 2's q2-feature-classifier plugin. *Microbiome* **2018**, *6*, 90. [CrossRef] [PubMed]
34. Quast, C.; Pruesse, E.; Yilmaz, P.; Gerken, J.; Schweer, T.; Yarza, P.; Peplies, J.; Glockner, F.O. The SILVA ribosomal RNA gene database project: Improved data processing and web-based tools. *Nucleic Acids Res.* **2013**, *41*, D590–D596. [CrossRef] [PubMed]
35. Segata, N.; Izard, J.; Waldron, L.; Gevers, D.; Miropolsky, L.; Garrett, W.S.; Huttenhower, C. Metagenomic biomarker discovery and explanation. *Genome Biol.* **2011**, *12*, R60. [CrossRef] [PubMed]
36. Williams, B.L.; Hornig, M.; Parekh, T.; Lipkin, W.I. Application of novel PCR-based methods for detection, quantification, and phylogenetic characterization of Sutterella species in intestinal biopsy samples from children with autism and gastrointestinal disturbances. *MBio* **2012**, *3*, e00261-11. [CrossRef] [PubMed]
37. Song, Y.L.; Liu, C.X.; Finegold, S.A. Real-time PCR quantitation of Clostridia in feces of autistic children. *Appl. Environ. Microbiol.* **2004**, *70*, 6459–6465. [CrossRef] [PubMed]
38. De Angelis, M.; Piccolo, M.; Vannini, L.; Siragusa, S.; De Giacomo, A.; Serrazzanetti, D.I.; Cristofori, F.; Guerzoni, M.E.; Gobbetti, M.; Francavilla, R. Fecal microbiota and metabolome of children with autism and pervasive developmental disorder not otherwise specified. *PLoS ONE* **2013**, *8*, 18. [CrossRef] [PubMed]
39. Wang, L.; Christophersen, C.T.; Sorich, M.J.; Gerber, J.P.; Angley, M.T.; Conlon, M.A. Increased abundance of *Sutterella* spp. and *Ruminococcus torques* in feces of children with autism spectrum disorder. *Mol. Autism.* **2013**, *4*, 42. [CrossRef]
40. Tomova, A.; Husarova, V.; Lakatosova, S.; Bakos, J.; Vlkova, B.; Babinska, K.; Ostatnikova, D. Gastrointestinal microbiota in children with autism in Slovakia. *Physiol. Behav.* **2015**, *138*, 179–187. [CrossRef]
41. Yuan, J.; Zhu, L.; Liu, X.; Li, T.; Zhang, Y.; Ying, T.; Wang, B.; Wang, J.; Dong, H.; Feng, E. A Proteome Reference Map and Proteomic Analysis of Bifidobacterium longum NCC2705. *Mol. Cell. Proteom.* **2006**, *5*, 11. [CrossRef]
42. Sokol, H.; Pigneur, B.; Watterlot, L.; Lakhdari, O.; Bermúdez-Humarán, L.G.; Gratadoux, J.J.; Blugeon, S.; Bridonneau, C.; Furet, J.P.; Corthier, G. *Faecalibacterium prausnitzii* is an anti-inflammatory commensal bacterium identified by gut microbiota analysis of Cron disease patients. *Proc. Natl. Acad. Sci. USA* **2008**, *105*, 16731–16736. [CrossRef] [PubMed]
43. Rossi, O.; van Berkel, L.A.; Chain, F.; Tanweer Khan, M.; Taverne, N.; Sokol, H.; Duncan, S.H.; Flint, H.J.; Harmsen, H.J.; Langella, P. *Faecalibacterium prausnitzii* A2-165 has a high capacity to induce IL-10 in human and murine dendritic cells and modulates T cell responses. *Sci. Rep.* **2016**, *6*, 18507. [CrossRef] [PubMed]
44. Zeamer, A.L.; Salive, M.C.; An, X.; Beaudoin, F.L.; House, S.L.; Stevens, J.S.; Zeng, D.; Neylan, T.C.; Clifford, G.D.; Linnstaedt, S.D. Association between microbiome and the development of adverse posttraumatic neuropsychiatric sequelae after traumatic stress exposure. *Transl. Psychiatry* **2023**, *13*, 354. [CrossRef] [PubMed]
45. Coello, K.; Hansen, T.H.; Sørensen, N.; Ottesen, N.M.; Miskowiak, K.W.; Pedersen, O.; Kessing, L.V.; Vinberg, M. Affective disorders impact prevalence of Flavonifractor and abundance of Christensenellaceae in gut microbiota. *Prog. Neuropsychopharmacol. Biol. Psychiatry* **2021**, *110*, 110300. [CrossRef] [PubMed]
46. Vinod, K.; Narendrakumar, M.; Suchismitha, I.; Chitra, D. Divergences in gene repertoire among the reference Prevotella genomes derived from distinct human body sites. *BMC Genom.* **2015**, *16*, 153.
47. Ley, R.E. Gut microbiota in 2015: Prevotella in the gut: Choose carefully. Nature Reviews. *Gastroenterol. Hepatol.* **2016**, *13*, 69–70.
48. Tett, A.; Pasolli, E.; Masetti, G.; Ercolini, D.; Segata, N. Prevotella diversity, niches and interactions with the human host. *Nat. Rev. Microbiol.* **2021**, *19*, 585–599. [CrossRef] [PubMed]
49. Agarwala, S.; Naik, B.; Ramachandra, N.B. Mucosa-associated specific bacterial species disrupt the intestinal epithelial barrier in the autism phenome. *Brain Behav. Immun. Health* **2021**, *15*, 100269. [CrossRef] [PubMed]
50. Abdelsalam, N.A.; Hegazy, S.M.; Aziz, R.K. The curious case of *Prevotella copri*. *Gut Microbes* **2023**, *15*, 2249152. [CrossRef]
51. Chang, C.J.; Lin, T.L.; Tsai, Y.L.; Wu, T.R.; Lai, W.F.; Lu, C.C.; Lai, H.C. Next generation probiotics in disease amelioration. *J. Food Drug Anal.* **2019**, *27*, 615–622. [CrossRef]
52. Tremaroli, V.; Bäckhed, F. Functional interactions between the gut microbiota and host metabolism. *Nature* **2012**, *489*, 242–249. [CrossRef] [PubMed]
53. García-Montero, C.; Fraile-Martínez, O.; Gómez-Lahoz, A.M.; Pekarek, L.; Castellanos, A.J.; Noguerales-Fraguas, F.; Coca, S.; Guijarro, L.G.; García-Honduvilla, N.; Asúnsolo, A. Nutritional Components in Western Diet Versus Mediterranean Diet at the Gut Microbiota-Immune System Interplay. Implications for Health and Disease. *Nutrients* **2021**, *13*, 699. [CrossRef] [PubMed]

54. Khine, W.W.T.; Rahayu, E.S.; See, T.Y.; Kuah, S.; Salminen, S.; Nakayama, J.; Lee, Y.K. Indonesian children fecal microbiome from birth until weaning was different from microbiomes of their mothers. *Gut Microbes* **2020**, *12*, 1761240. [CrossRef] [PubMed]
55. Kubasova, T.; Davidova-Gerzova, L.; Babak, V.; Cejkova, D.; Montagne, L.; Le-Floc'h, N.; Rychlik, I. Effects of host genetics and environmental conditions on fecal microbiota composition of pigs. *PLoS ONE* **2018**, *13*, e0201901. [CrossRef] [PubMed]
56. Karasova, D.; Crhanova, M.; Babak, V.; Jerabek, M.; Brzobohaty, L.; Matesova, Z.; Rychlik, I. Development of piglet gut microbiota at the time of weaning influences development of postweaning diarrhea—A field study. *Res. Vet. Sci.* **2021**, *135*, 59–65. [CrossRef] [PubMed]
57. Simopoulos, A.P. *Importance of the Omega-6/Omega-3 Balance in Health and Disease: Evolutionary Aspects of Diet, Healthy Agriculture, Healthy Nutrition, Healthy People*; Karger Publishers: Basel, Switzerland, 2011; pp. 10–21.
58. Yau, S.Y.; Yip, Y.S.L.; Formolo, D.A.; He, S.; Lee, T.H.Y.; Wen, C.; Hryciw, D.H. Chronic consumption of a high linoleic acid diet during pregnancy, lactation and post-weaning period increases depression-like behavior in male, but not female offspring. *Behav. Brain Res.* **2022**, *416*, 113538. [CrossRef] [PubMed]
59. Liu, Y.W.; Liong, M.T.; Chung, Y.E.; Huang, H.Y.; Peng, W.S.; Cheng, Y.F.; Lin, Y.S.; Wu, Y.Y.; Tsai, Y.C. Effects of *Lactobacillus plantarum* PS128 on Children with Autism Spectrum Disorder in Taiwan: A Randomized, Double-Blind, Placebo-Controlled Trial. *Nutrients* **2019**, *11*, 820. [CrossRef] [PubMed]
60. Critchfield, J.W.; van Hemert, S.; Ash, M.; Mulder, L.; Ashwood, P. The potential role of probiotics in the management of childhood autism spectrum disorders. *Gastroenterol. Res. Pract.* **2011**, *2011*, 161358. [CrossRef]
61. Santocchi, E.; Guiducci, L.; Fulceri, F.; Billeci, L.; Buzzigoli, E.; Apicella, F.; Calderoni, S.; Grossi, E.; Morales, M.A.; Muratori, F. Gut to brain interaction in Autism Spectrum Disorders: A randomized controlled trial on the role of probiotics on clinical, biochemical and neurophysiological parameters. *BMC Psychiatry* **2016**, *16*, 183. [CrossRef] [PubMed]
62. Arnold, L.E.; Luna, R.A.; Williams, K.; Chan, J.; Parker, R.A.; Wu, Q.; Hollway, J.A.; Jeffs, A.; Lu, F.; Coury, D.L.; et al. Probiotics for Gastrointestinal Symptoms and Quality of Life in Autism: A Placebo-Controlled Pilot Trial. *J. Child Adolesc. Psychopharmacol.* **2019**, *29*, 659–669. [CrossRef]
63. Sanctuary, M.R.; Kain, J.N.; Chen, S.Y.; Kalanetra, K.; Lemay, D.G.; Rose, D.R.; Yang, H.T.; Tancredi, D.J.; German, J.B.; Slupsky, C.M. Pilot study of probiotic/colostrum supplementation on gut function in children with autism and gastrointestinal symptoms. *PLoS ONE* **2019**, *14*, e0210064. [CrossRef] [PubMed]
64. Kang, D.W.; Adams, J.B.; Coleman, D.M.; Pollard, E.L.; Maldonado, J.; McDonough-Means, S.; Caporaso, J.G.; Krajmalnik-Brown, R. Long-term benefit of Microbiota Transfer Therapy on autism symptoms and gut microbiota. *Sci. Rep.* **2019**, *9*, 5821. [CrossRef] [PubMed]
65. Wang, Y.; Li, N.; Yang, J.J.; Zhao, D.M.; Chen, B.; Zhang, G.Q.; Chen, S.; Cao, R.F.; Yu, H.; Zhao, C.Y.; et al. Probiotics and fructo-oligosaccharide intervention modulate the microbiota-gut brain axis to improve autism spectrum reducing also the hyper-serotonergic state and the dopamine metabolism disorder. *Pharmacol. Res.* **2020**, *157*, 104784. [CrossRef] [PubMed]
66. Sarkar, A.; Lehto, S.M.; Harty, S.; Dinan, T.G.; Cryan, J.F.; Burnet, P.W.J. Psychobiotics and the Manipulation of Bacteria-Gut-Brain Signals. *Trends Neurosci.* **2016**, *39*, 763–781. [CrossRef]
67. Kong, X.J.; Liu, J.; Liu, K.; Koh, M.; Sherman, H.; Liu, S.; Tian, R.; Sukijthamapan, P.; Wang, J.; Fong, M. Probiotic and Oxytocin Combination Therapy in Patients with Autism Spectrum Disorder: A Randomized, Double-Blinded, Placebo-Controlled Pilot Trial. *Nutrients* **2021**, *13*, 1552. [CrossRef]

nutrients

MDPI

Article

Acute and Repeated Ashwagandha Supplementation Improves Markers of Cognitive Function and Mood

Megan Leonard [1], Broderick Dickerson [1], Landry Estes [1], Drew E. Gonzalez [1], Victoria Jenkins [1], Sarah Johnson [1], Dante Xing [1], Choongsung Yoo [1], Joungbo Ko [1], Martin Purpura [2], Ralf Jäger [2], Mark Faries [1,3], Wesley Kephart [4], Ryan Sowinski [1], Christopher J. Rasmussen [1] and Richard B. Kreider [1,*]

1 Exercise & Sport Nutrition Lab, Department of Kinesiology and Sports Management, Texas A&M University, College Station, TX 77843, USA; meganleonard10@tamu.edu (M.L.); dickersobl5@email.tamu.edu (B.D.); landry.estes@tamu.edu (L.E.); dg18@tamu.edu (D.E.G.); victoria.jenkins@tamu.edu (V.J.); sjohnson2216@tamu.edu (S.J.); dantexing@tamu.edu (D.X.); choongsungyoo@tamu.edu (C.Y.); joungboko10@tamu.edu (J.K.); mark.faries@ag.tamu.edu (M.F.); rjs370@tamu.edu (R.S.); crasmussen@tamu.edu (C.J.R.)

2 Increnovo LLC, Whitefish Bay, WI 53217, USA; martin.purpura@increnovo.com (M.P.); ralf.jaeger@increnovo.com (R.J.)

3 Texas A&M AgriLife Extension, Texas A&M University, College Station, TX 77843, USA

4 Department of Kinesiology, University of Wisconsin—Whitewater, Whitewater, WI 53190, USA; kephartw@uww.edu

* Correspondence: rbkreider@tamu.edu

Abstract: Background: Ashwagandha has been reported to reduce stress and attenuate cognitive decline associated with inflammation and neurodegeneration in clinical populations. However, the effects as a potential nootropic nutrient in younger populations are unclear. This study examined the effects of liposomal ashwagandha supplementation on cognitive function, mood, and markers of health and safety in healthy young men and women. Methods: 59 men and women (22.7 ± 7 yrs., 74.9 ± 16 kg, 26.2 ± 5 BMI) fasted for 12 h, donated a fasting blood sample, and were administered the COMPASS cognitive function test battery (Word Recall, Word recognition, Choice Reaction Time Task, Picture Recognition, Digit Vigilance Task, Corsi Block test, Stroop test) and profile of mood states (POMS). In a randomized and double-blind manner, participants were administered 225 mg of a placebo (Gum Arabic) or ashwagandha (*Withania somnifera*) root and leaf extract coated with a liposomal covering. After 60-min, participants repeated cognitive assessments. Participants continued supplementation (225 mg/d) for 30 days and then returned to the lab to repeat the experiment. Data were analyzed using a general linear model (GLM) univariate analysis with repeated measures and pairwise comparisons of mean changes from baseline with 95% confidence intervals (CI). Results: Ashwagandha supplementation improved acute and/or 30-day measures of Word Recall (correct and recalled attempts), Choice Reaction Time (targets identified), Picture Recognition ("yes" correct responses, correct and overall reaction time), Digit Vigilance (correct reaction time), Stroop Color-Word (congruent words identified, reaction time), and POMS (tension and fatigue) from baseline more consistently with several differences observed between groups. Conclusion: Results support contentions that ashwagandha supplementation (225 mg) may improve some measures of memory, attention, vigilance, attention, and executive function while decreasing perceptions of tension and fatigue in younger healthy individuals. Retrospectively registered clinical trial ISRCTN58680760.

Keywords: nootropic; executive function; cognition; memory; vigilance; attention; mood

1. Introduction

Ashwagandha (*Withania somnifera*) is a plant used in Ayurvedic medicine for over 3000 years as a naturally occurring adaptogen to help manage stress, anxiety, and inflammation [1–13]. Ashwagandha has antioxidant properties [9,14–16], influences endocrine function [17–21], and has immunomodulatory effects [10,22]. Ashwagandha supplementation has been reported to attenuate cognitive decline associated with inflammation [9,11,12] and neurodegeneration [8,10,15,23–26]. Since oral ingestion of ashwagandha is bioavailable and crosses the blood–brain barrier [24], ashwagandha has been considered a naturally occurring therapeutic agent for individuals with type 2 diabetes mellitus [9,10,25], mild cognitive impairment [8,11,14] and individuals with neurodegenerative diseases [14,15,24,27–29].

Basic research studies provide a solid theoretical rationale that ashwagandha supplementation may benefit health and cognition [25,27,30]. However, fewer studies have evaluated the effects of ashwagandha supplementation on cognitive function in clinical and healthy populations [31]. For example, Chengappa et al. [32] reported that ashwagandha supplementation (2 × 250 mg/d for eight weeks) in 60 patients with medically managed bipolar disorders improved measures of working memory, reaction time, and social cognition. Choudhary and associates [33] reported that ashwagandha supplementation (2 × 300 mg/d for eight weeks) in 50 individuals with mild cognitive impairment (MCI) improved measures of immediate and general memory, executive function, attention, and information processing speed. This group also reported that ashwagandha supplementation (2 × 300 mg/d for eight weeks) reduced cortisol and improved mental well-being in 52 patients suffering from chronic stress and related disorders [7]. Similarly, Remenapp et al. [34] reported that supplementation with a liposomal-coated ashwagandha (225 or 400 mg/d for 30 days) reduced markers of stress and improved cognitive function in individuals experiencing perceived stress. In healthy individuals, Pingali and collaborators [35] reported that ashwagandha supplementation (2 × 250 mg/d for 14 days) improved reaction times and cognitive and psychomotor performance in 20 healthy young males. Moreover, Baker and coworkers [36] reported that ashwagandha supplementation (2 × 350 mg/d for 30 days) improved college students' perceptions of well-being, energy, mental clarity, and sleep quality. Our group found that acute ingestion of ashwagandha (400 mg) enhanced measures of executive function, short-term/working memory, and the ability to sustain attention for up to six hours in healthy individuals [37]. Additionally, Bonilla and coworkers [38] conducted a Bayesian meta-analysis and systemic literature review and found evidence that ashwagandha supplementation improved physical performance-related variables. While these findings are promising, more research is needed to determine the acute and long-term effects of ashwagandha supplementation on cognitive function and health measures, particularly in healthy populations.

This study aimed to examine the effects of liposomal ashwagandha supplementation (acute dose and 225 mg/d for 30 days) on a comprehensive battery of cognitive function assessments, mood, and markers of health and safety in healthy young men and women. The primary outcomes were cognitive function and measures. Secondary outcomes included markers of health and safety. We hypothesized that acute supplementation with ashwagandha would increase markers of cognitive function and that 30-day supplementation of ashwagandha would result in additive benefits while well tolerated. The following describes the study's methods, procedures, and results, followed by a discussion and recommendations for additional work.

2. Methods

2.1. Experimental Design of Study

This clinical trial was conducted at a university clinical research facility in a double-blind, placebo-controlled manner. The independent variable was nutritional supplementation. Primary dependent variable outcomes included measures of cognitive function. Secondary outcomes were markers of health obtained from clinical chemistry panels and perceptions of side effects.

2.2. Participants of the Study

This clinical trial was retrospectively registered with the ISRCTN registry on 15 May 2024 (ISRCTN58680760). The study was approved by the Human Research Protection Program Review Board (IRB2022-0621, 21 September 2022) and adhered to the Declaration of Helsinki ethical standards for conducting human participant research. Participants were recruited for this study through emails and advertising on flyers, websites, and online flyers. Potential subjects were screened via questionnaires to determine initial eligibility and invited to attend a familiarization session if they appeared to meet eligibility criteria. Those interested in participating in the study signed consent statements, completed health history questionnaires, and underwent physical examination to confirm eligibility.

Healthy volunteers between 18 and 60 years of age were recruited for this study with the following inclusion criteria: They must (1) have no diagnosed cognitive deficits from a physician; (2) have no diagnosed sleep disorders from a physician; (3) have no history of cardiovascular, metabolic or pulmonary disease from a physician; (4) have no history of migraine headaches, hypertension, cardiac arrhythmias or anxiety; (5) have no ulcers or gastrointestinal reflux disease; (6) be willing to provide voluntary, written, and informed consent; (7) be willing to consume the investigational product daily for the duration of the study; and (8) have no allergies to the fiber Gum Arabic. Participants were excluded from the study if they (1) were pregnant or desired pregnancy during the study; (2) had a documented history of taking prescription medications in the prior month that might affect study testing. Individuals taking medications that the investigators deemed would not affect primary study outcomes and were taken throughout the study (e.g., glucose management, lipid-lowering, anti-hypertensive, thyroid medications, etc.) were permitted to participate in the study; and (3) were recently instructed by their physician (within the past month) to abstain or limit caffeine or other products containing stimulants.

A Consolidated Standards of Reporting Trials (CONSORT) illustration is provided in Figure 1. A total of 321 individuals underwent phone screening to assess general eligibility. Of these, 122 cleared the initial screening criteria and were invited to familiarization sessions. Sixty-seven people were familiarized with the study and provided informed consent. Of these, 7 subjects were unable to participate in the study due to scheduling conflicts, and 60 individuals were enrolled in the study and matched according to age, body mass index, and sex for random assignment into treatment groups. A total of 30 participants were randomized into the placebo (PLA) group and 30 into the ashwagandha (ASH) group. One was removed from the study due to non-compliance. Data from 59 participants were analyzed statistically.

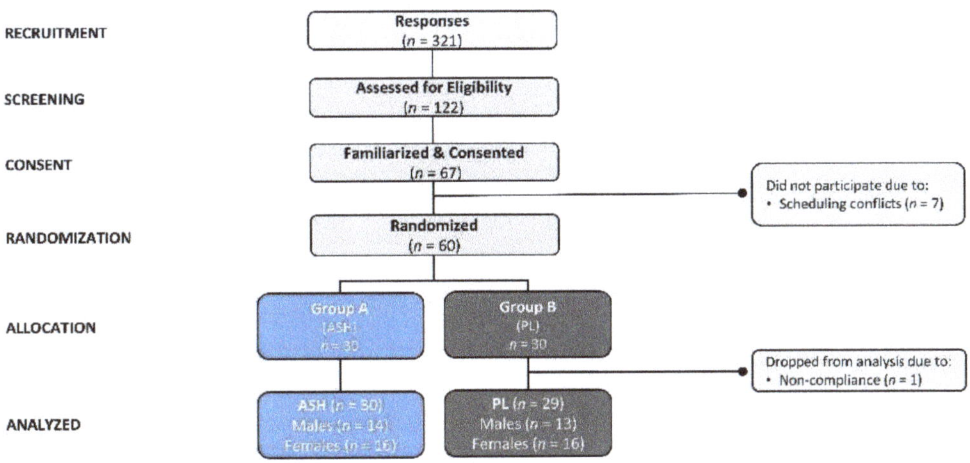

Figure 1. Consolidated Standards of Reporting Trials (CONSORT) flow chart for the ashwagandha (ASH) and placebo (PL) groups.

2.3. Testing Sequence

Figure 2 shows the experimental timeline. Participants attended one familiarization and two experimental testing sessions. At the familiarization session, participants completed health history questionnaires and had height, weight, and resting hemodynamics determined. The methods and expectations of the study were also described. Those eligible to participate practiced the cognitive function tests at least three times to familiarize themselves with the assessments and minimize the learning effects. Participants were instructed on recording diet and energy-containing beverage intake and provided a list of foods and beverages containing stimulants to avoid consuming before each testing session.

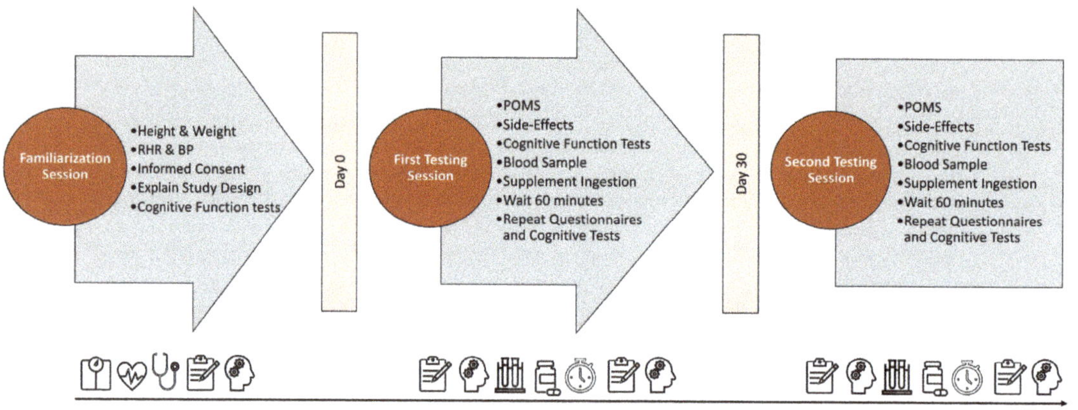

Figure 2. Testing timeline. RHR indicates resting heart rate, BP indicates resting blood pressure, and POMS represents Profile of Mood States inventory.

Before the first testing session, participants recorded their food and beverage intake for 4 days. They were also asked to refrain from consuming unusual amounts of caffeine and other stimulants for 48 h, fast for 12 h, and refrain from intense exercise for 24 h before testing. Upon reporting to the lab, participants completed a pre-supplementation (Pre) side-effect questionnaire and performed Computerized Mental Performance Assessment System (COMPASS) cognitive tests, including Word Recall, Word Recognition, Choice Reaction Time, Picture Recognition, Corsi Block, Digit Vigilance, and Stroop Color–Word test) as well as a Profile of Mood States (POMS) questionnaires. Participants then donated a fasted blood sample. Participants were randomly assigned to a placebo or ashwagandha supplement. One hour following ingestion, participants repeated cognitive testing through COMPASS, the POMS questionnaire, and a post-supplement side effect questionnaire. Participants were then given a 30-day supply of the supplement ingested during the visit and, in between testing sessions, were instructed to ingest daily in the morning with breakfast. Participants maintain their regular diet and physical activity levels and return to the lab after 30 days of supplementation to repeat the testing protocol from the first session and ingest the last dose of their randomly assigned supplement.

2.4. Supplementation Protocol

Participants consumed one capsule per day of either a placebo (PL) consisting of 225 mg of Gum Arabic (Spraygum Bai, Lot #190 057, Nexira Food, Rouen, France) or 225 mg of liposomal ashwagandha (*Withania somnifera*) root and leaf extract (ASH, NooGandha®, Specnova LLC, Lot #221221, Tysons Corner, VA, USA). This dosage and source of ashwagandha had been previously shown to reduce markers of stress and improve cognitive function in individuals with perceived stress [34]. The liposomal root and leaf extract was manufactured using sunflower lecithin consisting of a proprietary blend of phosphatidylcholine, phosphatidylserine, phosphatidylinositol, phosphatidylethanolamine, and phosphatidic acid, and a surface coating using gum Arabic-derived polysaccharides and ashwagandha plant fibers to improve the stability of the liposomes as they pass through the gastrointestinal tract. A homogenizer was used to mix the liposomal formulation with the ashwagandha extract in water/ethanol and then spray dried. The raw ingredients were encapsulated using Vcaps® Plus Capsules (Capsugel®, Lot #5412055, Colmer, France). Certificates of analysis were provided by the raw ingredient supplier and the company that encapsulated the supplements, verifying dosage and ensuring that the supplements were free from contaminants. Capsules were the same size and color and were shipped in labeled bags. Once received, the capsules were placed into individual participant supplementation containers and labeled as designated for double-blinded administration. Participants ingested the supplements after breakfast (or about 8:00 a.m.) for 29 days. Participants ingested the supplement on day 30 following pre-supplementation cognitive testing and obtaining a blood sample.

3. Procedures

3.1. Participant Descriptives

Weight and height were obtained from a Health-O-Meter Professional 500KL (Pelstar LLC, Alsip, IL, USA) digital scale. After sitting passively for 5 min, resting heart rate and blood pressure were obtained using standard procedures with a Connex® ProBP™ 3400 monitor (Welch Allyn, Tilburg, The Netherlands).

3.2. Diet Control

Volunteers documented beverage and food consumption for four days before baseline assessments using written food logs or a phone application (MyFitnessPal, Inc., Baltimore, MD, USA) [37]. This initial diet was replicated before the second testing session for diet consistency.

3.3. Cognitive Function Assessment

The Computerized Mental Performance Assessment (COMPASS) software version 6.0 (Northumbria University, Newcastle, UK) was used to evaluate measures of cognitive function. The assessments included (1) the Word Recall test that measures working memory and the transfer to long-term memory through secondary memory by recalling words shown 60 seconds (immediately) [39] after being seen and recalling them again at the end of cognitive testing 10–20 min following the initial recall (i.e., delayed recall [40]; (2) the Word Recognition test, which assesses secondary memory by indicating if the words displayed through the Word Recall test show familiarity, representing an accuracy percentage with the number of words recognized from the original list [41]; (3) the Choice Reaction Time (CRT) test that assesses reaction time vigilance towards a target stimulus with sustained attention by scoring the accuracy of a response that correlates with its respective target [42]; (4) The Picture Recognition test, which measures secondary memory by showing a series of pictures and following the initial presentation, showing another series of images and indicating YES/NO if the image was in the initial presentation and scoring the reaction time of recognition of correct responses [43]; (5) the Digit Vigilance test, which measures the overall accuracy of correct responses to a numbering scheme presented while assessing the number of false selections as well as measuring the reaction time of correct responses [44]; (6) the Corsi Block test that evaluates present working and spatial memory by replicating a display sequence of colored blocks and is scored on the number of correct trials accurately remembered [45]; and (7) the Stroop color-word test, which measures executive function, attention, and vigilance through accuracy of overall, congruent, and incongruent stimuli by responding to a series of color names written in different colored fonts and accurately identifying the color of the font [46]. The COMPASS cognitive tests have been used in several studies and have been shown as an accurate and reliable cognitive function assessment [47–49].

3.4. Profile of Mood States

The Profile of Mood States (POMS) 65-item questionnaire evaluated changes in mood states during supplementation throughout the study. Ratings to questions were categorized into six domains (i.e., anger, confusion, depression, fatigue, tension, and vigor). Total mood disturbance score (TMDS) was determined by summing scores from anger, confusion, depression, fatigue, and tension and subtracting the vigor score. The POMS is a commonly used and valid assessment of mood states [50,51].

3.5. Blood Collection

A fasted blood sample was obtained from a certified phlebotomist before supplement ingestion on days 0 and 30 of the study to assess clinical health markers and safety. This involved collecting approximately 20 mL of blood from an antecubital vein into two serum separator tubes (SST) and one chelated potassium (K2) ethylenediaminetetraacetic acid (EDTA) tube (BD Vacutainer®, Becton, Dickinson and Company, Franklin Lakes, NJ, USA). Serum SST tubes sat for 15–30 min at room temperature and were then centrifuged (3000× *g*) at 4 °C for 10 min (MegaFuge 40R Centrifuge, Thermo Electron North America LLC, West Palm Beach, FL, USA). Serum from one SST tube was aliquoted into 1.5 mL Eppendorf storage tubes (VWR, Radnor, PA, USA) and stored at −80 °C. One SST and the EDTA tube were refrigerated until transport to Clinical Pathology Labs, Inc. (Austin, TX, USA, CLIA #45D0505003, CAP Accreditation #21525-01) for whole blood cell blood count with percent differential and serum metabolic panel analysis.

3.6. Side Effects Questionnaire

Perceptions of symptoms related to dietary supplementation (i.e., dizziness, headache, tachycardia, heart palpitations, dyspnea, nervousness, blurred vision, and others) were assessed using Likert-type scales as previously described [52,53].

3.7. Statistical Analysis

Statistical analysis software (Version 29 SPSS®, IBM Corp., Armonk, NY, USA) was used for statistical analysis. The sample size was selected based on our previous work [54–57] and assumed a 5% improvement with a power of 80% in primary outcome variables. Our prior worked dose effectiveness study indicated that this sample size was sufficient to determine clinically significant differences [55–59]. Participants were randomized into treatment groups using a balanced Latin Square designer program [60]. Repeated measures multivariate and univariate General linear model (GLM) analyses using Wilks' Lambda and Greenhouse-Geisser statistical tests were performed on the data. Type I error was considered at a probability level of <0.05, while *p*-values between >0.05 and <0.10 are noted to identify a tendency toward significance. Pairwise comparisons were assessed using the Fisher's least significant difference statistic. Data were also analyzed using relative dose as a covariate, but it was not significant. Mean changes from baseline with confidence intervals (CIs) of 95% assessed the clinical significance of findings. Means and CIs completely above or below baseline were considered clinically significant findings [61]. Data are shown as means and standard deviations or mean changes from baseline with lower and upper CIs (mean [LL, UL]). Partial Eta squared (η_p^2) values were used to assess effect sizes. Effect sizes were considered small (0.01), medium (0.06), and large (0.14) [62]. Ch-squared analysis was used to assess differences in categorical ratings of side effects. This statistical approach is consistent with recommendations from Earnest and colleagues [63] on best practices in reporting sport nutrition-related research.

4. Results

4.1. Demographic Data

Participant demographic data are shown in Table S1. Participants were 22.7 ± 7.4 years (range 18–49), 166.8 ± 21.7 cm, 74.9 ± 16.5 kg, and had a body mass index (BMI) of 26.2 ± 5.1 kg/m^2. Sex differences were observed in body weight (females 68.8 ± 16.6 kg, males 82.1 ± 13.5 kg, $p = 0.001$) and resting blood pressure (females 112.3 ± 9.2 mmHg, males 125.8 ± 9.9 mmHg, $p < 0.001$). No sex differences were seen in age ($p = 0.889$), height ($p = 0.238$), BMI ($p = 0.626$), resting heart rate ($p = 0.143$), or diastolic blood pressure ($p = 0.962$).

4.2. Cognitive Function Assessment

4.2.1. Word Recall

Table S2 presents word recall test results. Overall, multivariate Wilk's Lambda analysis revealed a significant time effect ($p < 0.001$, $\eta_p^2 = 0.063$, moderate effect) with no interaction effects observed ($p = 0.736$, $\eta_p^2 = 0.017$, small effect). Univariate analysis revealed significant time effects for recall attempts ($p = 0.010$, $\eta_p^2 = 0.066$, medium effect), correct attempts ($p = 0.011$, $\eta_p^2 = 0.065$, medium effect), delayed recall attempts ($p < 0.001$, $\eta_p^2 = 0.155$, large effect), and delayed correctly recalled ($p < 0.001$, $\eta_p^2 = 0.129$, medium effect) with no significant group × time effects. Analysis of percent changes from baseline (Figure 3) revealed that after 30 days of supplementation, the number of correct attempts recalled attempts, and recalled correct attempts tended to increase above baseline with ASH supplementation, while no differences were observed in the PLA group.

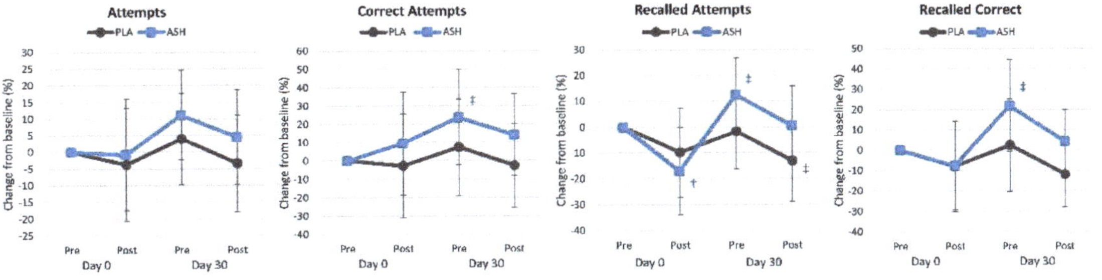

Figure 3. Word recall test results. † = $p < 0.05$ effect from baseline value while ‡ = $p > 0.05$ to $p < 0.10$ difference from baseline.

4.2.2. Word Recognition

Table S3 presents word recognition test results. Overall, multivariate Wilk's Lambda analysis revealed a significant time ($p = 0.008$, $\eta_p^2 = 0.076$, moderate effect) with no interaction effects observed ($p = 0.725$, $\eta_p^2 = 0.033$, small effect). Univariate analysis revealed significant time effects for words correct ($p = 0.080$, $\eta_p^2 = 0.039$, small effect), yes words correct ($p = 0.003$, $\eta_p^2 = 0.079$, medium effect), and no words reaction time ($p < 0.003$, $\eta_p^2 = 0.104$, moderate effect). No significant interaction effects were observed in words, yes, and no words correct or reaction time or overall reaction time between groups. Analysis of percent changes from baseline (Figure 4) revealed some changes from baseline values in both treatment groups after acute and 30 days of supplementation, with no significant differences observed between groups.

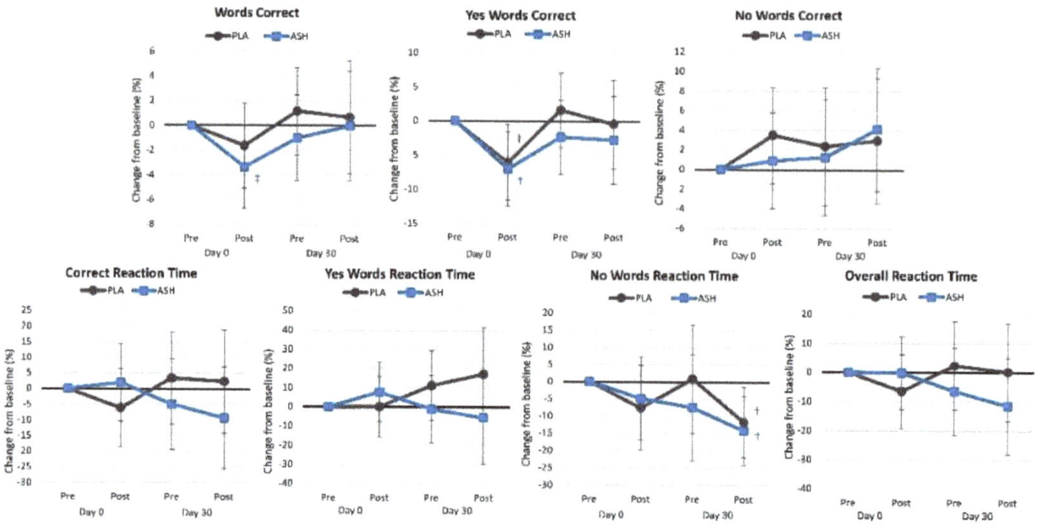

Figure 4. Word recognition test results. † = $p < 0.05$ difference from baseline value while ‡ = $p > 0.05$ to $p < 0.10$ difference from baseline.

4.2.3. Choice Reaction Time

Table S4 shows the choice reaction time test results. Overall, multivariate Wilk's Lambda analysis revealed no significant time ($p = 0.247$, $\eta_p^2 = 0.022$, small effect), while there tended to be an interaction effect ($p = 0.063$, $\eta_p^2 = 0.031$, small effect). Univariate analysis revealed that the percent targets correct tended to change over time ($p = 0.078$, $\eta_p^2 = 0.039$, small effect) while correct reaction time ($p = 0.183$, $\eta_p^2 = 0.029$, small effect) and overall reaction time ($p = 0.176$, $\eta_p^2 = 0.030$, small effect) were not significantly different from baseline values. Similarly, no significant interaction effects were observed in the percent targets correct ($p = 0.118$, $\eta_p^2 = 0.034$, small effect), correct reaction time ($p = 0.695$, $\eta_p^2 = 0.007$, small effect), or overall reaction time ($p = 0.682$, $\eta_p^2 = 0.007$, small effect) on this test. Figure 5 shows the percent changes from baseline in choice reaction variables. Acute supplementation with ASH tended to maintain the ability to correctly identify targets (−0.045% [−0.81, 0.72], $p = 0.906$) while those taking the PLA observed a significant decline in the ability to correctly identify targets (−1.096% [−1.88, −0.32], $p = 0.007$) with the difference tending to differ between treatment groups (1.051% [−0.044, 2.146], $p = 0.060$). No other differences were observed over time or between treatment groups.

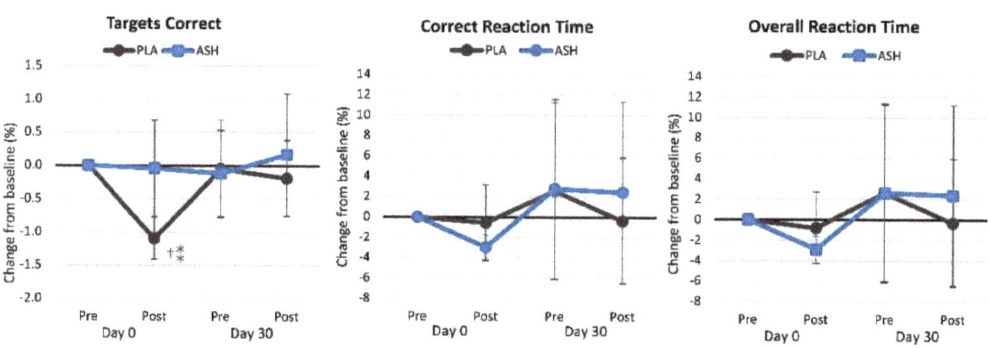

Figure 5. Choice reaction test results. ⁑ = $p > 0.05$ to $p < 0.10$ difference between treatment groups. † = $p < 0.05$ difference from baseline value.

4.2.4. Picture Recognition Test

Table S5 presents picture recognition test results. Overall, multivariate Wilk's Lambda analysis revealed a significant time (p = 0.006, η_p^2 = 0.077, moderate effect) while no significant group × time interaction effects were observed (p = 0.222, η_p^2 = 0.049, small effect). Univariate analysis revealed time effects in the percentage of pictures correctly identified (p = 0.006, η_p^2 = 0.086, moderate effect) while the percent of yes responses (p = 0.039, η_p^2 = 0.052, small effect) and no response (p = 0.051, η_p^2 = 0.054, small effect) tended to change over time. No significant group × time effects were observed in these and other measures of the picture recognition test. The pairwise comparison revealed some evidence that acute ASH ingestion improved correct reaction time (p = 0.006). Analysis of mean changes from baseline demonstrated that correct reaction time (p = 0.008) and overall reaction time (p = 0.023) were significantly faster in the ASH group. Figure 6 shows the percentage changes from baseline in picture recognition test variables. Results revealed that acute ASH supplementation significantly improved correct (p = 0.008), yes (p = 0.002), and overall reaction times (p = 0.018) and that 30 days of ASH supplementation better maintained correct picture recall, correct yes responses, correct reaction times, yes reaction time, no reaction time, and overall reaction time. Significant differences were seen between treatment groups after acute ingestion on Day 0 in correct reaction time and overall reaction time.

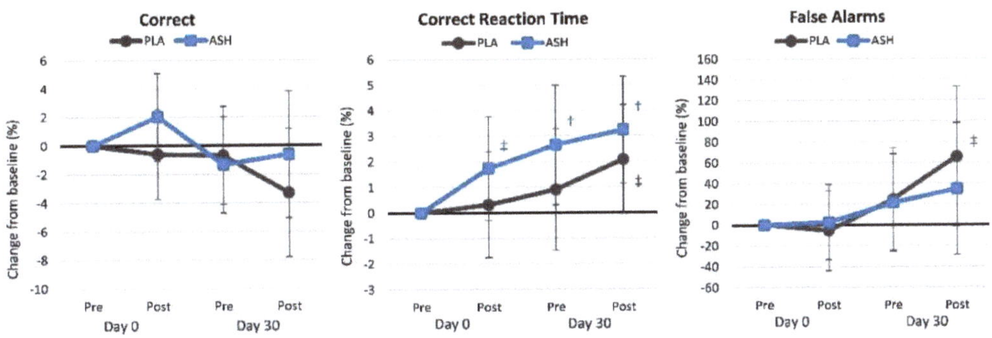

Figure 6. Picture recall test results. † = $p < 0.05$ difference from baseline value while ‡ = $p > 0.05$ to $p < 0.10$ difference from baseline.

4.2.5. Digit Vigilance

Digit vigilance test results are presented in Table S6. Overall, multivariate Wilk's Lambda analysis revealed a significant time (p = 0.006, η_p^2 = 0.044, small effect) but no interaction effect (p = 0.698, η_p^2 = 0.012, small effect). Univariate analysis revealed a significant time effect in correct response reaction time (p = 0.002, η_p^2 = 0.087, moderate effect), while no time or group × time effects were observed in the percent targets correct or false alarm responses. Figure 7 shows the percent changes from baseline in digit vigilance test results. Correct reaction time in the ASH group increased from baseline after 30 days of supplementation (p = 0.042), while false alarms tended to increase over time (p = 0.055) in the PLA group. No significant differences were observed between treatment groups.

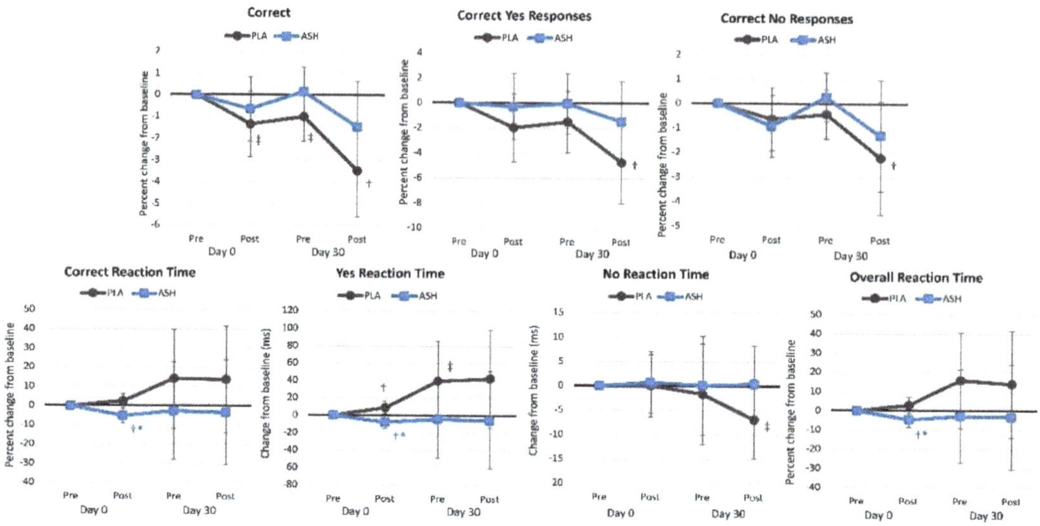

Figure 7. Digit vigilance test results. * = $p < 0.05$ difference between treatment groups. † = $p < 0.05$ difference from baseline value while ‡ = $p > 0.05$ to $p < 0.10$ difference from baseline.

4.2.6. Corsi Block

Table S7 shows Corsi Block test results. No significant time ($p = 0.592$, $\eta_p^2 = 0.034$, small effect) or group × time ($p = 0.754$, $\eta_p^2 = 0.021$, small effect) effects were observed in Corsi Block span score results. Likewise, no significant time or interaction effects were observed in the Corsi Block span score when analyzed as mean or percent changes from baseline (Figure 8).

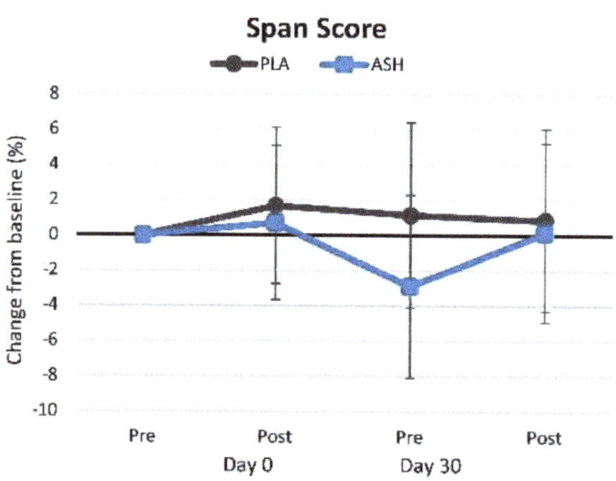

Figure 8. Corsi Block test results.

4.2.7. Stroop Test

Table S8 presents Stroop Color–Word test results. Overall, multivariate Wilk's Lambda analysis revealed a significant time ($p = 0.040$, $\eta_p^2 = 0.079$, moderate effect) but no interaction effect ($p = 0.514$, $\eta_p^2 = 0.051$, small effect). Univariate analysis revealed significant time effects in overall reaction time ($p < 0.001$, $\eta_p^2 = 0.124$, moderate effect), correct response reaction time ($p = 0.001$, $\eta_p^2 = 0.117$, moderate effect), congruent words overall reaction time ($p < 0.001$, $\eta_p^2 = 0.106$, moderate effect), incongruent words overall reaction time ($p = 0.003$, $\eta_p{}^2 = 0.099$, moderate effect), and correct congruent reaction time ($p = 0.001$, $\eta_p{}^2 = 0.104$, moderate effect) while the congruent words percent correct ($p = 0.087$, $\eta_p{}^2 = 0.039$, small effect) and correct incongruent reaction time ($p = 0.056$, $\eta_p{}^2 = 0.056$, small effect) tended to increase over time. However, no significant interaction effects were seen between treatments. Analysis of mean percent changes from baseline with 95% CIs (Figure 9) indicated that ASH tended to maintain the ability to identify congruent words correctly, while this ability declined in the PLA group. Reaction times improved from pre- to post-supplementation on days 0 and 30 of testing, with no differences observed between treatment groups.

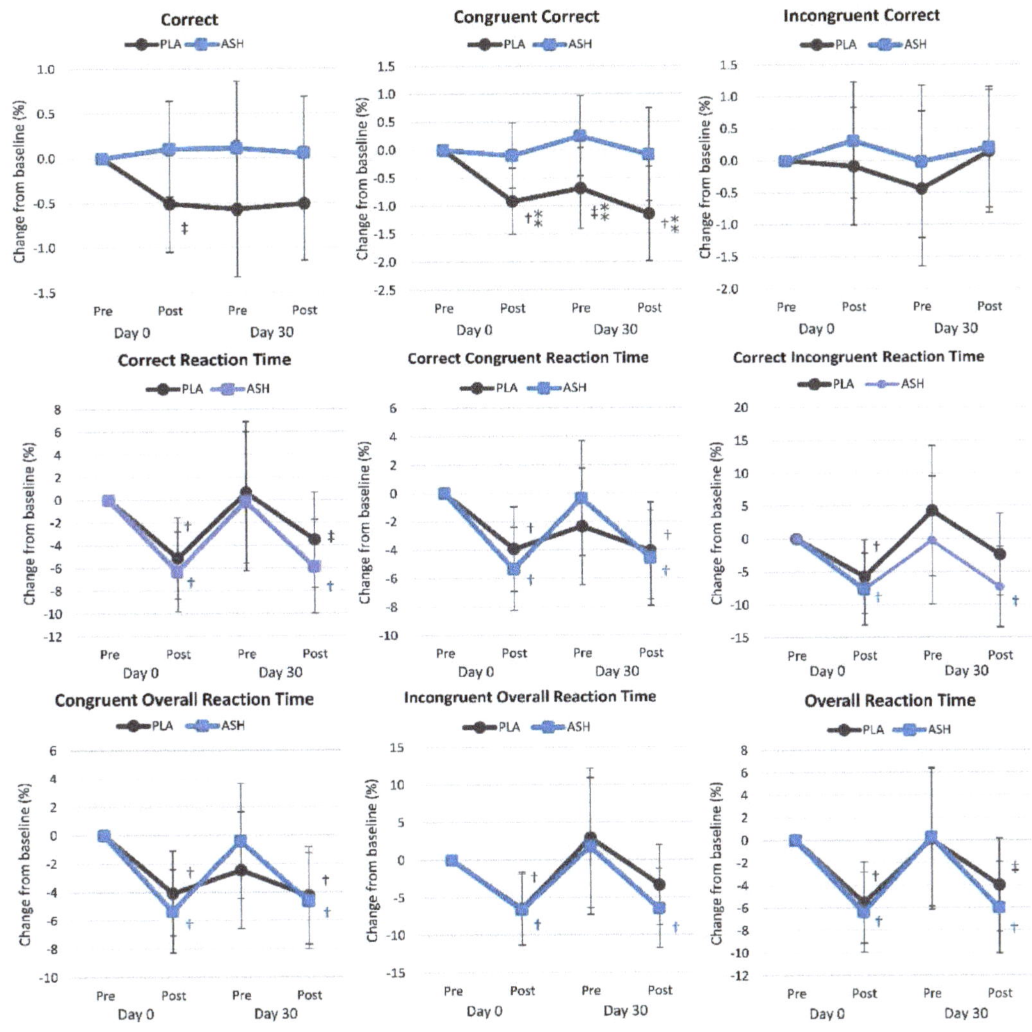

Figure 9. Stroop Color–Word test results. ⁑ = $p < 0.05$ difference between treatment groups. † = $p < 0.05$ difference from baseline value while ‡ = $p > 0.05$ to $p < 0.10$ difference from baseline.

4.3. *Profile of Mood States*

Table S9 shows the results of the POMS questionnaire. Overall, multivariate Wilk's Lambda analysis indicated a time ($p < 0.001$, $\eta_p{}^2 = 0.148$, large effect) but no interaction effects observed ($p = 0.148$, $\eta_p^2 = 0.047$, small effect). Univariate analysis revealed significant time effects in tension ($p < 0.001$, $\eta_p^2 = 0.189$, large effect), depression ($p < 0.001$, $\eta_p^2 = 0.120$, large effect), anger ($p < 0.001$, $\eta_p^2 = 0.139$, large effect), fatigue ($p < 0.001$, $\eta_p^2 = 0.200$, large effect), confusion ($p < 0.001$, $\eta_p^2 = 0.145$ large effect), vigor ($p = 0.004$, $\eta_p^2 = 0.098$, moderate effect, and TMDS ($p < 0.001$, $\eta_p^2 = 0.188$, large effect). Fatigue levels tended to interact ($p = 0.058$, $\eta_p^2 = 0.047$, small effect) with no other group × time effects observed. Analysis of changes from baseline (Figure 10) revealed evidence that participants in the ASH experienced a decrease in tension scores from baseline after acute ($p = 0.001$), 30 days ($p = 0.01$), and acute ingestion after 30 days ($p < 0.001$) of supplementation while tension scores were not significantly different from baseline values after 30 days of supplementation. Additionally, there was evidence that supplementation with ASH for 30 days reduced perceptions of fatigue after acute supplementation on Day 0 ($p = 0.005$) and Day 30 ($p = 0.002$) with 30 Day Pre values significantly lower in the ASH group compared to the PLA group (−1.91 [−5.6, −0.3], $p = 0.023$). No other significant differences were observed between treatment groups in the remaining POMS variables.

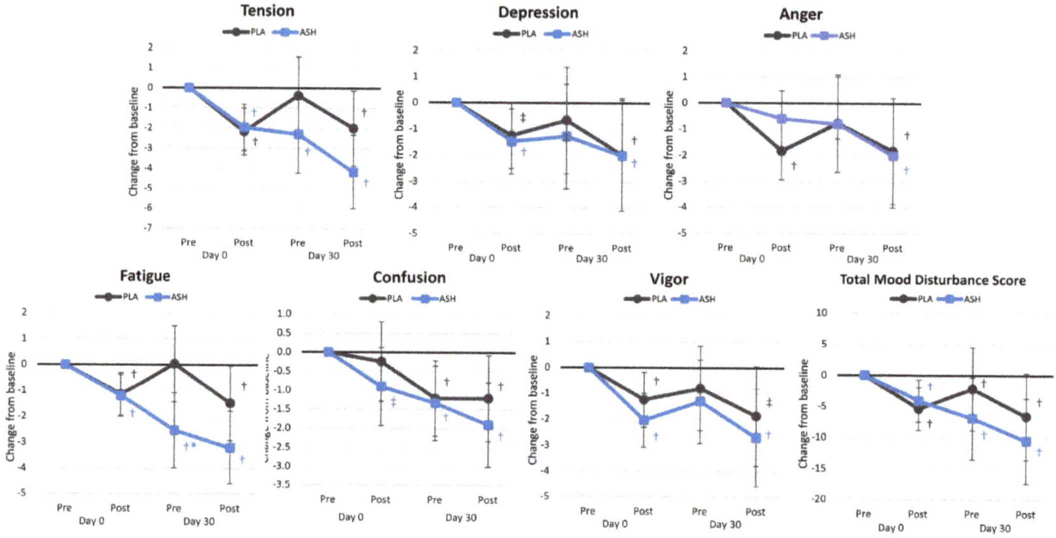

Figure 10. Profile of Mood States results. * = $p > 0.05$ to $p < 0.10$ difference between treatment groups. † = $p < 0.05$ difference from baseline value while ‡ = $p > 0.05$ to $p < 0.10$ difference from baseline.

4.4. Markers of Health and Safety

Overall GLM analysis of cell blood count data (Table S10) revealed no significant time ($p = 0.120$, $\eta_p^2 = 0.336$, large effect) or group × time effect ($p = 0.341$, $\eta_p^2 = 0.269$, large effect). Univariate analysis revealed a significant interaction effect in hematocrit ($p = 0.004$), where hematocrit decreased in the PLA group while maintained in the ASH group. The pairwise comparison also revealed no differences between treatment groups in baseline cell blood counts. However, lymphocytes in the ASH group were significantly lower than in the PLA group after 30 days of ASH supplementation. With ASH supplementation, red blood cells, neutrophils, and platelets tended to be higher. Overall GLM analysis of standard serum clinical markers (Table S11) revealed a trend toward a time effect ($p = 0.0065$, $\eta_p^2 = 0.554$, large effect) with no interaction effects ($p = 0.150$, $\eta_p^2 = 0.509$, large effect). Univariate analysis revealed significant time effects in several variables, with interaction effects observed in high-density lipoprotein ($p = 0.011$), protein ($p = 0.005$), albumin ($p = 0.027$), alkaline phosphatase ($p = 0.002$), and estimated glomerular filtration rate ($p = 0.037$). Participants in the ASH group observed a significant increase in high-density lipoprotein (HDL) cholesterol and a decrease in the HDL to total cholesterol ratio from baseline, while the blood urea nitrogen (BUN) to creatinine ratio increased significantly from baseline in the PLA group. There was also evidence that BUN and total bilirubin levels were lower and total protein and calcium levels higher after supplementation with ASH. Collectively, although differences over time and between treatment groups were small and within normal clinical values, results suggest that clinical blood profiles more favorably improved with ASH supplementation. Finally, Table S12 presents the side effects evaluated during the study. No significant differences in the frequency or severity of side effects evaluated or any other complaints. No participant withdrew from the study due to any perceived side effects.

5. Discussion

Ashwagandha has been reported to possess antioxidant [9,14–16], anti-inflammatory [9,11,12], neuroprotective [8,10,15,23–26], endocrinological [17–21], and immune-modulatory properties [10,22] that influence cognition [1,23,25,31]. Preliminary human clinical trials indicate that ashwagandha supplementation may have therapeutic benefits in individuals with medically managed bipolar disease [32], mild cognitive impairment [33], and chronic stress [7]. There is also evidence that ashwagandha supplementation improves memory, reaction times, and psychomotor performance in healthy participants [35,37,38]. Additionally, there is evidence that 30 days of ashwagandha supplementation improved perceptions of energy, sleep quality, and well-being in college students [36]. However, more research is needed to assess the potential nootropic effects of ashwagandha on cognitive function. Results from this study provide additional evidence that acute and 30 days of liposomal ashwagandha supplementation affect cognitive function and mood in healthy men and women. The following provides additional insight regarding the observed results.

5.1. Primary Outcomes

We evaluated the effects of ashwagandha supplementation on several cognitive function assessments to assess episodic memory (word presentation, picture recognition, delayed word recall, delayed picture recognition, delayed word recognition), attention and vigilance (simple reaction times, digit vigilance test, choice reaction time), and executive function (Stroop color-word test) [4]. Consistent with our prior work [37], we found evidence that acute ingestion of liposomal ashwagandha affected cognitive function in healthy individuals at a lower dose as previously established (225 mg compared to 400 mg). In this regard, present findings indicated that acute ashwagandha supplementation maintained the ability to identify correct targets on the choice reaction test, improved picture recognition reaction time, and prevented a decline in correct responses on the Stroop test. After 30 days of supplementation, there was some evidence that word recall correct attempts, correct reaction time in the digit vigilance task test, and Stroop test congruent correct responses improved to a greater degree from baseline with ashwagandha supplementation. Moreover, POMS tension and fatigue scores decreased from baseline, with Pre 30 Day fatigue scores in the ashwagandha group significantly lower than placebo responses. Collectively, these findings provide additional evidence that acute and longer periods of ashwagandha supplementation can improve measures of episodic memory, attention and vigilance, and executive function, as well as perceptions of tension and fatigue. These findings support reports that ashwagandha supplementation improved reaction times and cognitive performance in healthy young males [35], perceptions of energy and mental clarity in college students [36], and that 30 days of liposomal ashwagandha supplementation (225 or 400 mg/d) reduced stress-related markers and improved cognitive function in individuals with perceived stress [34].

While not directly assessed in the present study, there are several potential mechanisms described in the literature in which ashwagandha may influence cognition. First, ashwagandha has been reported to serve as an adaptogen, thereby improving response and resilience to stress [2,3,11,14,36,64,65]. The cognitive function tests produce a stressful state. Consequently, ashwagandha may have improved cognitive function by allowing the participants to perform better during stressful situations. Second, ashwagandha has been reported to inhibit acetylcholinesterase, increase neurotransmission [11,14,65], and provide neuroprotective effects [8,15,25,65,66]. Ashwagandha has also been reported to influence gamma-aminobutyric acid (GABA) activity [1,67–69]. GABA is an inhibitory neurotransmitter that has been a target in the treatment of anxiety [69]. Ashwagandha has been reported to have GABAergic activity and have similar effects as anxiolytic pharmaceutical interventions [69]. It has also been reported to promote the quality of sleep [68]. Thus, ashwagandha supplementation may help individuals in stressful conditions manage anxiety and/or improve the quality of sleep, thereby improving cognitive function. Finally, ashwagandha has also been reported to reduce oxidative stress [9,16] and inflammation [8,15,22,25]. Oxidative stress and inflammation in the brain have been implicated in the progression of cognitive impairment during aging. While we studied younger participants, it is possible that ashwagandha mediated oxidative stress and/or inflammation, thereby improving cognitive function. Additional research should examine the role of ashwagandha supplementation on these mechanisms, memory, and cognitive function in younger and older populations under stressful conditions who might benefit.

5.2. Secondary Outcomes

Since ashwagandha has antioxidant [9,14–16], endocrine [17–21], immune [10,22], and anti-inflammatory effects [9,11,12], there has been interest in determining the effects of ashwagandha supplementation on markers of health and safety. In this study, we assessed the effects of 30 days of ashwagandha supplementation on a comprehensive panel of whole blood and serum health markers and perceptions about the frequency and severity of common side effects. Results revealed that ashwagandha supplementation either had no effect or more favorably improved clinical blood panels with no differences in reported perceptions of side effects. These findings are consistent with results from Chandrasekhar et al. [6], who reported that ashwagandha supplementation (2 × 300 mg/d for 60 days) in individuals with a history of stress and anxiety had no clinically significant effects on clinical blood parameters or reported side effects. Additionally, results support findings from Smith and coworkers [18], who reported that 2 × 200 mg/d of ashwagandha root extract supplementation for 12 weeks in overweight men and women between the ages of 40–75 years experiencing high levels of fatigue and stress did not significantly affect blood glucose, HbA1c, cell blood counts, markers liver function, markers of renal function, or perceptions of side effects. Current findings also support other studies reporting that ashwagandha supplementation (500–700 mg/d for 2–8 weeks) in adults was well-tolerated and did not result in perceptions of side effects [33,64,70,71].

5.3. Limitations and Future Directions

Although positive results were observed, there are several limitations that should be considered. First, although participants maintained a similar diet, activity level, and fasting state before each testing session, diet and exercise were not controlled during the entire study. It is possible that variations in physical activity may have affected fatigue perception and/or ingestion of various stimulants in the participant's normal diet may have affected results. Second, we performed a single-dose effectiveness study and established that ingestion of 400 mg of liposomal ashwagandha improved some measures of memory, attention, and reaction time for several hours after ingestion [37]. Additionally, Remenapp et al. [34] reported that dietary supplementation of 225 mg/d with this source of liposomal ashwagandha improved cognitive function measures in individuals with perceived stress. Present findings indicate that younger healthy individuals may also benefit from supplementing their diet with 225 mg/d of liposomal ashwagandha. However, it is possible that ingesting higher amounts of ashwagandha, several doses a day as has been previously investigated, or a longer period of supplementation may have promoted greater effects. Consequently, additional research is warranted to (1) conduct studies on more frequent, higher doses, and/or a longer supplementation period; (2) evaluate the effects of ashwagandha supplementation during a standardized and supervised exercise and diet intervention; (3) examine the potential nootropic effects of ashwagandha supplementation in older individuals as a nutrition strategy to maintain cognitive function before experiencing clinically significant mild cognitive impairment, (4) assessing the effects of ashwagandha supplementation on managing stress and improving quality of life as people age; and, the potential role of ashwagandha supplementation in improving memory, attention, and executive function in individuals with signs of memory and cognitive impairment.

6. Conclusions

This study's results support that acute (225 mg) and 30 days of dietary supplementation of a liposomal form of ashwagandha (225 mg/d) can improve short-term memory, attention and vigilance, and reaction times in healthy younger men and women. While these results are promising, additional research is warranted to evaluate the potential nootropic effects of ashwagandha at different dosing strategies, longer supplementation periods, older individuals to help maintain cognitive function as people age, and potential clinical applications in individuals experiencing stressful conditions, memory and/or cognitive decline.

Supplementary Materials: The following supporting information can be downloaded at: https://www.mdpi.com/article/10.3390/nu16121813/s1, Table S1: Participant Demographic Data, Table S2: Word Recall test results, Table S3: Word Recognition test results, Table S4: Choice Reaction Time (CRT) test results, Table S5: Picture Recognition test results, Table S6: Digit Vigilance test results, Table S7: Corsi Block test results, Table S8: Stroop Color Word test results, Table S9: Profile of Mood States results, Table S10: Complete Blood Count data, Table S11: Serum Metabolic Panel, Table S12: Frequency and Severity of Side Effects.

Author Contributions: Conceptualization R.J., M.P., R.S. and R.B.K.; project management, R.B.K., C.J.R., M.L. and R.S.; data collection M.L., B.D., D.E.G., S.J., L.E., D.X., V.J., C.Y. and J.K.; data analysis R.B.K., R.S., M.F. and W.K.; writing—original draft preparation, R.B.K. and M.L.; writing—review and editing, R.B.K., R.S., M.L., R.J., M.P., M.F. and W.K.; funding acquisition, R.B.K. All authors have read and agreed to the published version of the manuscript.

Funding: The Human Clinical Research Facility at Texas A&M University received a fee-for-service award from Specnova LLC (Tysons Corner, VA, USA) in collaboration with Increnovo LLC (Whitefish Bay, WI, USA). Members of the Exercise & Sport Nutrition Lab collected and independently analyzed the data. Specnova was not involved in data collection, analysis, or manuscript writing.

Institutional Review Board Statement: This study was conducted with authorization by Texas A&M University's Institutional Review Board IRB2022-0621 (approved 26 September 2022).

Informed Consent Statement: The sponsor approved the publication of this paper.

Data Availability Statement: Data and statistical analyses are available upon request on a case-by-case basis for non-commercial scientific inquiry and educational use if IRB restrictions and research agreement terms are not violated.

Acknowledgments: We thank the study participants, J.P. Bramhall, for providing medical oversight and Peter Murano as an external safety and quality assurance monitor for this study.

Conflicts of Interest: R.J. and M.P. are principals and researchers affiliated with Increnovo. They were not involved in the data collection or analysis. R.B.K. served as Principal Investigator at the study location. He has conducted sponsored grants and contracts awarded to the universities with which he has been affiliated, received an honorarium for making scientific presentations, served as an expert on legal cases, and consulted with industry on product development unrelated to the product studied. He has no financial or intellectual property-related conflict of interest with the study sponsor or nutrient examined. Other authors report no conflicts.

References

1. D'Cruz, M.; Andrade, C. Potential clinical applications of Ashwagandha (*Withania somnifera*) in medicine and neuropsychiatry. *Expert Rev. Clin. Pharmacol.* **2022**, *15*, 1067–1080. [CrossRef]
2. Akhgarjand, C.; Asoudeh, F.; Bagheri, A.; Kalantar, Z.; Vahabi, Z.; Shab-Bidar, S.; Rezvani, H.; Djafarian, K. Does Ashwagandha supplementation have a beneficial effect on the management of anxiety and stress? A systematic review and meta-analysis of randomized controlled trials. *Phytother. Res.* **2022**, *36*, 4115–4124. [CrossRef]
3. Salve, J.; Pate, S.; Debnath, K.; Langade, D. Adaptogenic and Anxiolytic Effects of Ashwagandha Root Extract in Healthy Adults: A Double-blind, Randomized, Placebo-controlled Clinical Study. *Cureus* **2019**, *11*, e6466. [CrossRef]
4. Pratte, M.A.; Nanavati, K.B.; Young, V.; Morley, C.P. An alternative treatment for anxiety: A systematic review of human trial results reported for the Ayurvedic herb ashwagandha (*Withania somnifera*). *J. Altern. Complement. Med.* **2014**, *20*, 901–908. [CrossRef]
5. Liao, L.-Y.; He, Y.-F.; Li, L.; Meng, H.; Dong, Y.-M.; Yi, F.; Xiao, P.-G. A preliminary review of studies on adaptogens: Comparison of their bioactivity in TCM with that of ginseng-like herbs used worldwide. *Chin. Med.* **2018**, *13*, 57. [CrossRef]
6. Chandrasekhar, K.; Kapoor, J.; Anishetty, S. A Prospective, Randomized Double-Blind, Placebo-Controlled Study of Safety and Efficacy of a High-Concentration Full-Spectrum Extract of Ashwagandha Root in Reducing Stress and Anxiety in Adults. *Indian J. Psychol. Med.* **2012**, *34*, 255–262. [CrossRef]
7. Choudhary, D.; Bhattacharyya, S.; Joshi, K. Body Weight Management in Adults Under Chronic Stress Through Treatment With Ashwagandha Root Extract. *J. Evid.-Based Complement. Altern. Med.* **2017**, *22*, 96–106. [CrossRef]
8. Gupta, M.; Kaur, G. *Withania somnifera* (L.) Dunal ameliorates neurodegeneration and cognitive impairments associated with systemic inflammation. *BMC Complement. Altern. Med.* **2019**, *19*, 217. [CrossRef]
9. Gomez Afonso, A.; Fernandez-Lazaro, D.; Adams, D.P.; Monserda-Vilaro, A.; Fernandez-Lazaro, C.I. Effects of Withania somnifera (Ashwagandha) on Hematological and Biochemical Markers, Hormonal Behavior, and Oxidant Response in Healthy Adults: A Systematic Review. *Curr. Nutr. Rep.* **2023**, *12*, 465–477. [CrossRef] [PubMed]
10. Alanazi, H.H.; Elfaki, E. The immunomodulatory role of *Withania somnifera* (L.) dunal in inflammatory diseases. *Front. Pharmacol.* **2023**, *14*, 1084757. [CrossRef] [PubMed]
11. Kurapati, K.R.; Samikkannu, T.; Atluri, V.S.; Kaftanovskaya, E.; Yndart, A.; Nair, M.P. beta-Amyloid1-42, HIV-1Ba-L (clade B) infection and drugs of abuse induced degeneration in human neuronal cells and protective effects of ashwagandha (*Withania somnifera*) and its constituent Withanolide A. *PLoS ONE* **2014**, *9*, e112818. [CrossRef] [PubMed]
12. Schliebs, R.; Liebmann, A.; Bhattacharya, S.K.; Kumar, A.; Ghosal, S.; Bigl, V. Systemic administration of defined extracts from *Withania somnifera* (Indian Ginseng) and Shilajit differentially affects cholinergic but not glutamatergic and GABAergic markers in rat brain. *Neurochem. Int.* **1997**, *30*, 181–190. [CrossRef] [PubMed]
13. Sood, A.; Kumar, A.; Dhawan, D.K.; Sandhir, R. Propensity of *Withania somnifera* to Attenuate Behavioural, Biochemical, and Histological Alterations in Experimental Model of Stroke. *Cell. Mol. Neurobiol.* **2016**, *36*, 1123–1138. [CrossRef]
14. Thakral, S.; Yadav, A.; Singh, V.; Kumar, M.; Kumar, P.; Narang, R.; Sudhakar, K.; Verma, A.; Khalilullah, H.; Jaremko, M.; et al. Alzheimer's disease: Molecular aspects and treatment opportunities using herbal drugs. *Ageing Res. Rev.* **2023**, *88*, 101960. [CrossRef] [PubMed]
15. Dar, N.J.; MuzamilAhmad. Neurodegenerative diseases and *Withania somnifera* (L.): An update. *J. Ethnopharmacol.* **2020**, *256*, 112769. [CrossRef]
16. Durg, S.; Dhadde, S.B.; Vandal, R.; Shivakumar, B.S.; Charan, C.S. *Withania somnifera* (Ashwagandha) in neurobehavioural disorders induced by brain oxidative stress in rodents: A systematic review and meta-analysis. *J. Pharm. Pharmacol.* **2015**, *67*, 879–899. [CrossRef] [PubMed]
17. Wicinski, M.; Fajkiel-Madajczyk, A.; Kurant, Z.; Kurant, D.; Gryczka, K.; Falkowski, M.; Wisniewska, M.; Slupski, M.; Ohla, J.; Zabrzynski, J. Can Ashwagandha Benefit the Endocrine System?—A Review. *Int. J. Mol. Sci.* **2023**, *24*, 16513. [CrossRef] [PubMed]
18. Smith, S.J.; Lopresti, A.L.; Fairchild, T.J. Exploring the efficacy and safety of a novel standardized ashwagandha (*Withania somnifera*) root extract (Witholytin(R)) in adults experiencing high stress and fatigue in a randomized, double-blind, placebo-controlled trial. *J. Psychopharmacol.* **2023**, *37*, 1091–1104. [CrossRef]
19. Santos, H.O.; Cadegiani, F.A.; Forbes, S.C. Nonpharmacological Interventions for the Management of Testosterone and Sperm Parameters: A Scoping Review. *Clin. Ther.* **2022**, *44*, 1129–1149. [CrossRef]
20. Smith, S.J.; Lopresti, A.L.; Teo, S.Y.M.; Fairchild, T.J. Examining the Effects of Herbs on Testosterone Concentrations in Men: A Systematic Review. *Adv. Nutr.* **2021**, *12*, 744–765. [CrossRef]
21. Lopresti, A.L.; Drummond, P.D.; Smith, S.J. A Randomized, Double-Blind, Placebo-Controlled, Crossover Study Examining the Hormonal and Vitality Effects of Ashwagandha (*Withania somnifera*) in Aging, Overweight Males. *Am. J. Mens. Health* **2019**, *13*, 1557988319835985. [CrossRef]
22. Agarwal, R.; Diwanay, S.; Patki, P.; Patwardhan, B. Studies on immunomodulatory activity of *Withania somnifera* (Ashwagandha) extracts in experimental immune inflammation. *J. Ethnopharmacol.* **1999**, *67*, 27–35. [CrossRef]
23. Ng, Q.X.; Loke, W.; Foo, N.X.; Tan, W.J.; Chan, H.W.; Lim, D.Y.; Yeo, W.S. A systematic review of the clinical use of *Withania somnifera* (Ashwagandha) to ameliorate cognitive dysfunction. *Phytother. Res.* **2020**, *34*, 583–590. [CrossRef]

24. Gregory, J.; Vengalasetti, Y.V.; Bredesen, D.E.; Rao, R.V. Neuroprotective Herbs for the Management of Alzheimer's Disease. *Biomolecules* **2021**, *11*, 543. [CrossRef]
25. de la Pena, I.; Afable, T.; Dahilig-Talan, V.R.; Cruz, P. Review of Plant Extracts and Active Components: Mechanisms of Action for the Treatment of Obesity-Induced Cognitive Impairment. *Brain Sci.* **2023**, *13*, 929. [CrossRef]
26. Lerose, V.; Ponticelli, M.; Benedetto, N.; Carlucci, V.; Lela, L.; Tzvetkov, N.T.; Milella, L. *Withania somnifera* (L.) Dunal, a Potential Source of Phytochemicals for Treating Neurodegenerative Diseases: A Systematic Review. *Plants* **2024**, *13*, 771. [CrossRef]
27. Kushwah, S.; Maurya, N.S.; Kushwaha, S.; Scotti, L.; Chawade, A.; Mani, A. Herbal Therapeutics for Alzheimer's Disease: Ancient Indian Medicine System from the Modern Viewpoint. *Curr. Neuropharmacol.* **2023**, *21*, 764–776. [CrossRef]
28. Das, R.; Rauf, A.; Akhter, S.; Islam, M.N.; Emran, T.B.; Mitra, S.; Khan, I.N.; Mubarak, M.S. Role of Withaferin A and Its Derivatives in the Management of Alzheimer's Disease: Recent Trends and Future Perspectives. *Molecules* **2021**, *26*, 3696. [CrossRef]
29. Kurapati, K.R.; Atluri, V.S.; Samikkannu, T.; Nair, M.P. Ashwagandha (*Withania somnifera*) reverses beta-amyloid1-42 induced toxicity in human neuronal cells: Implications in HIV-associated neurocognitive disorders (HAND). *PLoS ONE* **2013**, *8*, e77624. [CrossRef] [PubMed]
30. Sen, D.; Rathee, S.; Pandey, V.; Jain, S.K.; Patil, U.K. Comprehensive Insights into Pathophysiology of Alzheimer's Disease: Herbal Approaches for Mitigating Neurodegeneration. *Curr. Alzheimer Res.* **2024**. [CrossRef] [PubMed]
31. Hersant, H.; He, S.; Maliha, P.; Grossberg, G. Over the Counter Supplements for Memory: A Review of Available Evidence. *CNS Drugs* **2023**, *37*, 797–817. [CrossRef] [PubMed]
32. Chengappa, K.N.; Bowie, C.R.; Schlicht, P.J.; Fleet, D.; Brar, J.S.; Jindal, R. Randomized placebo-controlled adjunctive study of an extract of *Withania somnifera* for cognitive dysfunction in bipolar disorder. *J. Clin. Psychiatry* **2013**, *74*, 1076–1083. [CrossRef] [PubMed]
33. Choudhary, D.; Bhattacharyya, S.; Bose, S. Efficacy and Safety of Ashwagandha (*Withania somnifera* (L.) Dunal) Root Extract in Improving Memory and Cognitive Functions. *J. Diet. Suppl.* **2017**, *14*, 599–612. [CrossRef] [PubMed]
34. Remenapp, A.; Coyle, K.; Orange, T.; Lynch, T.; Hooper, D.; Hooper, S.; Conway, K.; Hausenblas, H.A. Efficacy of *Withania somnifera* supplementation on adult's cognition and mood. *J. Ayurveda Integr. Med.* **2022**, *13*, 100510. [CrossRef] [PubMed]
35. Pingali, U.; Pilli, R.; Fatima, N. Effect of standardized aqueous extract of *Withania somnifera* on tests of cognitive and psychomotor performance in healthy human participants. *Pharmacogn. Res.* **2014**, *6*, 12–18. [CrossRef] [PubMed]
36. Baker, C.; Kirby, J.B.; O'Connor, J.; Lindsay, K.G.; Hutchins, A.; Harris, M. The Perceived Impact of Ashwagandha on Stress, Sleep Quality, Energy, and Mental Clarity for College Students: Qualitative Analysis of a Double-Blind Randomized Control Trial. *J. Med. Food* **2022**, *25*, 1095–1101. [CrossRef] [PubMed]
37. Xing, D.; Yoo, C.; Gonzalez, D.; Jenkins, V.; Nottingham, K.; Dickerson, B.; Leonard, M.; Ko, J.; Faries, M.; Kephart, W.; et al. Effects of Acute Ashwagandha Ingestion on Cognitive Function. *Int. J. Environ. Res. Public Health* **2022**, *19*, 11852. [CrossRef] [PubMed]
38. Bonilla, D.A.; Moreno, Y.; Gho, C.; Petro, J.L.; Odriozola-Martinez, A.; Kreider, R.B. Effects of Ashwagandha (*Withania somnifera*) on Physical Performance: Systematic Review and Bayesian Meta-Analysis. *J. Funct. Morphol. Kinesiol.* **2021**, *6*, 20. [CrossRef] [PubMed]
39. Buchner, A.; Wippich, W. On the Reliability of Implicit and Explicit Memory Measures. *Cogn. Psychol.* **2000**, *40*, 227–259. [CrossRef]
40. Tsoi, K.K.F.; Chan, J.Y.C.; Hirai, H.W.; Wong, A.; Mok, V.C.T.; Lam, L.C.W.; Kwok, T.C.Y.; Wong, S.Y.S. Recall Tests Are Effective to Detect Mild Cognitive Impairment: A Systematic Review and Meta-analysis of 108 Diagnostic Studies. *J. Am. Med. Dir. Assoc.* **2017**, *18*, 807.e17–807.e29. [CrossRef]
41. Cutting, L.E.; Scarborough, H.S. Prediction of Reading Comprehension: Relative Contributions of Word Recognition, Language Proficiency, and Other Cognitive Skills Can Depend on How Comprehension Is Measured. *Sci. Stud. Read.* **2006**, *10*, 277–299. [CrossRef]
42. Lange, B.; Hunfalvay, M.; Murray, N.; Roberts, C.-M.; Bolte, T. Reliability of computerized eye-tracking reaction time tests in non-athletes, athletes, and individuals with traumatic brain injury. *Optom. Vis. Perform.* **2018**, *6*, 119–129.
43. Levie, W.H.; Hathaway, S.N. Picture recognition memory: A review of research and theory. *J. Vis. Verbal Languaging* **1988**, *8*, 6–45. [CrossRef]
44. Wesnes, K.A.; Brooker, H.; Ballard, C.; McCambridge, L.; Stenton, R.; Corbett, A. Utility, reliability, sensitivity and validity of an online test system designed to monitor changes in cognitive function in clinical trials. *Int. J. Geriatr. Psychiatry* **2017**, *32*, e83–e92. [CrossRef] [PubMed]
45. de Paula, J.J.; Malloy-Diniz, L.F.; Romano-Silva, M.A. Reliability of working memory assessment in neurocognitive disorders: A study of the Digit Span and Corsi Block-Tapping tasks. *Braz. J. Psychiatry* **2016**, *38*, 262–263. [CrossRef]
46. Siegrist, M. Test-Retest Reliability of Different Versions of the Stroop Test. *J. Psychol.* **1997**, *131*, 299–306. [CrossRef]
47. Mumme, K.D.; Conlon, C.A.; von Hurst, P.R.; Jones, B.; Haskell-Ramsay, C.F.; de Seymour, J.V.; Stonehouse, W.; Heath, A.-L.M.; Coad, J.; Mugridge, O.; et al. Dietary patterns and cognitive function in older New Zealand adults: The REACH study. *Eur. J. Nutr.* **2022**, *61*, 1943–1956. [CrossRef]
48. Patan, M.J.; Kennedy, D.O.; Husberg, C.; Hustvedt, S.O.; Calder, P.C.; Middleton, B.; Khan, J.; Forster, J.; Jackson, P.A. Differential Effects of DHA- and EPA-Rich Oils on Sleep in Healthy Young Adults: A Randomized Controlled Trial. *Nutrients* **2021**, *13*, 248. [CrossRef]

49. Wightman, E.L.; Jackson, P.A.; Spittlehouse, B.; Heffernan, T.; Guillemet, D.; Kennedy, D.O. The Acute and Chronic Cognitive Effects of a Sage Extract: A Randomized, Placebo Controlled Study in Healthy Humans. *Nutrients* **2021**, *13*, 218. [CrossRef]
50. Chen, K.-M.; Snyder, M.; Krichbaum, K. Translation and equivalence: The Profile of Mood States Short Form in English and Chinese. *Int. J. Nurs. Stud.* **2002**, *39*, 619–624. [CrossRef]
51. Shacham, S. A Shortened Version of the Profile of Mood States. *J. Personal. Assess.* **1983**, *47*, 305–306. [CrossRef] [PubMed]
52. Leonard, M.; Maury, J.; Dickerson, B.; Gonzalez, D.E.; Kendra, J.; Jenkins, V.; Nottingham, K.; Yoo, C.; Xing, D.; Ko, J.; et al. Effects of Dietary Supplementation of a Microalgae Extract Containing Fucoxanthin Combined with Guarana on Cognitive Function and Gaming Performance. *Nutrients* **2023**, *15*, 1918. [CrossRef] [PubMed]
53. Sowinski, R.J.; Grubic, T.J.; Dalton, R.L.; Schlaffer, J.; Reyes-Elrod, A.G.; Jenkins, V.M.; Williamson, S.; Rasmussen, C.; Murano, P.S.; Earnest, C.P.; et al. An Examination of a Novel Weight Loss Supplement on Anthropometry and Indices of Cardiovascular Disease Risk. *J. Diet. Suppl.* **2021**, *18*, 478–506. [CrossRef] [PubMed]
54. Yoo, C.; Xing, D.; Gonzalez, D.; Jenkins, V.; Nottingham, K.; Dickerson, B.; Leonard, M.; Ko, J.; Faries, M.; Kephart, W.; et al. Acute Paraxanthine Ingestion Improves Cognition and Short-Term Memory and Helps Sustain Attention in a Double-Blind, Placebo-Controlled, Crossover Trial. *Nutrients* **2021**, *13*, 3980. [CrossRef] [PubMed]
55. Sowinski, R.; Gonzalez, D.; Xing, D.; Yoo, C.; Jenkins, V.; Nottingham, K.; Dickerson, B.; Humphries, M.; Leonard, M.; Ko, J.; et al. Effects of Inositol-Enhanced Bonded Arginine Silicate Ingestion on Cognitive and Executive Function in Gamers. *Nutrients* **2021**, *13*, 3758. [CrossRef] [PubMed]
56. Jung, Y.P.; Earnest, C.P.; Koozehchian, M.; Galvan, E.; Dalton, R.; Walker, D.; Rasmussen, C.; Murano, P.S.; Greenwood, M.; Kreider, R.B. Effects of acute ingestion of a pre-workout dietary supplement with and without p-synephrine on resting energy expenditure, cognitive function and exercise performance. *J. Int. Soc. Sports Nutr.* **2017**, *14*, 3. [CrossRef] [PubMed]
57. Collins, P.B.; Earnest, C.P.; Dalton, R.L.; Sowinski, R.J.; Grubic, T.J.; Favot, C.J.; Coletta, A.M.; Rasmussen, C.; Greenwood, M.; Kreider, R.B. Short-Term Effects of a Ready-to-Drink Pre-Workout Beverage on Exercise Performance and Recovery. *Nutrients* **2017**, *9*, 823. [CrossRef] [PubMed]
58. Dalton, R.L.; Sowinski, R.J.; Grubic, T.J.; Collins, P.B.; Coletta, A.M.; Reyes, A.G.; Sanchez, B.; Koozehchian, M.; Jung, Y.P.; Rasmussen, C.; et al. Hematological and Hemodynamic Responses to Acute and Short-Term Creatine Nitrate Supplementation. *Nutrients* **2017**, *9*, 1359. [CrossRef]
59. Galvan, E.; Walker, D.K.; Simbo, S.Y.; Dalton, R.; Levers, K.; O'Connor, A.; Goodenough, C.; Barringer, N.D.; Greenwood, M.; Rasmussen, C.; et al. Acute and chronic safety and efficacy of dose dependent creatine nitrate supplementation and exercise performance. *J. Int. Soc. Sports Nutr.* **2016**, *13*, 12. [CrossRef]
60. Kim, B.; Stein, H. A spreadsheet program for making a balanced Latin Square design. *Rev. Colomb. Cienc. Pecu.* **2009**, *22*, 591–596.
61. Page, P. Beyond statistical significance: Clinical interpretation of rehabilitation research literature. *Int. J. Sports Phys. Ther.* **2014**, *9*, 726–736.
62. Cohen, J. *Statistical Power Analysis for the Social Sciences*, 2nd ed.; Lawrence Erlbaum Associates: Hillsdale, NJ, USA, 1988.
63. Earnest, C.P.; Roberts, B.M.; Harnish, C.R.; Kutz, J.L.; Cholewa, J.M.; Johannsen, N.M. Reporting Characteristics in Sports Nutrition. *Sports* **2018**, *6*, 139. [CrossRef]
64. Gopukumar, K.; Thanawala, S.; Somepalli, V.; Rao, T.S.S.; Thamatam, V.B.; Chauhan, S. Efficacy and Safety of Ashwagandha Root Extract on Cognitive Functions in Healthy, Stressed Adults: A Randomized, Double-Blind, Placebo-Controlled Study. *Evid.-Based Complement. Altern. Med.* **2021**, *2021*, 8254344. [CrossRef]
65. Konar, A.; Gupta, R.; Shukla, R.K.; Maloney, B.; Khanna, V.K.; Wadhwa, R.; Lahiri, D.K.; Thakur, M.K. M1 muscarinic receptor is a key target of neuroprotection, neuroregeneration and memory recovery by i-Extract from *Withania somnifera*. *Sci. Rep.* **2019**, *9*, 13990. [CrossRef]
66. Wongtrakul, J.; Thongtan, T.; Kumrapich, B.; Saisawang, C.; Ketterman, A.J. Neuroprotective effects of *Withania somnifera* in the SH-SY5Y Parkinson cell model. *Heliyon* **2021**, *7*, e08172. [CrossRef]
67. Candelario, M.; Cuellar, E.; Reyes-Ruiz, J.M.; Darabedian, N.; Feimeng, Z.; Miledi, R.; Russo-Neustadt, A.; Limon, A. Direct evidence for GABAergic activity of *Withania somnifera* on mammalian ionotropic GABAA and GABArho receptors. *J. Ethnopharmacol.* **2015**, *171*, 264–272. [CrossRef]
68. Park, C.W.; Hong, K.B.; Suh, H.J.; Ahn, Y. Sleep-promoting activity of amylase-treated Ashwagandha (*Withania somnifera* L. Dunal) root extract via GABA receptors. *J. Food Drug Anal.* **2023**, *31*, 278–288. [CrossRef]
69. Savage, K.; Firth, J.; Stough, C.; Sarris, J. GABA-modulating phytomedicines for anxiety: A systematic review of preclinical and clinical evidence. *Phytother. Res.* **2018**, *32*, 3–18. [CrossRef] [PubMed]
70. Kelgane, S.B.; Salve, J.; Sampara, P.; Debnath, K. Efficacy and Tolerability of Ashwagandha Root Extract in the Elderly for Improvement of General Well-being and Sleep: A Prospective, Randomized, Double-blind, Placebo-controlled Study. *Cureus* **2020**, *12*, e7083. [CrossRef] [PubMed]
71. Deshpande, A.; Irani, N.; Balkrishnan, R.; Benny, I.R. A randomized, double blind, placebo controlled study to evaluate the effects of ashwagandha (*Withania somnifera*) extract on sleep quality in healthy adults. *Sleep Med.* **2020**, *72*, 28–36. [CrossRef] [PubMed]

nutrients

MDPI

Article

Pediatric Onset Multiple Sclerosis and Obesity: Defining the Silhouette of Disease Features in Overweight Patients

Laura Papetti [1,*], Elena Panella [2], Gabriele Monte [1], Michela Ada Noris Ferilli [1], Samuela Tarantino [1], Martina Proietti Checchi [1] and Massimiliano Valeriani [1,3]

[1] Developmental Neurology Unit, Bambino Gesù Children's Hospital, IRCCS, 00165 Rome, Italy; gabriele.monte@opbg.net (G.M.); michela.ferilli@opbg.net (M.A.N.F.); samuela.tarantino@opbg.net (S.T.); martina.proietti@opbg.net (M.P.C.); massimiliano.valeriani@opbg.net (M.V.)

[2] Child Neurology and Psychiatry Unit, Systems Medicine Department, Hospital of Rome, Tor Vergata University, 00133 Rome, Italy; elena.panella@opbg.net

[3] Center for Sensory Motor Interaction, Aalborg University, DK-9220 Aalborg, Denmark

* Correspondence: laura.papetti@opbg.net; Tel.: +39-066-859-2865

Abstract: Obesity has been suggested as an environmental risk factor for multiple sclerosis (MS) and may negatively effect the progression of the disease. The aim of this study is to determine any correlation between overweight/obesity and the clinical and neuroradiological features at the onset of pediatric onset multiple sclerosis (POMS). Were included patients referred to the POMS Unit of the Bambino Gesù Children's Hospital between June 2012 and June 2021. The diagnosis of MS with an onset of less than 18 years was required. For all included subjects, we considered for the analysis the following data at the onset of symptoms: general data (age, sex, functional system compromised by neurological signs, weight and height), brain and spinal magnetic resonance imaging (MRI), cerebrospinal fluid exams. We identified 55 pediatric cases of POMS and divided them into two groups according to the body mass index (BMI): 60% were healthy weight (HW) and 40% were overweight/obese (OW/O). OW/O patients experienced a two-year age difference in disease onset compared to the HW patients (12.7 ± 3.8 years vs. 14.6 ± 4.1 years; $p < 0.05$). Onset of polyfocal symptoms was seen more frequently in OW/O patients than in HW (72.7% vs. 21.2%; $p < 0.05$). The pyramidal functions were involved more frequently in the OW/O group than in the HW group (50% vs. 25%; $p < 0.005$). Black holes were detected more frequently in OW/O patients in onset MRI scans compared to the HW group (50% vs. 15.5%; $p < 0.05$). Our findings suggest that being overweight/obese affects the risk of developing MS at an earlier age and is associated with an unfavorable clinical–radiological features at onset. Weight control can be considered as a preventive/therapeutic treatment.

Keywords: pediatric onset multiple sclerosis; obesity; onset; risk factor

1. Introduction

Multiple sclerosis (MS) is a chronic inflammatory autoimmune disease of the central nervous system (CNS) that is most commonly diagnosed in young adults; 3–5% of patients have disease onset under the age of 18, and less than 2% of patients are younger than 10 years [1–4]. Pediatric MS has some differences from adults in the clinical presentation: the pediatric population presents a higher relapse rate, despite better recoveries after relapses, and a slower evolution toward a secondarily progressive disease [1]. About half to two-thirds of pediatric patients present with multiple symptoms [4].

Obesity is a major public health problem, as the prevalence of obesity in Italian school-aged children is 9.4%, while the prevalence of overweight is 20.4% [5]. Furthermore, the prevalence of overweight and obesity persists, with an increasing trend, over the years, following the trend of MS diagnosis in children [6].

The etiology of MS is still unknown, although current evidence suggests that it is the result of autoimmune injury triggered by a combination of an environmental stimulus in genetically susceptible individuals, both adults and children. Low serum vitamin D levels, passive or active smoking, remote infection with Epstein–Barr virus, and obesity are factors that may increase environmental risks [6–8].

Adipose tissue, which is involved in many metabolic functions, regulates the immune system and endocrine function. These functions are mediated by various components of adipose tissue: adipocytes, which exert an endocrine function through the secretion of various adipokines (adiponectin, leptin, and resistin), innate and adaptive immune cells (macrophages, neutrophils, eosinophils, mast cells, and various T and B cells), and fibroblasts. This is closely connected to how adipose tissue regulates inflammatory status [5]. An excessive amount of adipose tissue leads to an increased secretion of inflammatory substances, such as pro-inflammatory cytokines like tumor necrosis factor-α (TNF-α), monocyte chemoattractant protein-1 (MCP-1), and interleukin-6 (IL-6); at the same time, the secretion of anti-inflammatory adipokines is decreased in the over-weight/obese body. Obesity is associated with chronic inflammatory status as a result of these factors [5]. Indeed, obesity in adults is associated with increased cerebrospinal fluid (CSF) levels of proinflammatory molecules, including IL -6 and leptin, and a de-crease in anti-inflammatory mediators such as IL-13 [9,10]. Similar data have been found in the pediatric population, particularly in prepubertal patients, suggesting an interaction between excess body fat, sex hormones, and the occurrence of pediatric onset multiple sclerosis (POMS) [11]. An hypothesis is that children are more susceptible to inflammatory damage due to blood–brain barrier (BBB) permeability, resulting in more pronounced acute axonal damage in inflammatory demyelinating lesions than in adults [12], with a greater number and volume of new T2 lesions on brain magnetic resonance imaging (MRI) [13], which may affect outcomes, especially if they are abundant on baseline neuroimaging [14,15]. It is particularly relevant to identify the risk factors involved in the development of the disease during the early stages of life, considering this time as a window to potentially interfere: in this sense many efforts have been spent and some studies focused on the role for childhood obesity in risk of developing MS.

The presence of high BMI values in pediatric multiple sclerosis patients in their childhood and before the onset of neurological symptoms is a common occurrence [6,11,16,17]. Despite considering environmental and genetic risk factors, the connection between high BMI and increased risk of MS remains valid [18]. Regarding genetic factors, a recent study performing mendelian randomization could confirm the positive causal association between high BMI and MS, and detected as well common risk genes shared between MS and obesity, suggesting a potential pathogenetic mechanism to justify this comorbidity [16]. High BMI may be considered not only a risk factor but also a negative predictive fac-tor, as it appears to be associated with higher rates of relapse [10], development of disability [19], and negative response to disease-modifying drugs in terms of relapses under treatment [20]. Meanwhile, as evidence increases about this correlation, the prevalence of obesity in the MS pediatric population is growing at a rate of 25–50% compared to the past [21].

The objective of this research is to examine whether overweight/obesity is associated with the clinical and neuroradiological presentation of POMS.

2. Materials and Methods

2.1. Participants and Inclusion Criteria

We performed a retrospective study that includes patients that referred to the POMS Unit of the Bambino Gesù Children's Hospital, between June 2012 and June 2021. The 2013 International Pediatric MS Study Group (IP-MSSG) criteria were used to diagnose MS, and these patients were included in the analysis [22]. Pediatric MS diagnosis was made based on these criteria if the patient had a history of:

(1) Two or more non-encephalopathic clinical CNS events with presumed inflammation, separated by more than 30 days, involving more than one area of the CNS;

(2) One non encephalopathic episode typical of MS which was associated with MRI findings consistent with 2010 Revised McDonald criteria for dissemination in space (DIS) and in which a follow up MRI shows at least one new enhancing or non enhancing lesion consistent with dissemination in time (DIT);
(3) One ADEM attack followed by a non encephalopathic clinical event, three or more months after symptom onset, that was associated with new MRI lesions that fulfill 2010 Revised McDonald for DIS criteria;
(4) A first, single, acute event that does not meet ADEM criteria and whose MRI findings are consistent with the 2010 Revised McDonald criteria for DIS and DIT (this last criterion was valid only for children over 12 years of age.

Other inclusion criteria for the studied are: age younger than 18 years at clinical onset; the other inclusion criteria included: availability of an examination of the liquor taken by rachicentes (physical chemical examination, determination of oligoclonal bands on liquor and serum and microbiological examinations); examination of cerebral and spinal MRI with and without injection of gadolinium and carried out 30 days after possible steroid therapy and without the initiation of therapy modifying the course of disease.

Patients with other intercurrent chronic pathologies (endocrine, tumoral, cardiac, respiratory) that could affect BMI were excluded.

The study was approved by the Ethical Committee of Bambino Gesù Children's Hospital (date 19 October 2023). All patients enrolled and their parents provided consent for the publication of the results.

2.2. Data Collection

Medical records performed at the moment of disease onset, before starting steroid treatment and disease modifying therapy were retrospectively reviewed for each patient. The collected data included: (1) demographic variables: sex, age; (2) clinical variables: measurement of weight (kg), height (cm), BMI calculated as weight in kilograms divided by height in meters squared, expanded disability status scale (EDSS) score, presenting clinical symptoms; (3) laboratory data: presence of oligoclonal bands (OCBs) in CSF, determined by isoelectric focusing, combined with immunoblotting of matched serum, and CSF sample pairs, presence of pleocytosis in CSF, defined as >5 white blood cells/mm^3, previous Epstein–Barr virus (EBV) infection, defined by measuring serum viral antibodies (IgM and IgG by ELISA) and performing quantitative real-time PCR for the EBV; (4) characteristics of brain and spine MRI, performed at the time of clinical onset with a 3T scanner, acquiring axial and sagittal T2-weighted, fluid-attenuated in-version recovery (FLAIR)- weighted, T1-weighted MRI sequences, and T1-weighted MRI images after administration of gadolinium.

Clinical symptoms were later classified according to the involvement of various functional systems, related to nervous system activity. In details the following symptoms or signs had been considered: Visual function (visual acuity, visual fields, and scotoma); Brainstem Functions (extraocular movement impairment, nystagmus, trigeminal damage, facial weakness, hearing loss, dysarthria, dysphagia, and other cranial nerve functions); Pyramidal functions (reflexes, limb strength, and spasticity); Cerebellar Functions (head tremor, truncal ataxia, tremor or dysmetria of limbs, and gai ataxia); Sensory Functions (superficial sensation of trunk and limbs, vibration sense of limbs, position sense of limbs, and paranesthesia); Bowel and Bladder Function (urinary hesitancy or urgency, and bowel disfunction).

We ensured that the brain MRI, performed at the onset, included axial and sagittal T2-weighted, fluid-attenuated inversion recovery (FLAIR)-weighted, T1-weighted MRI sequences, and T1-weighted MRI images after administration of gadolinium. All patients included in the study also underwent a spinal MRI at the onset of symptoms and before the start of corticosteroid therapy; dual-echo (proton-density and T2-weighted) conventional and/or fast spin-echo, STIR (as alternative to proton-density weighted) and contrast-enhanced T1-weighted spin-echo (in case of presence of T2 lesions) sequences were acquired. The MRI scan revision was centralized and carried out by two operators (LP and MANF) blinded to clinical outcome, at the Bambino Gesù children's hospital. Lesion characteristics were recorded, including the location, distribution, border outline, symmetry, and number, as well as size and gadolinium capture. Tumefactive lesions were defined as such if larger than two cortical gyri. The presence or absence of black holes (non-enhancing hypointense lesions visible on T1-weighted sequences) and post-gadolinium enhancement were analyzed.

2.3. Subgroups Classification

The population was divided in two groups, according to the BMI, using the Center for Disease Control (CDC) metrics [6]: the first group (healthy weight, HW) includes people with BMI under 25, and the second group (overweight/obese, OW/O) included people with BMI of 25–29.9, considered overweight, and people with a BMI over 30, categorized as obese.

2.4. Statistical Analysis

We used the chi-squared test to compare the distribution of categorical variables in the two groups. The categorical variables included sex, presence of encephalopathy at clinical onset, symptoms at onset in various functional systems, presence of oligoclonal bands in CSF, distribution of lesions in MRI in the different areas (periventricular, subcortical, subtentorial, and spine), presence of gadolinium-enhancing lesions, black holes and swelling lesions.

The Mann–Whitney U-test was used to compare continuous variables like age, the number of the relapses before diagnosis and the number of lesions visible in MRI scans taken at onset.

Statistical analysis was conducted with SPSS software version 22. The level of significance was set for p value values lower than 0.05.

3. Results

Among 65 patients selected, 10 of them were excluded for unavailability of data required (3 no onset MRI images available, 2 no data from CSF collected at onset available, 5 no measurement of weight and height performed at onset).

The statistical analysis included 55 patients with a diagnosis of relapsing-remitting MS with pediatric onset. This population was composed of 34 girls (61.8%) and 21 boys (38.2%), with a mean age of 13.5 years (Table 1).

Table 1. Demographic features of the patients included in the study. SD: standard deviation.

	POMS Subjects			
Sex	**No.**	**%**	**Age, y**	
			Mean	SD
Male	21	38.2%	13.1	3.2
Female	34	61.8%	13.7	3.06
Total	55	100%	13.5	3.1

Patients were divided into two groups: HW (33 patients, 60%) and OW/O (22 patients, 40%). The HW group consisted of 57.6% girls and 42.4% boys while the OW/O population was composed of 68.2% girls and 31.8% boys ($p > 0.05$) (Table 2).

Table 2. Demographic features of the healthy weight and overweight/obese patients.

	Healthy Weight				**Obese/Overweight**			
	No.	**%**	**Age, y**		**No.**	**%**	**Age, y**	
			Mean	DS			Mean	DS
Total	33	60	14.6	4.1	22	40	12.7	3.8
Male	14	42.4	12.4	3.5	7	31.8	14.5	1.8
Female	19	57.6	13	3.8	15	68.2	14.6	1.2

3.1. *Age at Disease Onset*

OW/O patients had an age of disease onset approximately two years lower than HW patients (12.7 ± 3.8 years vs. 14.6 ± 4.1 years; $p < 0.05$).

3.2. *Clinical Features at Disease Onset*

An acute disseminated encephalomyelitis (ADEM)-like with encephalopathy at onset occurred with a similar frequency in the OW/O group versus HW patients ($p > 0.05$). Excluding patients with ADEM-like, an onset with polyfocal symptoms was more often seen in OW/O patients than in the HW group (72.7% vs. 21.2%; $p < 0.05$). In order of frequency, the onset symptoms occurred in patients with polyfocal presentation involved superficial/proprioceptive sensation (No. 16, 69.5%), brainstem deficit (No. 12, 52.1%), pyramidal functions (No. 12, 52.1%), cerebellar functions (No. 11, 47.8%), visual deficit (No. 8, 34.7%), bowel and bladder functions (No. 2, 8.6%).

On the contrary, monofocal onset was seen more often in HW patients than in the OW/O group (66.6% vs. 18.1%; $p < 0.05$). In order of frequency, the onset symptom occurred in patients with monofocal presentation involved visual deficit (No. 9, 34.6%), brainstem deficit (No. 7, 26.9%), superficial/proprioceptive sensation (No. 6, 23%), pyramidal function (No.4, 15.3%).

Involvement of the pyramidal functions was more often detected in the OW/O group than in the HW group (50% vs. 25.4%; $p < 0.005$), (Table 3). The analysis of the total number of relapses that occurred before diagnosis did not reveal a statistically significant difference between the two groups.

Table 3. Clinical features at onset of the healthy weight and overweight/obese patients.

	Healthy Weight		Obese/Overweight		*p*-Value
	Mean	SD	Mean	SD	
Age at onset, y	14.6	4.1	12.7	3.8	<0.05
Expanded disability status scale (EDSS) at onset	1.9	0.5	2.2	0.6	0.07
	No.	%	No.	%	
Monofocal onset	22	66.6	4	18.1	<0.05
Polifocal onset	7	21.2	16	72.7	<0.05
Encephalopathy at onset	4	12.1	2	9.1	0.08
Pyramidal functions	8	25.4	11	50	<0.05
Superficial sensation (light, touch and pain)	9	27.3	11	50	0.07
Proprioceptive sensation	9	27.3	6	27.3	0.08
Cerebellar functions	9	27.3	3	13.6	0.19
Brainstem functions	16	48.5	9	40.9	0.39
Visual deficit	10	30.3	9	40.8	0.3
Bowel and bladder functions	2	6.1	2	9.1	0.52
Oligoclonal band (intrathecal IgG synthesis)	26	83.9	19	95	0.23
Pleocytosis (>5 cell/mmc)	13	50	14	70	0.14
Ig G anti-Epstein–Bar virus	30	91	22	100	0.2

3.3. *MRI Features at Disease Onset*

Regarding MRI scans performed at onset, a statistically significant difference was revealed analyzing the number of black holes: they were more frequently detected in OW/O patients in onset MRI compared to the HW group (50% vs. 15.5%; $p < 0.05$) (Figure 1).

When analyzing the presence of other MRI features such as periventricular lesions, juxtacortical/cortical lesions, infratentorial lesions, spinal cord lesions, gadolinium-enhanced lesions, and tumefactive lesions, no statistically significant differences were found between the two groups (Table 4).

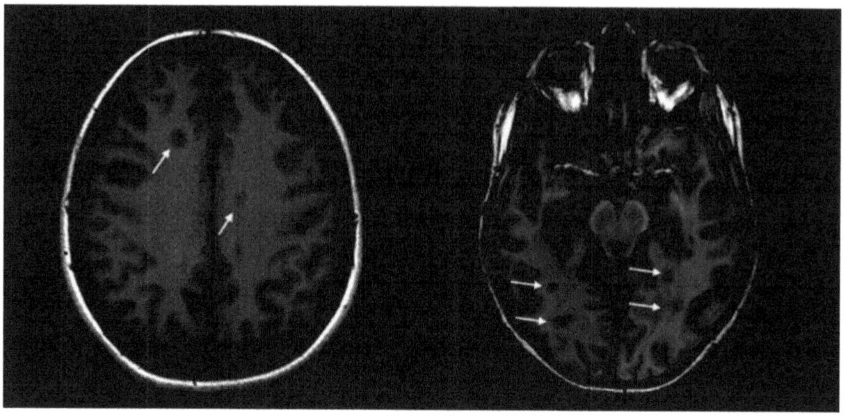

Figure 1. Onset MRI scans of two enrolled patients. The white arrows indicate T1 hypointense lesions, known as black holes.

Table 4. Onset Magnetic Resonance Imaging features for the healthy weight and overweight/obese patients.

	Healthy Weight		Obese/Overweight		*p*-Value
	No.	%	No.	%	
Black holes	5	15.5	12	54.5	<0.05
Periventricular lesion	29	87.9	18	81.8	0.4
Juxtacortical/cortical	25	75.8	19	86.4	0.27
Infratentorial	23	69.7	13	59.1	0.3
Optic nerve	13	39.4	9	40.9	0.5
Spinal cord	20	60.6	13	59.1	0.09
Gadolinium enhancing lesions	25	75.8	19	86.4	0.27
Tumefactive lesions	8	24.2	6	27.3	0.52

We found no statistically significant difference between the two groups about the total number of T2 hyperintense cerebral lesions at onset.

3.4. Laboratorial Features at Disease Onset

Oligoclonal bands (OCBs) in CSF were found in 26 HW patients (83.9%) and in 19 (95%) OW/O patients, pleocytosis were found in 13 HW patients (50%) and in 14 OW/O patients (70%), previous EBV infection were found in 30 HW patients (91%) and in 22 OW/O patients (100%).

We found no statistically significant difference between the two groups analyzing the presence of oligoclonal bands (OCBs) in CSF, the presence of pleocytosis in CSF, previous EBV infection.

4. Discussion

Our study shows that patients with POMS who are OW/O experience clinical onset at a younger age than those who are HW. At the onset of the disease, OW/O individuals exhibit a worse clinical picture and less favorable MRI findings than HW individuals. POMS is a disease that is caused by genetic and environmental factors, which increase the risk of development [23].

Susceptibility to SM has been identified in several genes within the major histocompatibility complex (MHC) loci. Around one-third of them have been associated with POMS, which suggests that there is a shared genetic inheritance. The HLA-DRB1 gene is the gene responsible for the most significant genetic contribution, which is associated with changes in HLA in general and specifically [24].

Epstein–Barr virus (EBV) infection is a determinant of POMS that has been extensively studied [25,26]. Vitamin D deficiency, not breastfeeding infants, pesticide exposure, smoking, air quality, and hormonal influences [18] were among the other factors [18]. Furthermore, certain risk factors seem to have a more significant impact during a particular time frame. BMI and obesity in adolescence, not during childhood, is associated with an increased risk of developing MS [27–30].

Authors, in particular, attempted to establish if there was a correlation between obesity age and the risk of MS [17,23]. Identifying the age range in which obesity may increase the future risk of developing MS is important for implementing prevention measures for being overweight. Some studies showed that in obese subjects, correcting their body weight reduces the risk of developing MS during the life period [17,27].

Recently a Mendelian randomization study was performed to evaluate whether childhood BMI has a causal influence on MS, and whether this putative effect is independent of early adult obesity and pubertal timing. This study found that a higher genetically predicted childhood BMI was associated with increased odds of MS. The association between childhood obesity and MS susceptibility was mediated by the persistence of obesity into early adulthood, but independent of the timing of puberty [31].

A Danish longitudinal study conducted in school children found that among girls, at each age from 7 to 13, a one-unit increase in a the BMI z-score was associated with significant hazard ratios of MS. The risk of MS increased by 1.61–1.95 times for girls in the 95th percentile for BMI compared to girls in the 85th percentile. The associations were weaker in boys. A hazard ratio of 1.17 was found for a one-unit increase in BMI z-score at age 7, and 1.15 was found at age 13 [17].

Munger et al. reported a study in which some women were questioned, using a self-reported representative pictogram about their body silhouettes at the ages of 5, 10, and 20. In this study, the author found that the women who reported having a larger body size at age 20 had a two-fold increased risk of MS compared to women who reported a thinner body size. There was also a suggestion that having a larger body size during childhood at ages 5 or 10 may increase the risk of MS. Furthermore, after adjusting for body size at age 20, there was no increased risk of MS associated with having a large body size during childhood. The twofold risk of MS associated with large body size at age 20 remained unchanged [28].

These studies support the previous findings that overweight individuals in late adolescence/early adulthood have a 40% increased risk of MS [32]. Unfortunately, we do not have information about the BMI trajectory of our cohort, but our data show that a high BMI at the onset of MS is associated with an earlier age of onset of MS, polyfocal symptoms at onset, and early hypointense MRI lesions.

The onset of MS in childhood has significant implications for the prognosis. Compared to adults, children and adolescents with MS experience a higher relapse rate and more commonly affect the cerebellar and brainstem regions [33]. Pediatric MS patients take about 10 years longer than adults to reach irreversible levels of disability. Nevertheless, these levels are attained by a final age that is 10 years younger than in adult-onset patients [34]. Furthermore, POMS can impact the cognitive function and development of children. Early-onset MS patients have a faster decline in cognitive performances compared to patients with adult-onset disease, resulting in a higher risk of cognitive impairment and psychiatric comorbidity in adulthood, even when adjusted for disease duration [35–40]. POMS is linked to psychiatric comorbidity in adulthood. Overall, it is estimated that approximately 25–30% of patients with POMS experience mild cognitive changes [40], mainly related to attention and processing speed [41]. Long-term repercussions on the cognitive performance of POMS patients have also been described [36]. Therefore, efforts towards early diagnosis, the discovery of early predictors of long-term outcomes, and appropriate early drug intervention are highly warranted.

Regarding clinical features at onset, a recent review estimated that approximately half to two-thirds of pediatric MS patients have a polysymptomatic presentation [42] and this is more frequent in patients with onset at a younger age, which may reflect a greater susceptibility of an immature brain to the inflammatory insult [43]. Children are most commonly diagnosed with motor dysfunction (30%), sensory symptoms (15–30%), brainstem symptoms (25%), optic neuritis (10–22%), and ataxia (5–15%) [38]. The results of our analysis show that a clinical onset with polyfocal signs is more frequently observed in overweight/obese patients than in healthy weight patients, who tend to present with monofocal clinical manifestations. In addition, in our study, overweight/obese patients are more likely to have pyramidal domain involvement (50% OW/O) than healthy weight patients (25.4% HW).

The presence of polyfocal symptoms at the clinical onset of MS has been associated in both children and adults, with an increased risk of moderate or severe disability [2,44–46] and a decreased response to disease-modifying treatments [47]. A higher EDSS score is linked to the involvement of the pyramidal system [48]. Langer-Gould et al. analyzing a multiethnic population of 75 new diagnoses of pediatric clinically isolated Syndrome (CIS) and POMS, found that moderately and extremely obese children were more likely to present with motor/sensory symptoms of transverse myelitis [16].

Regarding MRI findings at the onset of the disease, our analysis shows that hypointense T1 lesions (also known as Black Holes) [49], are found more frequently in the overweight/obese group than in the healthy weight group. The presence of black holes is a sign of chronic inflammatory damage and tissue damage due to axonal loss [50].

It is known in adults that the black hole burden is related to cerebral atrophy [51–54], and both reflect and are a negative prognostic factor of physical disability [49,55,56] and poor cognitive performance [57]. Furthermore, children with MS are more likely to present black holes already at onset of the disease compared to AOMS, probably due to a more aggressive disease early on and the susceptibility to axonal damage [12], and tend to have more rapid loss of brain parenchyma during the course of the disease [58].

Studies in the adult MS population on the effects of over weight on neuroradiological factors that are associated with progression and persistent disability have yielded conflicting results [10,59–62]. Some studies have documented a relationship between an increase in BMI and a reduction in cerebral gray matter over time, which leads to a greater burden of T1 lesions [59,61,62]. This type of relationship [10,60] has not been found by others, however. Other factors, such as the duration of the disease, may affect the reduction in brain volume or lesion load in the adult population [53].

The study of these phenomena in the pediatric population allows us to analyze the disease at an earlier stage, which therefore gives more value to the effect of weight. Few studies have investigated this topic in POMS, and those that have, have found that patient BMI did not affect the probability of presenting numerous T2 lesions or contrast enhancing lesions at the onset [63]. However, a second study did not observe significant differences in the clinical characteristics of POMS between normal weight and overweight subjects [20].

The correlation between overweight and MS can also have implications in the therapeutic field. In adults with MS, the risk of developing persistent disability has gradually decreased through the early use of effective disease-modifying treatments [64,65]. However, the therapeutic management of POMS has greater limitations than that of adults because we have fewer pharmacological strategies available than in adults [66,67]. Therapeutic targets can be focused on preventing and controlling overweight. Prior studies have demonstrated the protective effect of vigorous physical exercise during adolescence and childhood against the development of MS [14,68,69] and the protective effect of physical exercise on disease outcome, in terms of accumulation of disability, relapse rate, neurocognitive performance and MS-related MRI lesions accrual [70].

Diet may also play a role in susceptibility to developing MS and may affect the course of the disease in both adults [71,72] and in pediatric populations with MS [73,74]. In recent years, MS treatment strategies have included other therapeutic dietary interventions, in addition to vitamin D supplementation [53]. Omega-3 and omega-6 acid supplements [75] and the ketogenic diet [76,77] are included. Dietary interventions should be considered as a potential therapeutic strategies that may impact pathophysiological mechanisms and the well-being of patients with MS.

In conclusion, although our study highlighted and presented more evidence about how negatively obesity influences the onset features of POMS [53], it is possible, even after the disease onset, to remove risk factors, ameliorate the progression of disability and comorbidities, and improve the neurological reserve, able to repair and compensate for neuronal damage [19,58,78].

Our study has a series of limitations, which consist of the retrospective analysis of the data and the failure to observe changes in BMI over time.

5. Conclusions

Early diagnosis and treatment of MS have a very strong impact on the prognosis of multiple sclerosis. The identification of risk factors for the development of the disease is of great importance. Our study strengthens the thesis that being overweight may have an unfavorable prognostic role in MS patients. The onset of MS in childhood offers disadvantages in terms of impact on cognition, future disability, and reduced availability of drugs to reduce disease progression. To these, the addition of overweight in childhood may have a further unfavorable prognostic factor. Controlling weight during adolescence, rather than childhood or adulthood, is critical in determining the risk of MS.

Author Contributions: Conceptualization, L.P.; Validation, L.P. and M.V.; Investigation, M.A.N.F., S.T. and M.P.C.; Data curation, L.P. and G.M.; Writing—original draft, E.P.; Writing—review & editing, L.P.; Visualization, M.V.; Supervision, M.V. All authors have read and agreed to the published version of the manuscript.

Funding: This work was supported also by the Italian Ministry of Health with Current Research funds.

Institutional Review Board Statement: The study was approved by the Ethical Committee of Bambino Gesù Children's Hospital (date 19 October 2023).

Informed Consent Statement: All patients enrolled and their parents provided consent for the publication of the results.

Data Availability Statement: The data presented in this study are available on request from the corresponding author. The data are not publicly available due to ethical reasons.

Conflicts of Interest: The authors declare no conflict of interest.

References

1. Boiko, A.; Vorobeychik, G.; Paty, D.; Devonshire, V.; Sadovnick, D. University of British Columbia MSCN. Early onset multiple sclerosis: A longitudinal study. *Neurology* **2002**, *59*, 1006–1010. [CrossRef] [PubMed]
2. Bigi, S.; Banwell, B. Pediatric Multiple Sclerosis. *J. Child. Neurol.* **2012**, *27*, 1378–1383. [CrossRef] [PubMed]
3. Gordon-Lipkin, E.; Banwell, B. An update on multiple sclerosis in children: Diagnosis, therapies, and prospects for the future. *Expert. Rev. Clin. Immunol.* **2017**, *13*, 975–989. [CrossRef]
4. Tenembaum, S. Multiple sclerosis in childhood and adolescence. *J. Neurol. Sci.* **2011**, *311*, S53–S57. [CrossRef] [PubMed]
5. Rasul, T.; Frederiksen, J.L. Link between overweight/obese in children and youngsters and occurrence of multiple sclerosis. *J. Neurol.* **2018**, *265*, 2755–2763. [CrossRef] [PubMed]
6. Brenton, J.N.; Woolbright, E.; Briscoe-Abath, C.; Qureshi, A.; Conaway, M.; Goldman, M.D. Body mass index trajectories in pediatric multiple sclerosis. *Dev. Med. Child. Neurol.* **2019**, *61*, 1289–1294. [CrossRef]
7. Mowry, E.M.; Krupp, L.B.; Milazzo, M.; Chabas, D.; Strober, J.B.; Belman, A.L.; McDonald, J.C.; Oksenberg, J.R.; Bacchetti, P.; Waubant, E. Vitamin D status is associated with relapse rate in pediatric-onset multiple sclerosis. *Ann. Neurol.* **2010**, *67*, 618–624. [CrossRef]
8. Pohl, D.; Krone, B.; Rostasy, K.; Kahler, E.; Brunner, E.; Lehnert, M.; Wagner, H.-J.; Gärtner, J.; Hanefeld, F. High seroprevalence of Epstein–Barr virus in children with multiple sclerosis. *Neurology* **2006**, *67*, 2063–2065. [CrossRef]
9. Stampanoni Bassi, M.; Iezzi, E.; Buttari, F.; Gilio, L.; Simonelli, I.; Carbone, F.; Micillo, T.; De Rosa, V.; Sica, F.; Furlan, R. Obesity worsens central inflammation and disability in multiple sclerosis. *Mult. Scler. J.* **2020**, *26*, 1237–1246. [CrossRef]

10. Manuel Escobar, J.; Cortese, M.; Edan, G.; Freedman, M.S.; Hartung, H.-P.; Montalbán, X.; Sandbrink, R.; Radü, E.-W.; Barkhof, F.; Wicklein, E.-M.; et al. Body mass index as a predictor of MS activity and progression among participants in BENEFIT. *Mult. Scler. J.* **2022**, *28*, 1277–1285. [CrossRef]
11. Milles, P.; De Filippo, G.; Maurey, H.; Tully, T.; Deiva, K. Obesity in Pediatric-Onset Multiple Sclerosis: A French Cohort Study. *Neurol. Neuroimmunol. Neuroinflamm.* **2021**, *8*, e1044. [CrossRef]
12. Pfeifenbring, S.; Bunyan, R.F.; Metz, I.; Röver, C.; Huppke, P.; Gärtner, J.; Lucchinetti, C.F.; Brück, W. Extensive acute axonal damage in pediatric multiple sclerosis lesions: Axonal Damage in Pediatric MS. *Ann. Neurol.* **2015**, *77*, 655–667. [CrossRef] [PubMed]
13. Huppke, P.; Huppke, B.; Ellenberger, D.; Rostasy, K.; Hummel, H.; Stark, W.; Brück, W.; Gärtner, J. Therapy of highly active pediatric multiple sclerosis. *Mult. Scler. J.* **2019**, *25*, 72–80. [CrossRef] [PubMed]
14. Stephens, S.; Berenbaum, T.; Finlayson, M.; Motl, R.W.; Yeh, E.A. Youth with multiple sclerosis have low levels of fitness. *Mult. Scler. J.* **2021**, *27*, 1597–1605. [CrossRef]
15. Di Filippo, M.; Anderson, V.M.; Altmann, D.R.; Swanton, J.K.; Plant, G.T.; Thompson, A.J.; Miller, D.H. Brain atrophy and lesion load measures over 1 year relate to clinical status after 6 years in patients with clinically isolated syndromes. *J. Neurol. Neurosurg. Psychiatry* **2010**, *81*, 204–208. [CrossRef]
16. Langer-Gould, A.; Brara, S.M.; Beaber, B.E.; Koebnick, C. Childhood obesity and risk of pediatric multiple sclerosis and clinically isolated syndrome. *Neurology* **2013**, *80*, 548–552. [CrossRef]
17. Munger, K.L.; Bentzen, J.; Laursen, B.; Stenager, E.; Koch-Henriksen, N.; Sørensen, T.I.; Baker, J.L. Childhood body mass index and multiple sclerosis risk: A long-term cohort study. *Mult. Scler. J.* **2013**, *19*, 1323–1329. [CrossRef] [PubMed]
18. Teleanu, R.I.; Niculescu, A.G.; Vladacenco, O.A.; Roza, E.; Perjoc, R.S.; Teleanu, D.M. The State of the Art of Pediatric Multiple Sclerosis. *Int. J. Mol. Sci.* **2023**, *24*, 8251. [CrossRef] [PubMed]
19. Van Hijfte, L.; Loret, G.; Bachmann, H.; Reynders, T.; Breuls, M.; Deschepper, E.; Kuhle, J.; Willekens, B.; Laureys, G. Lifestyle factors in multiple sclerosis disability progression and silent brain damage: A cross-sectional study. *Mult. Scler. Relat. Disord.* **2022**, *65*, 104016. [CrossRef]
20. Huppke, B.; Ellenberger, D.; Hummel, H.; Stark, W.; Röbl, M.; Gärtner, J.; Huppke, P. Association of Obesity with Multiple Sclerosis Risk and Response to First-line Disease Modifying Drugs in Children. *JAMA Neurol.* **2019**, *76*, 1157. [CrossRef]
21. Woolbright, E.; Koshiya, H.; Brenton, J.N. Body size perceptions & diet modification in youth with multiple sclerosis. *Mult. Scler. Relat. Disord.* **2022**, *58*, 103402. [CrossRef] [PubMed]
22. Krupp, L.B.; Tardieu, M.; Amato, M.P.; Banwell, B.; Chitnis, T.; Dale, R.C.; Ghezzi, A.; Hintzen, R.; Kornberg, A.; Pohl, D.; et al. International Pediatric Multiple Sclerosis Study Group criteria for pediatric multiple sclerosis and immune-mediated central nervous system demyelinating disorders: Revisions to the 2007 definitions. *Mult. Scler. J.* **2013**, *19*, 1261–1267. [CrossRef] [PubMed]
23. Yan, K.; Balijepalli, C.; Desai, K.; Gullapalli, L.; Druyts, E. Epidemiology of pediatric multiple sclerosis: A systematic literature review and meta-analysis. *Mult. Scler. Relat. Disord.* **2020**, *44*, 102260. [CrossRef] [PubMed]
24. Boiko, A.N.; Gusev, E.I.; Sudomoina, M.A.; Alekseenkov, A.D.; Kulakova, O.G.; Bikova, O.V.; Maslova, O.I.; Guseva, M.R.; Boiko, S.Y.; Guseva, M.E.; et al. Association and linkage of juvenile MS with HLA-DR2(15) in Russians. *Neurology* **2002**, *58*, 658. [CrossRef]
25. Balfour, H.H.; Schmeling, D.O.; Grimm-Geris, J.M. The promise of a prophylactic Epstein–Barr virus vaccine. *Pediatr. Res.* **2020**, *87*, 345–352. [CrossRef] [PubMed]
26. Moreno, M.A.; Or-Geva, N.; Aftab, B.T.; Khanna, R.; Croze, E.; Steinman, L.; Han, M.H. Molecular signature of Epstein-Barr virus infection in MS brain lesions. *Neurol. Neuroimmunol. Neuroinflamm.* **2018**, *5*, e466. [CrossRef] [PubMed]
27. Gianfrancesco, M.A.; Acuna, B.; Shen, L.; Briggs, F.B.S.; Quach, H.; Bellesis, K.H.; Bernstein, A.; Hedstrom, A.K.; Kockum, I.; Alfredsson, L.; et al. Obesity during childhood and adolescence increases susceptibility to multiple sclerosis after accounting for established genetic and environmental risk factors. *Obes. Res. Clin. Pract.* **2014**, *8*, e435–e447. [CrossRef]
28. Munger, K.L.; Chitnis, T.; Ascherio, A. Body size and risk of MS in two cohorts of US women. *Neurology* **2009**, *73*, 1543–1550. [CrossRef]
29. Hedström, A.; Olsson, T.; Alfredsson, L. Body mass index during adolescence, rather than childhood, is critical in determining MS risk. *Mult. Scler. J.* **2016**, *22*, 878–883. [CrossRef]
30. Wesnes, K.; Riise, T.; Casetta, I.; Drulovic, J.; Granieri, E.; Holmøy, T.; Kampman, M.T.; Landtblom, A.-M.; Lauer, K.; Lossius, A.; et al. Body size and the risk of multiple sclerosis in Norway and Italy: The EnvIMS study. *Mult. Scler. J.* **2015**, *21*, 388–395. [CrossRef]
31. Harroud, A.; Mitchell, R.E.; Richardson, T.G.; Morris, J.A.; Forgetta, V.; Davey Smith, G.; Baranzini, S.E.; Richards, J.B. Childhood obesity and multiple sclerosis: A Mendelian randomization study. *Mult. Scler. J.* **2021**, *27*, 2150–2158. [CrossRef] [PubMed]
32. Hedström, A.K.; Olsson, T.; Alfredsson, L. High body mass index before age 20 is associated with increased risk for multiple sclerosis in both men and women. *Mult. Scler. J.* **2012**, *18*, 1334–1336. [CrossRef] [PubMed]
33. Huppke, B.; Ellenberger, D.; Rosewich, H.; Friede, T.; Gärtner, J.; Huppke, P. Clinical presentation of pediatric multiple sclerosis before puberty. *Eur. J. Neurol.* **2014**, *21*, 441–446. [CrossRef] [PubMed]
34. Renoux, C.; Vukusic, S.; Mikaeloff, Y.; Edan, G.; Clanet, M.; Dubois, B.; Debouverie, M.; Brochet, B.; Lebrun-Frenay, C.; Pelletier, J. Natural history of multiple sclerosis with childhood onset. *N. Engl. J. Med.* **2007**, *356*, 2603–2613. [CrossRef] [PubMed]

35. Boesen, M.S.; Blinkenberg, M.; Thygesen, L.C.; Eriksson, F.; Magyari, M. School performance, psychiatric comorbidity, and healthcare utilization in pediatric multiple sclerosis: A nationwide population-based observational study. *Mult. Scler. J.* **2021**, *27*, 259–267. [CrossRef] [PubMed]
36. McKay, K.A.; Hillert, J.; Manouchehrinia, A. Long-term disability progression of pediatric-onset multiple sclerosis. *Neurology* **2019**, *92*, e2764–e2773. [CrossRef] [PubMed]
37. Baruch, N.F.; O'Donnell, E.H.; Glanz, B.I.; Benedict, R.H.; Musallam, A.J.; Healy, B.C.; Rintell, D.; Chitnis, T. Cognitive and patient-reported outcomes in adults with pediatric-onset multiple sclerosis. *Mult. Scler. J.* **2016**, *22*, 354–361. [CrossRef]
38. Ruano, L.; Branco, M.; Portaccio, E.; Goretti, B.; Niccolai, C.; Patti, F.; Chisari, C.; Gallo, P.; Grossi, P.; Ghezzi, A.; et al. Patients with paediatric-onset multiple sclerosis are at higher risk of cognitive impairment in adulthood: An Italian collaborative study. *Mult. Scler. J.* **2018**, *24*, 1234–1242. [CrossRef]
39. Ozakbas, S.; Kaya, D.; Idiman, E. Early Onset Multiple Sclerosis Has Worse Prognosis Than Adult Onset Multiple Sclerosis Based on Cognition and Magnetic Resonance Imaging. *Autoimmune Dis.* **2012**, *2012*, 563989. [CrossRef]
40. Amato, M.P.; Krupp, L.B.; Charvet, L.E.; Penner, I.; Till, C. Pediatric multiple sclerosis: Cognition and mood. *Neurology* **2016**, *87*, S82–S87. [CrossRef]
41. Amato, M.P.; Goretti, B.; Ghezzi, A.; Hakiki, B.; Niccolai, C.; Lori, S.; Moiola, L.; Falautano, M.; Viterbo, R.G.; Patti, F.; et al. Neuropsychological features in childhood and juvenile multiple sclerosis: Five-year follow-up. *Neurology* **2014**, *83*, 1432–1438. [CrossRef] [PubMed]
42. Reinhardt, K.; Weiss, S.; Rosenbauer, J.; Gärtner, J.; Von Kries, R. Multiple sclerosis in children and adolescents: Incidence and clinical picture–new insights from the nationwide G erman surveillance (2009–2011). *Eur. J. Neurol.* **2014**, *21*, 654–659. [CrossRef] [PubMed]
43. Banwell, B.; Ghezzi, A.; Bar-Or, A.; Mikaeloff, Y.; Tardieu, M. Multiple sclerosis in children: Clinical diagnosis, therapeutic strategies, and future directions. *Lancet Neurol.* **2007**, *6*, 887–902. [CrossRef]
44. Hammond, S.R.; McLeod, J.G.; Macaskill, P.; English, D.R. Multiple sclerosis in Australia: Prognostic factors. *J. Clin. Neurosci.* **2000**, *7*, 16–19. [CrossRef]
45. Bergamaschi, R. Prognostic Factors in Multiple Sclerosis. In *International Review of Neurobiology*; Elsevier: Amsterdam, The Netherlands, 2007; Volume 79, pp. 423–447. ISBN 978-0-12-373736-6.
46. Mikaeloff, Y.; Suissa, S.; Vallée, L.; Lubetzki, C.; Ponsot, G.; Confavreux, C.; Tardieu, M.; KIDMUS Study Group. First episode of acute CNS inflammatory demyelination in childhood: Prognostic factors for multiple sclerosis and disability. *J. Pediatr.* **2004**, *144*, 246–252. [CrossRef] [PubMed]
47. Kopp, T.I.; Blinkenberg, M.; Chalmer, T.A.; Petersen, T.; Ravnborg, M.H.; Soelberg Sørensen, P.; Magyari, M. Predictors of treatment outcome in patients with paediatric onset multiple sclerosis. *Mult. Scler. J.* **2020**, *26*, 964–975. [CrossRef] [PubMed]
48. Briggs, F.B.S.; Yu, J.C.; Davis, M.F.; Jiangyang, J.; Fu, S.; Parrotta, E.; Gunzler, D.D.; Ontaneda, D. Multiple sclerosis risk factors contribute to onset heterogeneity. *Mult. Scler. Relat. Disord.* **2019**, *28*, 11–16. [CrossRef]
49. Thaler, C.; Faizy, T.; Sedlacik, J.; Holst, B.; Stellmann, J.-P.; Young, K.L.; Heesen, C.; Fiehler, J.; Siemonsen, S. T1- Thresholds in Black Holes Increase Clinical-Radiological Correlation in Multiple Sclerosis Patients. *PLoS ONE* **2015**, *10*, e0144693. [CrossRef]
50. Bitsch, A.; Kuhlmann, T.; Stadelmann, C.; Lassmann, H.; Lucchinetti, C.; Brück, W. A longitudinal MRI study of histopathologically defined hypointense multiple sclerosis lesions: Longitudinal Study of Hypointense T1 Lesions. *Ann. Neurol.* **2001**, *49*, 793–796. [CrossRef]
51. Verhey, L.H.; Branson, H.M.; Shroff, M.M.; Callen, D.J.; Sled, J.G.; Narayanan, S.; Sadovnick, A.D.; Bar-Or, A.; Arnold, D.L.; Marrie, R.A.; et al. MRI parameters for prediction of multiple sclerosis diagnosis in children with acute CNS demyelination: A prospective national cohort study. *Lancet. Neurol.* **2011**, *10*, 1065–1073. [CrossRef]
52. Sailer, M.; Losseff, N.A.; Wang, L.; Gawne-Cain, M.L.; Thompson, A.J.; Miller, D.H. T1 lesion load and cerebral atrophy as a marker for clinical progression in patients with multiple sclerosis. A prospective 18 months follow-up study. *Eur. J. Neurol.* **2001**, *8*, 37–42. [CrossRef] [PubMed]
53. Paolillo, A.; Pozzilli, C.; Gasperini, C.; Giugni, E.; Mainero, C.; Giuliani, S.; Tomassini, V.; Millefiorini, E.; Bastianello, S. Brain atrophy in relapsing-remitting multiple sclerosis: Relationship with "black holes", disease duration and clinical disability. *J. Neurol. Sci.* **2000**, *174*, 85–91. [CrossRef] [PubMed]
54. Radue, E.-W.; Barkhof, F.; Kappos, L.; Sprenger, T.; Haring, D.A.; de Vera, A.; von Rosenstiel, P.; Bright, J.R.; Francis, G.; Cohen, J.A. Correlation between brain volume loss and clinical and MRI outcomes in multiple sclerosis. *Neurology* **2015**, *84*, 784–793. [CrossRef] [PubMed]
55. Giorgio, A.; Stromillo, M.L.; Bartolozzi, M.L.; Rossi, F.; Battaglini, M.; De Leucio, A.; Guidi, L.; Maritato, P.; Portaccio, E.; Sormani, M.P.; et al. Relevance of hypointense brain MRI lesions for long-term worsening of clinical disability in relapsing multiple sclerosis. *Mult. Scler. J.* **2014**, *20*, 214–219. [CrossRef] [PubMed]
56. Tam, R.; Traboulsee, A.; Riddehough, A.; Sheikhzadeh, F.; Li, D.K.B. The impact of intensity variations in T1-hypointense lesions on clinical correlations in multiple sclerosis. *Mult. Scler. J.* **2011**, *17*, 949–957. [CrossRef] [PubMed]
57. Vollmer, T.; Huynh, L.; Kelley, C.; Galebach, P.; Signorovitch, J.; DiBernardo, A.; Sasane, R. Relationship between brain volume loss and cognitive outcomes among patients with multiple sclerosis: A systematic literature review. *Neurol. Sci.* **2016**, *37*, 165–179. [CrossRef]

58. Yeh, E.A.; Weinstock-Guttman, B.; Ramanathan, M.; Ramasamy, D.P.; Willis, L.; Cox, J.L.; Zivadinov, R. Magnetic resonance imaging characteristics of children and adults with paediatric-onset multiple sclerosis. *Brain* **2009**, *132*, 3392–3400. [CrossRef]
59. Mowry, E.M.; Azevedo, C.J.; McCulloch, C.E.; Okuda, D.T.; Lincoln, R.R.; Waubant, E.; Hauser, S.L.; Pelletier, D. Body mass index, but not vitamin D status, is associated with brain volume change in MS. *Neurology* **2018**, *91*, e2256–e2264. [CrossRef]
60. Lutfullin, I.; Eveslage, M.; Bittner, S.; Antony, G.; Flaskamp, M.; Luessi, F.; Salmen, A.; Gisevius, B.; Klotz, L.; Korsukewitz, C.; et al. Association of obesity with disease outcome in multiple sclerosis. *J. Neurol. Neurosurg. Psychiatry* **2023**, *94*, 57–61. [CrossRef]
61. Kappus, N.; Weinstock-Guttman, B.; Hagemeier, J.; Kennedy, C.; Melia, R.; Carl, E.; Ramasamy, D.P.; Cherneva, M.; Durfee, J.; Bergsland, N.; et al. Cardiovascular risk factors are associated with increased lesion burden and brain atrophy in multiple sclerosis. *J. Neurol. Neurosurg. Psychiatry* **2015**, *87*, 181–187. [CrossRef]
62. Castro, K.; Ntranos, A.; Amatruda, M.; Petracca, M.; Kosa, P.; Chen, E.Y.; Morstein, J.; Trauner, D.; Watson, C.T.; Kiebish, M.A.; et al. Body Mass Index in Multiple Sclerosis modulates ceramide-induced DNA methylation and disease course. *EBioMedicine* **2019**, *43*, 392–410. [CrossRef] [PubMed]
63. Alkolfat, F.; Said, S.; Mekky, J.; Eldeeb, H. What an adult multiple sclerosis registry can tell us about pediatric onset multiple sclerosis? *Mult. Scler. Relat. Disord.* **2023**, *79*, 104962. [CrossRef] [PubMed]
64. Beiki, O.; Frumento, P.; Bottai, M.; Manouchehrinia, A.; Hillert, J. Changes in the Risk of Reaching Multiple Sclerosis Disability Milestones in Recent Decades: A Nationwide Population-Based Cohort Study in Sweden. *JAMA Neurol.* **2019**, *76*, 665. [CrossRef]
65. Capra, R.; Cordioli, C.; Rasia, S.; Gallo, F.; Signori, A.; Sormani, M.P. Assessing long-term prognosis improvement as a consequence of treatment pattern changes in MS. *Mult. Scler. J.* **2017**, *23*, 1757–1761. [CrossRef] [PubMed]
66. Narula, S.; Hopkins, S.E.; Banwell, B. Treatment of Pediatric Multiple Sclerosis. *Curr. Treat. Options Neurol.* **2015**, *17*, 10. [CrossRef]
67. Papetti, L.; Figà Talamanca, L.; Spalice, A.; Vigevano, F.; Centonze, D.; Valeriani, M. Predictors of Evolution into Multiple Sclerosis after a First Acute Demyelinating Syndrome in Children and Adolescents. *Front. Neurol.* **2019**, *9*, 1156. [CrossRef]
68. Wesnes, K.; Myhr, K.-M.; Riise, T.; Cortese, M.; Pugliatti, M.; Boström, I.; Landtblom, A.-M.; Wolfson, C.; Bjørnevik, K. Physical activity is associated with a decreased multiple sclerosis risk: The EnvIMS study. *Mult. Scler. J.* **2018**, *24*, 150–157. [CrossRef]
69. Dorans, K.S.; Massa, J.; Chitnis, T.; Ascherio, A.; Munger, K.L. Physical activity and the incidence of multiple sclerosis. *Neurology* **2016**, *87*, 1770–1776. [CrossRef]
70. Motl, R.W.; Pilutti, L.A. Is physical exercise a multiple sclerosis disease modifying treatment? *Expert. Rev. Neurother.* **2016**, *16*, 951–960. [CrossRef]
71. Lopez-Garcia, E.; Schulze, M.B.; Meigs, J.B.; Manson, J.E.; Rifai, N.; Stampfer, M.J.; Willett, W.C.; Hu, F.B. Consumption of Trans Fatty Acids Is Related to Plasma Biomarkers of Inflammation and Endothelial Dysfunction. *J. Nutr.* **2005**, *135*, 562–566. [CrossRef]
72. Mozaffarian, D.; Pischon, T.; Hankinson, S.E.; Rifai, N.; Joshipura, K.; Willett, W.C.; Rimm, E.B. Dietary intake of trans fatty acids and systemic inflammation in women. *Am. J. Clin. Nutr.* **2004**, *79*, 606–612. [CrossRef] [PubMed]
73. Timmermans, S.; Bogie, J.F.J.; Vanmierlo, T.; Lütjohann, D.; Stinissen, P.; Hellings, N.; Hendriks, J.J.A. High Fat Diet Exacerbates Neuroinflammation in an Animal Model of Multiple Sclerosis by Activation of the Renin Angiotensin System. *J. Neuroimmune Pharmacol.* **2014**, *9*, 209–217. [CrossRef] [PubMed]
74. Azary, S.; Schreiner, T.; Graves, J.; Waldman, A.; Belman, A.; Guttman, B.W.; Aaen, G.; Tillema, J.-M.; Mar, S.; Hart, J.; et al. Contribution of dietary intake to relapse rate in early paediatric multiple sclerosis. *J. Neurol. Neurosurg. Psychiatry* **2018**, *89*, 28–33. [CrossRef] [PubMed]
75. Aristotelous, P.; Stefanakis, M.; Pantzaris, M.; Pattichis, C.S.; Calder, P.C.; Patrikios, I.S.; Sakkas, G.K.; Giannaki, C.D. The Effects of Specific Omega-3 and Omega-6 Polyunsaturated Fatty Acids and Antioxidant Vitamins on Gait and Functional Capacity Parameters in Patients with Relapsing-Remitting Multiple Sclerosis. *Nutrients* **2021**, *13*, 3661. [CrossRef] [PubMed]
76. Atabilen, B.; Akdevelioğlu, Y. Effects of different dietary interventions in multiple sclerosis: A systematic review of evidence from 2018 to 2022. *Nutr. Neurosci.* **2022**, *26*, 1279–1291. [CrossRef]
77. Brenton, J.N.; Lehner-Gulotta, D.; Woolbright, E.; Banwell, B.; Bergqvist, A.G.C.; Chen, S.; Coleman, R.; Conaway, M.; Goldman, M.D. Phase II study of ketogenic diets in relapsing multiple sclerosis: Safety, tolerability and potential clinical benefits. *J. Neurol. Neurosurg. Psychiatry* **2022**, *93*, 637–644. [CrossRef] [PubMed]
78. Giovannoni, G.; Butzkueven, H.; Dhib-Jalbut, S.; Hobart, J.; Kobelt, G.; Pepper, G.; Sormani, M.P.; Thalheim, C.; Traboulsee, A.; Vollmer, T. Brain health: Time matters in multiple sclerosis. *Mult. Scler. Relat. Disord.* **2016**, *9*, S5–S48. [CrossRef]

nutrients

Article

Sex Specificity in the Mixed Effects of Blood Heavy Metals and Cognitive Function on Elderly: Evidence from NHANES

Shuaixing Song [1,2], Nan Liu [3], Guoxu Wang [2], Yulin Wang [2], Xiaoan Zhang [4], Xin Zhao [4], Hui Chang [4], Zengli Yu [2] and Xiaozhuan Liu [1,*]

1 Center for Clinical Single-Cell Biomedicine, Henan Provincial People's Hospital, People's Hospital of Zhengzhou University, Zhengzhou 450003, China
2 College of Public Health, Zhengzhou University, Zhengzhou 450001, China
3 Institute of Environment and Health, South China Hospital of Shenzhen University, Shenzhen 518000, China
4 The Third Affiliated Hospital of Zhengzhou University, Zhengzhou 450052, China
* Correspondence: zhuan@zzu.edu.cn

Abstract: The way that males and females react to environmental exposures and negative impacts on their neurological systems is often different. Although previous research has examined the cognitively impairing effects of solitary metal exposures, the relationship between metal mixtures and cognitive function, particularly when considering an individual's sex, remains elusive. This study aimed to investigate the sex differences in the association between multiple metal combinations and cognitive function in older Americans. This research employed the 2011–2014 NHANES survey of elderly Americans. The association between five mixed metals and four cognitive tests (the animal fluency test (AFT), the digit symbol substitution test (DSST), the instant recall test (IRT), and the delayed recall test (DRT)) were investigated with generalized linear regression model (GLM), Bayesian kernel machine regression model (BKMR), weighted quantile sum regression model (WQS), and quantile g-computation regression model (Qgcomp). A total of 1833 people, including 883 males and 950 females, enrolled in this cross-sectional study. We discovered that blood lead and blood cadmium were negatively associated with cognitive performance, while blood selenium demonstrated a positive association with cognitive function in older people. The negative relationship of heavy metal combinations on cognitive function might be somewhat reduced or even reversed via selenium. The IRT, AFT, and DSST are three of the four cognitive tests where men had more dramatic positive or negative results. There was a sex-specific connection between blood metal ratios and cognitive function among older Americans, as evidenced by the more significant relationship between mixed metals and cognitive performance in men (either positively or negatively). These results emphasize the impacts of ambient heavy metal exposure on cognitive function by employing sex-specific methods.

Keywords: heavy metals; mixture; cognitive function; joint effect; sex-specific

1. Introduction

Older people frequently experience cognitive impairment, which harms their quality of life and social interactions. Declining cognitive function significantly increases both personal and public health burdens as the percentage of the elderly population rises quickly [1]. According to reports, several physical, psychological, social, and living variables and health factors combine with inherited and external factors to trigger cognitive decline [2]. Current evidence indicates that environmental factors, such as heavy metals, may impact the cognitive function of older people [3,4]. Although metals are naturally existing elements, diffuse air pollution, cigarette smoke, and contaminated food and water are the main artificial sources of exposure [5]. Individuals with elevated levels of heavy metals face a risk of developing renal injury, neuropathy, coronary disease, and other serious illnesses [6].

When examining illness burden, gender differences in metal toxicity should be taken into account. Metals can cause cellular reactive stress after entering the body through many different paths, leading to various physiological, metabolic, and behavioral dysfunctions. Differences in the dopaminergic system and neuroimmune axis make males more vulnerable to exposure to metals such as mercury [7]. Evidence for sex-specific neurotoxic effects of manganese may also derive in part from the metal's differential assimilation, absorption, and storage [8]. It has been documented that exposure to dangerous metals, including lead, cadmium, tungsten, and manganese, is related to lowered cognitive function [7,9–11]. However, it has also been discovered that some heavy metals, such as selenium and zinc, can benefit cognitive function [12,13]. Laboratory data suggest that there might be sex disparities regarding how heavy metals affect cognitive performance due to variations in hormone levels, body makeup, and brain architecture [14,15]. Given that there are sex disparities in both the prevalence and manifestations of cognitive-related diseases, as well as in how well these disorders respond to treatment, it is crucial to examine how sex differences in changes in cognitive ability relate to these differences. The sex-specific relationship between mental exposure and cognitive function [16,17], however, is still poorly understood since there lacks sufficient population-based research that concentrates on such. The findings of this study regarding sex variations in cognitive function may be crucial for understanding cognitive disease, but this epidemiological study has no individual predictive value.

Humans are exposed to multiple environmental chemicals concurrently, which can contribute to interactions between co-managed chemicals and confound the study of effects. Individuals are inevitably exposed to multiple metals at once, and these metals can interact with one another. The majority of previous epidemiological research has been on the relationship between individual metal exposures and cognitive function. Only single pollutant models have been used in studies on the effects of heavy metal exposure on cognitive function; this may have an impact on effect estimates by neglecting mixed data [9,18]. Combinations of metals might function either effectively or antagonistically, as different metals may facilitate or prevent the absorption of other metals [3,4]. According to the research conducted on teenagers in Bangladesh, selenium is favorably correlated with cognition, whereas manganese, arsenic, and cadmium are negatively correlated with working memory, visual recognition, and memory [19]. Prenatal metal combination exposure has been linked to neurocognitive development in children, according to studies on neonates [20–22]. In this study, we investigated the correlation between cognitive function and five heavy blood metals in senior people participating in the National Health and Nutrition Examination Survey (NHANES) between 2011 and 2014. Additionally, we looked into correlations between sex and cognitive function, pinpointing individual metals with the most important mixtures. We expect that the academic and clinical communities will benefit greatly from the information this study will contribute, which should result in more efficient prevention for both men and women.

2. Materials and Methods

2.1. Study Design and Population

The National Health and Nutrition Examination Survey (NHANES) is a cross-sectional population health survey conducted by the National Center for Health Statistics (NCHS) of the Centers for Disease Control and Prevention (CDC) among the non-institutionalized U.S. population [23]. The representative samples for this investigation were selected through a complex multi-stage, hierarchical sampling approach. The protocol for NHANES has been approved by a review committee affiliated with CDC. All participants completed informed consent forms. This study combined and evaluated demographic, examination, laboratory, and questionnaire data from participants enrolled in the NHANES. Our study was limited to 3632 participants aged 60 or older out of 19,931 respondents who participated in the NHANES between 2011 and 2014. Throughout the investigation, we removed respondents with missing sociological characteristics and laboratory test results for serum heavy metals (n = 1486), as well as older persons who lacked four complete cognitive assessments (n = 313). Finally, this study comprised a total of 1833 individuals aged 60 or older, including 883 males and 950 women. The specific selection process for inclusion in the study is shown in Figure 1.

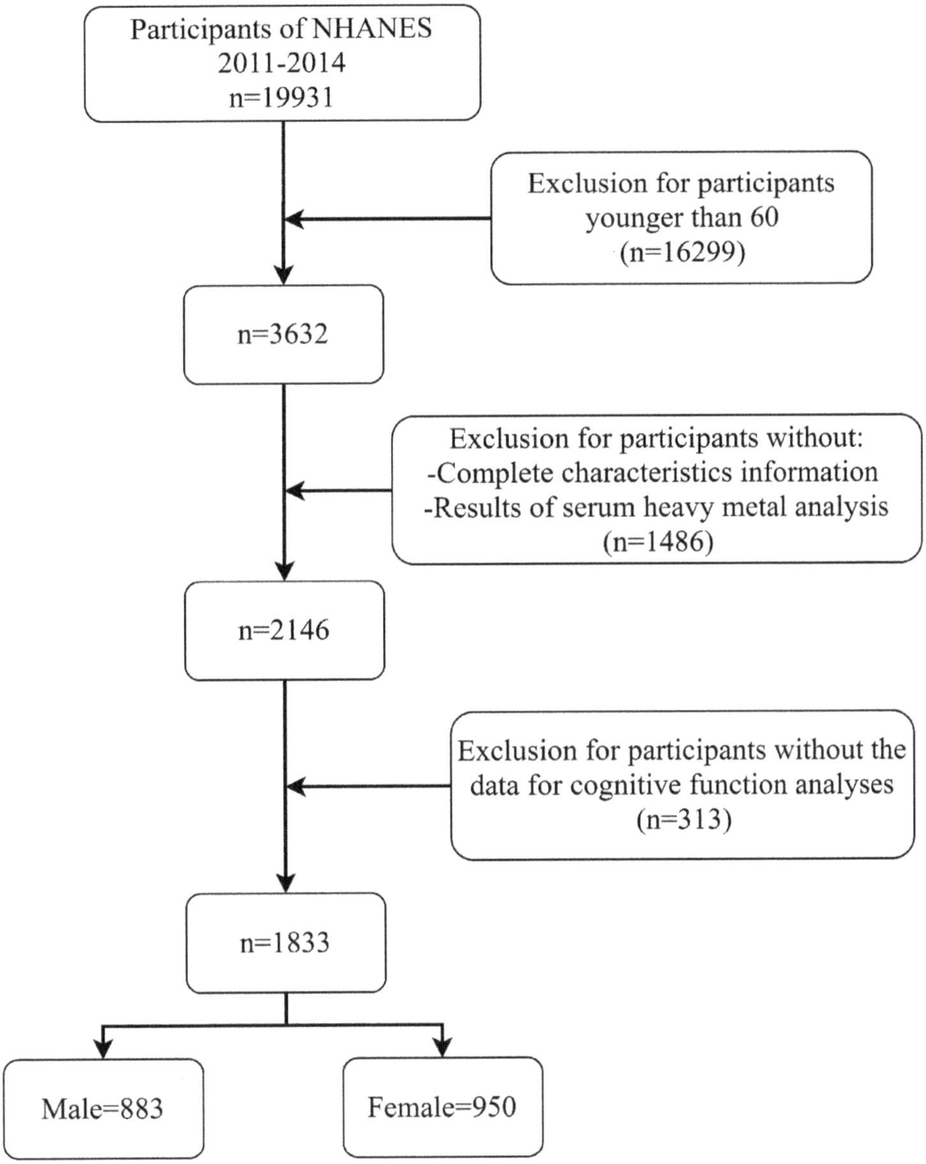

Figure 1. The specific selection process for inclusion in the study.

2.2. Measurements of Blood Heavy Metals

The processing, storage, and shipment of whole blood specimens to the NCHS, CDC for study. Serum concentrations of lead (Pb), cadmium (Cd), mercury (Hg), manganese (Mn), and selenium (Se) were measured employing inductively coupled plasma mass spectrometry with quadrupole ICPu20MS technology. The serum heavy metal concentrations of the participants in this investigation are provided in Table 1. In order to comply with the standards of statistical analysis, the serum heavy metal assay values were log-transformed during subsequent data processing according to the results of the statistical description.

Table 1. Descriptive statistical results of serum heavy metal content in the population were included.

		Percentile						
	Mean	25th	50th	75th	Skew.2SE [a]	Kurt.2SE [b]	Normtest.W [c]	Normtest.p
Blood lead	1.90	1.03	1.49	2.24	50.96	265.31	0.60	<0.001
Blood cadmium	0.55	0.27	0.40	0.65	23.98	43.02	0.72	<0.001
Blood mercury	1.87	0.56	1.04	2.11	42.98	176.52	0.57	<0.001
Blood selenium	195.25	177.90	193.50	208.30	52.00	364.67	0.72	<0.001
Blood manganese	9.48	7.04	8.79	11.18	37.83	177.17	0.75	<0.001

[a] The skewness coefficient g1 (skewness), its significant criterium (skew. 2SE, that is, g1/2. SEg1; if skew. 2SE > 1, then skewness is significantly different than zero). [b] The kurtosis coefficient g2 (kurtosis), its significant criterium (kurt. 2SE, same remark than for skew.2SE). [c] The statistic of a Shapiro–Wilk test of normality (normtest.W) and its associated probability (normtest.p).

2.3. Measurement of Cognitive Performance

The animal fluency test (AFT), the digit symbol substitution test (DSST), and the Consortium to Establish a Registry for Alzheimer's Disease (CERAD) word list learning test were employed to evaluate the cognitive function of participants. The reliability of these tests in determining the cognitive level of the individuals is quite considerable [24]. The instant recall test (IRT) and delayed recall test (DRT) are part of the CERAD word learning tests utilized to evaluate the immediate and delayed acquisition of new linguistic material. AFT was used to measure executive function, wherein participants had 60 s to identify as many animals as they could, with the total number of adequately identified animals counting toward the final score. Finally, the researchers employed the DSST to evaluate reaction speed, sustained attention, and working memory among participants, with the total number of adequately matched numbers and symbols serving as the score. There is currently no golden standard for determining poor cognitive performance on the four previous cognitive tests, with higher scores on all tests indicating superior cognitive performance.

2.4. Covariates

In addition to the five serum heavy metals described above, we investigated a number of potentially confounding variables, including age (60–69 years; 70–79 years; 80+ years), sex (male; female), race/ethnicity (non-Hispanic White; non-Hispanic Black; other Hispanic; other/Multi-racial; Mexican; Other Race, Including Multi-Racial), education level (less than 9th grade; 9–11th grade; high school grad/GED; some college or AA degree; college graduate or above), smoking status (never smoker; current smoker; former smoker), alcohol intake (1–5 drinks/month; 5–10 drinks/month; 10+ drinks/month; non-drinker).

2.5. Statistical Analysis

The data analysis in this study was performed with R (version 4.2.2, R Core Team, Vienna, Austria). Categorical variables are expressed as the number of instances (n) and frequency (%), whereas non-normally distributed continuous variables are expressed as the median (IQR = $Q_{75} - Q_{25}$). For group comparisons of data with a normal distribution, we utilize Student's *t*-test, or for group comparisons of skewed variables, we used the Wilcoxon rank sum test. The chi-square test was performed to examine the variations in rates for categorical variables amongst groups. $p < 0.05$ for a two-sided test was considered statistically significant. In this investigation, a subgroup analysis was conducted to investigate the sex differences in the relationship between serum heavy metals and cognitive function in older persons, taking into account the substantial variations between the four cognitive tests in the sex stratification. In addition, Pearson correlation analysis was employed to examine relationships between blood heavy metals.

2.5.1. Statistical Model 1: Generalized Linear Regression Model (GLM)

During the initial stage, this research employed weighted generalized linear regression models in a complex sampling environment to investigate the connection between four cognitive scores and serum heavy metal concentrations separately, adjusting for relevant demographic and behavioral confounders. Serum heavy metals were included in the analysis of generalized linear regression models with continuous and categorical variables due to the likelihood of nonlinear correlation between serum heavy metals and outcome. Dominance ratios and related 95% confidence intervals summarize the statistical results. Model was adjusted for age, sex, race/ethnicity, education level, smoking status, and alcohol intake.

2.5.2. Statistical Model 2: Bayesian Kernel Machine Regression (BKMR) Model

The combined effects and potential interactions between serum heavy metals and cognitive function were then examined via BKMR statistical modeling [22,25]. The nonlinear relationship between exposure and outcome was investigated in this research using exposure–response cross-sections for a single variable and outcome while other variables were held constant at the median. Bivariate exposure–response profiles represent how mixture compositions interact, which could be understood as potential interactions between the slope of the curve for one chemical at the 10th, 50th, and 90th modifications of another chemical (with the remaining variables fixed at the median). The association plot of the overall effect of the mixture with outcome shows the change in estimated outcome when all exposure variables are set at different percentiles simultaneously compared to when they are fixed at the median. Using the Markov Chain Monte Carlo method, iteration was set at 30,000.

2.5.3. Statistical Model 3: Weighted Quantile Sum (WQS) Regression Model

The cumulative impact of metal mixture components on cognitive function was estimated by employing WQS regression. The WQS statistical model for multiple regression in high-dimensional data sets calculates the effects of all exposure factors on outcomes by constructing a weighted index and determining whether that index is related to outcomes [26]. The relative intensity of the weights given to each variable by the model allows the researcher to subsequently evaluate the contribution of each environmental chemical to the overall index impact, allowing for the identification of significant substances in the mixture.

2.5.4. Statistical Model 4: Quantile g-Computation (Qgcomp) Regression Model

We further employed the Qgcomp model to overcome the limitations of the WQS regression model on the direction of association. The G-computation procedure has some advantages relative to traditional regression, including the decoupling of confounding adjustment and effect estimation and the causal parameter interpretation [27]. Qgcomp combines the inferential simplicity of weighted quantile sum regression with the flexibility of g-computation without the requirement of homogeneity assumption and the linearity and additivity of exposure. Qgcomp is a straightforward and computationally efficient method for estimating the association between a combination of exposures and the desired health outcome. Qgcomp can be used to consistently estimate effects of the exposure mixture in settings in which WQS regression may be biased or inconsistent but also yield equivalent estimates with WQS regression in large samples when its assumptions hold [28]. Using the qgcomp.noboot function, a linear model of cognitive function was fitted to evaluate the total effect through allocating positive or negative weighting indices to each blood heavy metal by segmenting each metal into quartiles. In order to determine the mixing effect's linearity and display it using a g computation to show the mixed effect, the qgcomp.boot function (R package, "qgcomp") was used.

3. Results

3.1. Characteristics of the Study Participants

This study enrolled 1833 eligible participants, including 883 men and 950 women. As shown in Table 2, the results of the sex subgroup analysis differed in terms of age, education, smoking, and alcohol consumption. Additionally, males were far more inclined to smoke and consume alcohol than women. There were significant sex differences on three of the four cognitive assessments, with males achieving lower cognitive scores than females (IRT, DRT, DSST). All five serum heavy metal concentrations exhibited a skewed distribution, as presented in Table 1 of the statistical description of serum heavy metal concentrations. Analysis of correlation revealed no correlation between serum heavy metal concentrations included in the study (Figures S1–S3).

Table 2. Characteristics of the study population.

Characteristic	Overall, N = 1833 (100%) [1]	Male, N = 883 (45%) [1]	Female, N = 950 (55%) [1]	*p*-Value [2]
Age				0.031
60–69 years	921 (51%)	460 (55%)	461 (48%)	
70–79 years	516 (29%)	247 (28%)	269 (29%)	
80+ years	396 (20%)	176 (17%)	220 (23%)	
Race/ethnicity				0.200
Non-Hispanic White	862 (80%)	389 (80%)	473 (81%)	
Non-Hispanic Black	436 (7.8%)	220 (7.2%)	216 (8.4%)	
Other Hispanic	203 (3.9%)	100 (4.0%)	103 (3.9%)	
Other/Multi-Racial	166 (3.5%)	82 (3.4%)	84 (3.5%)	
Mexican American	138 (2.9%)	74 (3.2%)	64 (2.6%)	
Other Race, Including Multi-Racial	28 (1.7%)	18 (2.2%)	10 (0.6%)	
Education				<0.001
Less Than 9th Grade	219 (6.0%)	124 (7.0%)	95 (5.1%)	
9–11th Grade	246 (9.4%)	116 (9.4%)	130 (9.3%)	
High School Grad/GED	421 (22%)	188 (18%)	233 (25%)	
Some College or AA degree	514 (31%)	222 (27%)	292 (34%)	
College Graduate or above	433 (32%)	233 (39%)	200 (27%)	
Smoking status				<0.001
Never smoker	894 (48%)	297 (34%)	597 (60%)	
Current smoker	234 (11%)	147 (13%)	87 (8.3%)	
Former smoker	705 (41%)	439 (53%)	266 (31%)	
Alcohol intake				<0.001
1–5 drinks/month	864 (44%)	497 (49%)	367 (40%)	
5–10 drinks/month	89 (6.5%)	51 (6.9%)	38 (6.1%)	
10+ drinks/month	307 (23%)	198 (31%)	109 (16%)	
Non-drinker	573 (27%)	137 (13%)	436 (38%)	
IRT	20.0 (17.0, 23.0)	19.0 (16.0, 22.0)	21.0 (17.0, 23.0)	<0.001
DRT	6.00 (5.00, 8.00)	6.00 (4.00, 7.00)	7.00 (5.00, 8.00)	<0.001
AFT	18.0 (14.0, 22.0)	19.0 (14.0, 22.0)	18.0 (14.0, 22.0)	0.100
DSST	54 (42, 65)	50 (40, 62)	56 (43, 67)	<0.001

Notes: [1] n (unweighted) (weighted%); Median (IQR). [2] chi-squared tests with Rao and Scott's second-order correction; Wilcoxon rank-sum test for complex survey samples.

3.2. Single Metal Exposures and Cognitive Function

According to the results provided by Tables S1–S6, a correlation existed between blood selenium and cognition. Blood selenium at Q2 equates was found to have the most positive effect on cognitive function according to the analysis of GLM based on quartiles of exposure variables. Additionally, blood lead was negatively correlated with DSST scores across all variables ((β(95% CI): -0.52 (-0.93, -0.11), $p < 0.01$); blood cadmium was negatively associated with IRT ((β(95% CI): −0.73 (−1.50, −0.03), $p = 0.03$); blood cadmium was negatively correlated with DSST (β(95% CI): −3.00 (−5.10, −0.99), $p = 0.03$); blood manganese was negatively correlated with AFT ((β(95% CI): −0.05 (−0.11, −0.01), $p = 0.04$).

3.3. Multi-Metal Exposures and Cognitive Function

3.3.1. Multi-Metal Exposures and Cognitive Function: BKMR Model

We examined the relationship between five blood heavy metal co-exposures and cognitive function through the BKMR model (Figures 2, 3, S4 and S5). The results of four cognitive assessments were positively correlated with blood heavy metal co-exposure among the male participants. The general impact of blood heavy metal in the female subset was positively correlated with DRT and DSST scores but negatively correlated with AFT. Other blood heavy metals were set at their 50th percentile exposure amounts to evaluate unilateral impacts in univariate exposure–response functions (Figures S6 and S7; Tables 3 and S7–S9). Blood selenium was found to be a significant component for improved cognitive performance in the study, and its association with cognitive performance was more remarkable in males (IRT: 0.99; DRT: 0.71; AFT: 0.81; DSST: 0.99) than in women (IRT: 0.56; DRT: 0.63; AFT: 0.25; DSST: 0.71). Additionally, there was a difference between male and female groups in the negative association between blood lead, blood cadmium, and blood manganese and cognitive capacity. For instance, the DRT scores of blood manganese among males showed an "inverted U-shaped" curve; however, the relationship with blood manganese in females was DSST. The DSST scores of blood lead and females exhibited an "inverted U-shaped" curve, yet the connection was falling in males. Notably, when the other five metals were set at the 10th, 50th, and 90th percentiles, blood manganese and blood selenium may have possible associations with other metal concentrations (Figures S8–S10).

Table 3. Summary results from BKMR and Qgcomp analysis in the whole population.

Variable	BKMR PIP				Qgcomp			
	IRT	DRT	AFT	DSST	IRT	DRT	AFT	DSST
Blood Lead	0.15	0.29	0.26	0.49	−0.42	−0.37	0.42	−0.12
Blood Cadmium	0.36	0.37	0.33	0.83	−0.58	−0.63	−0.58	−0.88
Blood Mercury	0.05	0.17	0.06	0.21	0.21	0.08	−0.14	0.02
Blood Selenium	0.99	0.94	0.47	1.00	0.78	0.52	0.58	0.57
Blood Manganese	0.21	0.75	0.25	0.93	0.054	0.40	−0.28	0.41

Models adjusted for gender, age, race/ethnicity, education level, alcohol intake, and smoking status.

3.3.2. Multi-Metal Exposures and Cognitive Function: WQS Model

We initially examined the combined effect of serum heavy metals on cognitive function. As observed in Figure 4, the IRT, AFT, and DSST are three of the four cognitive tests where men had more dramatic positive or negative results. Likewise, the beneficial effects of five blood heavy metals on cognitive performance in males were further supported by favorable WQS model analysis results (IRT: 3.00 (0.01, 6.00); DRT: 1.49 (0.24, 2.74); DSST: 13.80 (4.61, 22.99). No statistically meaningful variations were found in the negative WQS model. According to the weighing study of all the demographic factors, blood selenium had the most significant protective impact on brain performance, while blood cadmium and blood lead had the opposite effects (Figure 5, Table 4). Similar tendencies were observed in the sex subgroups. (Figures S11 and S12, Tables S10–S12). Remarkably, blood lead produced a protective effect in the female AFT test results but not blood selenium.

Table 4. Summary results from WQS analysis in the whole population.

Variable	IRT Positive	IRT Negative	DRT Positive	DRT Negative	AFT Positive	AFT Negative	DSST Positive	DSST Negative
Blood Selenium	0.94	0.00	0.68	0.02	0.80	0.00	0.82	0.00
Blood Manganese	0.01	0.34	0.25	0.02	0.04	0.17	0.14	0.01
Blood Mercury	0.04	0.02	0.04	0.07	0.08	0.04	0.03	0.07
Blood Lead	0.00	0.38	0.01	0.41	0.06	0.12	0.00	0.40
Blood Cadmium	0.00	0.25	0.02	0.48	0.02	0.66	0.00	0.52

Models adjusted for gender, age, race/ethnicity, education level, alcohol intake, and smoking status.

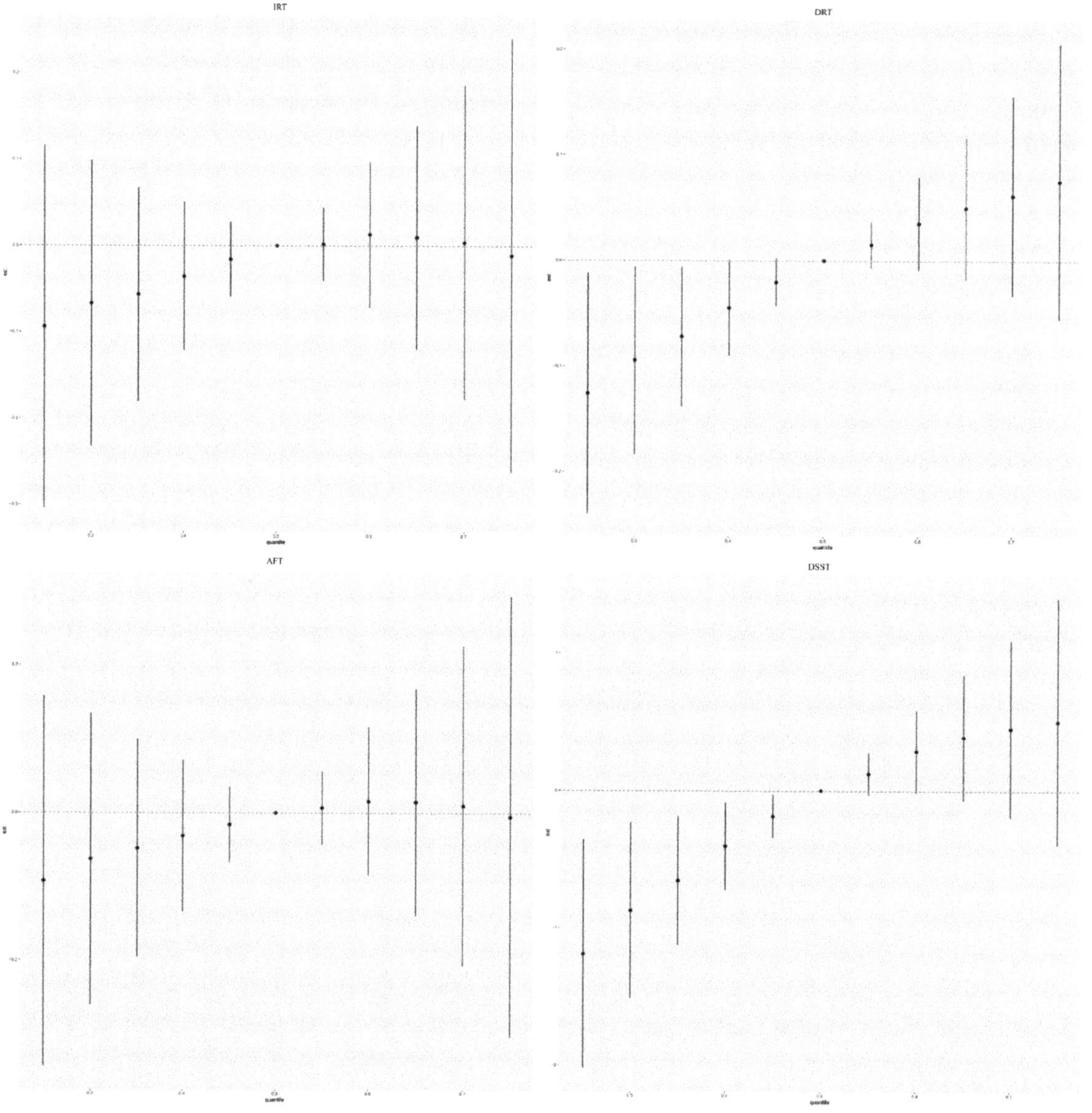

Figure 2. Combined effects of the metal as a mixture on cognitive function in elderly people.

Figure 3. Univariate exposure–response functions and 95% confidence interval for each heavy metal with the other metals fixed at the median in elderly people.

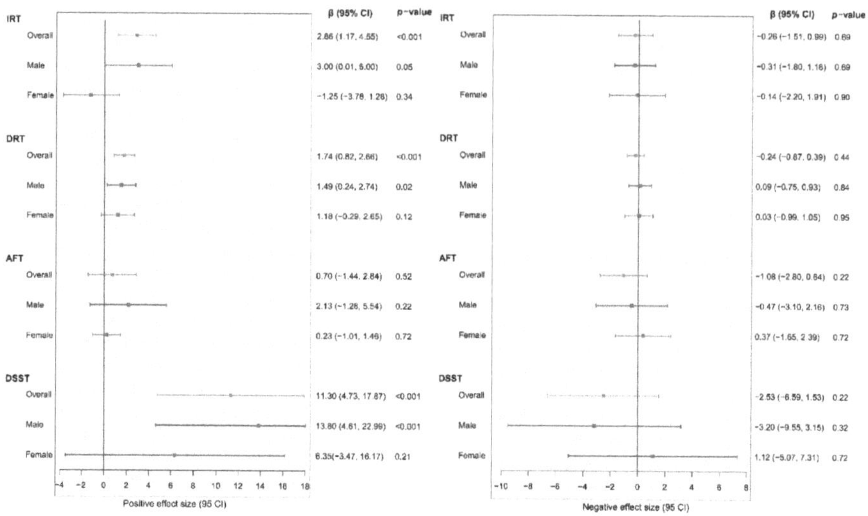

Figure 4. Associations between blood heavy metals and cognitive function by WQS regression model in NHANES 2011–2014. Cyan: whole population; Red: male; Blue: female.

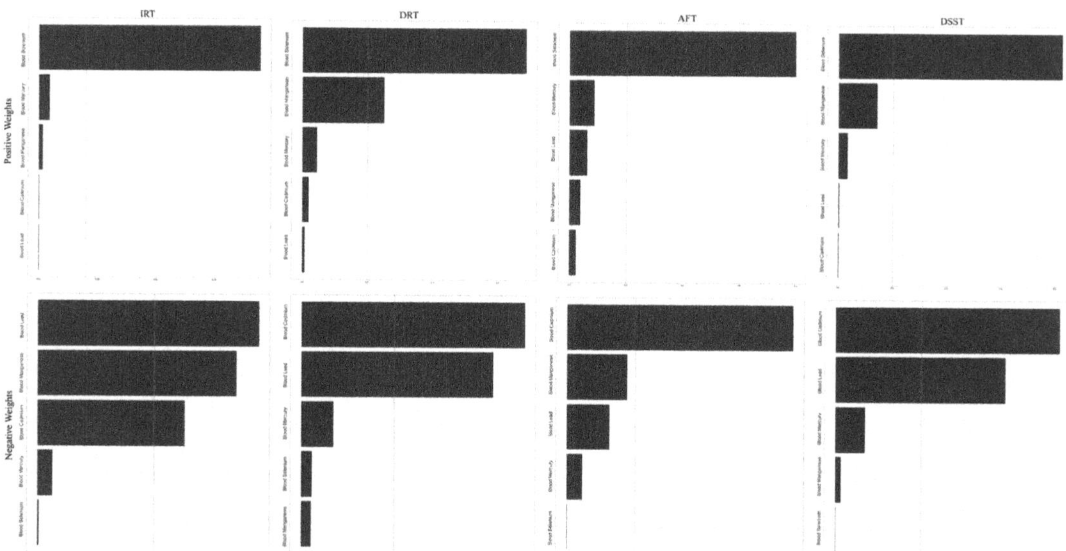

Figure 5. WQS model regression index weights for blood heavy metals and cognitive function in elderly people. Models were adjusted by age, race/ethnicity, education level, alcohol intake, and smoking status.

3.3.3. Multi-Metal Exposures and Cognitive Function: Qgcomp Model

Compared to the WQS model, the Qgcomp model allows for direct evaluation of the effects of different input variables on the dependent variable as it refrains from making assertions about the combined impacts before testing. The findings of Figure S13 and Table S13 demonstrate that in the whole population, the cumulative effects of blood heavy metal levels exhibited a favorable tendency with DRT and DSST. In the male elderly population, a positive trend within blood heavy metal levels was connected with DRT and AFT; in the female population, a negative trend was associated with IRT and AFT; instead, a positive trend was associated with DSST. Blood cadmium had a significant negative impact on the DRT test findings (DRT: −0.56), but blood selenium had the most important positive effect in the IRT, AFT, and DSST tests in the male group (IRT: 0.95; AFT: 0.70; DSST: 0.83) (Figures S14–S17; Table S14). Results for the female group revealed that blood cadmium had the greatest negative effects in IRT, DRT, and DSST (IRT: −0.66; DRT: −0.78; DSST: 1.00), while a substantial positive impact of blood lead (AFT: 1.00) could be identified in the AFT test (Figure S17, Table S15). The combined impact of serum heavy metals in the evidence for the entire community was similar to the earlier findings (positive: blood selenium; negative: blood cadmium) (Figure 6, Table S15).

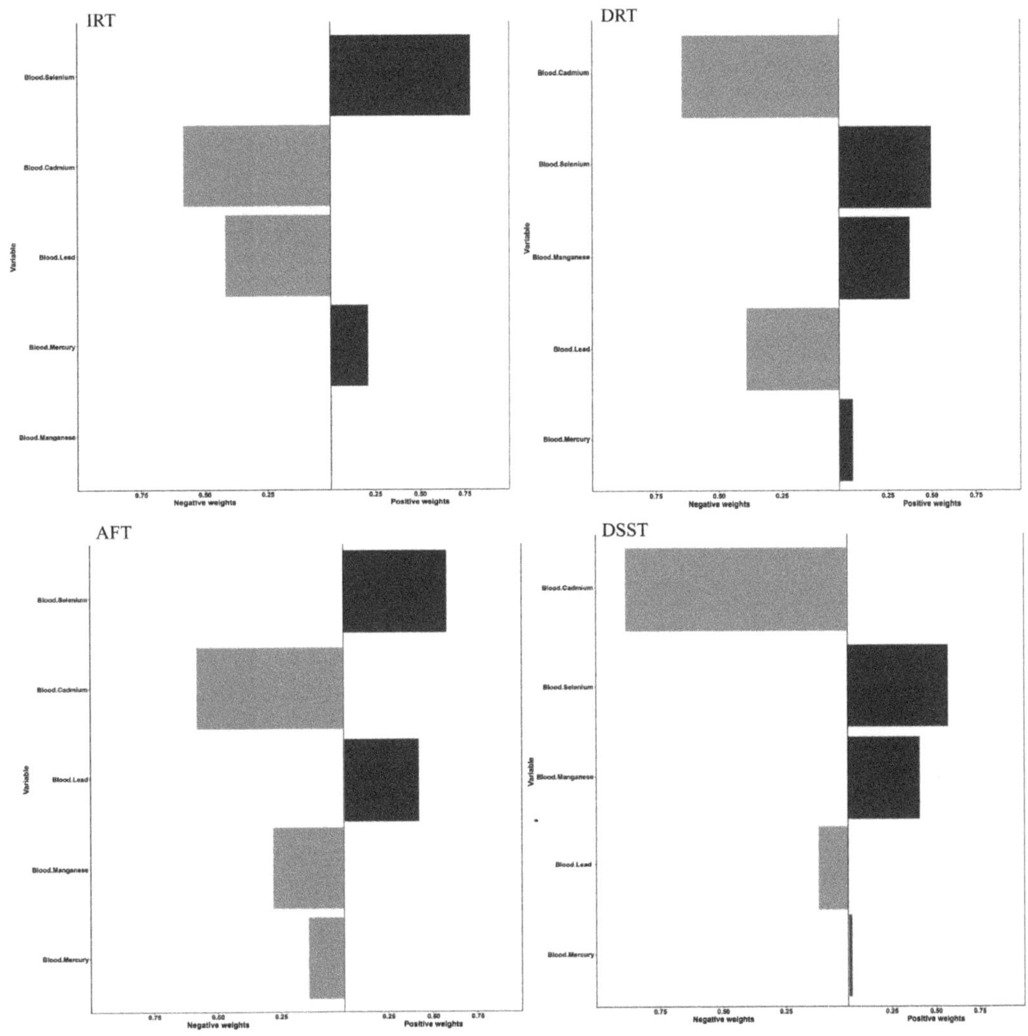

Figure 6. Qgcomp model regression index weights for blood heavy metals and four cognitive tests in older people. The Qgcomp models were adjusted by sex, age, race/ethnicity, education level, alcohol intake, and smoking status.

4. Discussion

Our study attempted to determine whether there is a relationship between cognitive function and blood heavy metals levels in older people and whether there are sex differences within this association. In this investigation, the GLM, BKMR, WQS, and Qgcomp models were utilized to evaluate the complex effects of five serum heavy metals on cognitive function. We identified a positive correlation between selenium levels in the blood and cognitive function in the elderly. Cadmium and lead were significantly and negatively associated with cognitive function. Men, on average, had weaker cognitive skills than women, according to studies that considered age into view. Moreover, both in the positive and negative related directions of the study, older male participants demonstrated a more marked reaction in cognitive capacity to blood heavy metals. Our main results imply that the negative correlation of other heavy metals with cognitive function may be reduced or even reversed by selenium. This study emphasizes the significance of determining the typical effects of metals in blood on cognitive function using various statistical techniques and contrasting the outcomes while considering the advantages and limitations of multiple methods.

Organs prone to metal ion enrichment and elevated metabolic activity include the nervous system in general and the brain in specific. Nerve deterioration and reactive stress may result from disturbances in metal balance [29]. This study discovered that in both monometallic and polymetallic models, selenium was favorably linked with brain performance. Due to its distinct neurophysiological characteristics and beneficial qualities, selenium is generally considered necessary [12,30]. The redox activity of metal elements and reactive stress constitute important biochemical signaling pathways for cognitive function [31]. Selenium, in the shape of selenoproteins, plays an assortment of roles in normal metabolic processes and metabolism. Glutathione peroxidase, a typical selenoprotein, has an antioxidant impact and may protect against cellular damage from reactive oxygen species [12]. Food sources are a substantial contribution to blood heavy metal levels [5], and dietary patterns have a considerable impact on serum elemental metal concentrations [28]. Through the correction of metabolic imbalances and the reduction of inflammation and oxidative stress, moderate selenium supplementation may enhance cognitive performance [32]. Notably, although the association of blood selenium on cognitive function was robust in this study, the strongest positive association was observed at moderate levels. Consideration should also be given to the health concerns associated with consuming too many foods high in selenium [33,34].

Concerns have been expressed with the increasing proportion of older individuals in the global population and their potential increased susceptibility to multiple metal exposures due to physiological factors [35]. According to this study, blood levels of lead and cadmium are negatively correlated with cognitive ability in the elderly. The same trends have been discovered in health investigations conducted in other areas [10,11]. People are continuously subjected to numerous elements at the exact moment rather than just one. Metals can have varying health impacts in situations of mixed exposure due to combined or adverse interactions [34,36]. Our core findings suggest that the protective effect of selenium might mitigate or even reverse the negative association between other heavy metals and cognitive function. All elements of the growth, operation, aging, and illness of the central nervous system depend on redox balance, and an imbalance causes neurodegeneration [31]. Redox homeostasis is readily impacted by abnormalities in metal homeostasis. Reduced selenium levels result in neural malfunction, which has been linked to gender in animal research [37]. Testes may make males more susceptible to the physiological effects of selenium antioxidants, offering neuroprotection [37]. According to several studies, oxidative stress caused by heavy metals like lead and cadmium causes a variety of physiological, biochemical, and behavioral dysfunctions in people [29,38,39]. This may be one of the possible mechanisms by which selenium counteracts the cognitive impairment associated with other heavy metals. We are, to our understanding, one of the few studies to have examined the impact of lead and cadmium, as well as selenium alone, on older people's cognitive function. For the variables affecting this joint result to be confirmed, more in vivo research is required.

Subgroup studies offered distinctive perspectives on metal exposure in various groups. According to the results of the mixed exposure study, males and females may be affected differently by exposure to mixed metals; a more excellent relationship (positive or negative) between mixed metals and cognitive performance in males was also shown. The disparities in the bodily loads of metal buildup between men and women may explain these sex-specific correlations [40]. Selenium metabolism and expression of selenoproteins are sexually dimorphic [30]. The liver and kidney of male and female rats differ in producing the proteins selenoprotein GPx1, selenoprotein P, and iodothyronine deiodinase 1, with females producing more selenoprotein when given identical amounts of selenium [30]. Males have a more significant load of heavy metals and are more vulnerable to the deleterious impacts of heavy metals, including impaired cognitive development, focus impairments, and behavioral issues, according to epidemiological and laboratory research [40,41]. Neurotoxic effects that are particular to one sex may be caused by sex variations impacting these processes, such as differences in accumulation, antioxidant capacity, nutritional needs, sex hormones, and methylation/gene expression. There are several interrelated variables that impact metal-related neurotoxicity, ranging from molecular, hormonal, and epigenetic processes influencing gene regulation, expression, and function to sexually dimorphic changes in absorption and metabolism [14,17,40,42]. The relationship between mental and physical makeup may also partially contribute to sex variations in the cumulative impacts of heavy metals [15]. In addition, additional hypotheses, such as those involving endocrine, genetic, biochemical, structural, and environmental variables, may explain the sex-based unequal vulnerability [16,43].

Previous research on the effects of blood heavy metals on cognitive function has only used single pollutant models, which may result in biased impact evaluations due to the neglect of mixed effects [9,18]. Therefore, it is crucial to employ specialized techniques to investigate the mixed implications of multiple pollutants. Determining which statistical method is most appropriate for this research is challenging as there is no a priori information. To investigate the impact of serum heavy metals on cognitive function in the elderly, this study employed a variety of statistical models, including the Bayesian kernel machine regression (BKMR) model, the weighted quantile sum (WQS) regression model, and the Quantile g-computation (Qgcomp) regression model [34,44]. The findings will vary depending on particular approaches because these statistical techniques place varying emphasis on handling distinct statistical characteristics (high dimensionality, multicollinearity, interaction, and nonlinear effects). Combining different statistical approaches can help provide reasonably comprehensive information and avoids the one-sidedness of a single approach [45]. Nevertheless, additional study is required to address mixture and interaction effects using statistical methods and to clarify biological processes due to the intricacy of the real environmental effects of multiple metals.

The application of multiple statistical models in this study, which enabled us to thoroughly evaluate the relationship between metal mixtures and individual metals with cognitive abilities and ensure the validity of our results, is one of its major advantages. Additionally, we determined the weight between each metal pair metal mixture and cognitive ability, which was infrequently evaluated in earlier research. Moreover, NHANES is broadly representative, so our findings could be applied to various American groups. However, some limitations should be acknowledged. The data used in this research originates from a cross-sectional survey. There is no longitudinal follow-up of cognitive status nor of heavy metals concentrations, which means that a causal relationship cannot be envisaged, as there may be reverse causality. Additionally, a single measurement might not accurately reflect exposure over the day because each metal has a distinct distribution and half-life in the human body. The physiological operation of the brain depends on the homeostasis of metal ions [38]; through a variety of biochemical mechanisms, trace mineral components, including iron, copper, zinc, and manganese, may have an impact on brain health [43]. Base metals such as copper, zinc, and iron were not included in this study, which limits the usable scope of the findings. The chance of unmeasured factors distorting the outcomes cannot be ruled out, although important predictors were included in our model. The chance of unmeasured factors distorting the outcomes cannot be ruled out, such as covariates vascular risk factors, although important predictors were included in our model. Although we adjusted for potential confounders to the best of our ability, residual confounders are inevitable in the observation environment, including age, education, lifestyle habits, etc. BKMR, WQS, and Qgcomp do not currently have weight-based statistical analyses, which may impact the validity of the study results.

5. Conclusions

In conclusion, we discovered that blood selenium was positively correlated with cognitive performance in older people. In contrast, levels of cadmium and lead in the blood were negatively associated with cognitive ability. Selenium might partially mitigate or even reverse the negative correlation of heavy metal mixtures on cognitive function. In addition, a significant result of our research was that males performed cognitively, on average, worse than women. It appears that males are more susceptible to the impacts of exposure to mixed metals, either positively or unfavorably, and the relationship between mixed metals and cognitive function has been demonstrated to be stronger in men. These findings need to be confirmed by additional cohort studies, nevertheless, given the cross-sectional approach of our research.

Supplementary Materials: The following supporting information can be downloaded at https://www.mdpi.com/article/10.3390/nu15132874/s1. Table S1. Associations of single blood heavy metals and cognitive function on elderly. Table S2. Associations of single blood heavy metals and cognitive function on male elderly. Table S3. Associations of single blood heavy metals and cognitive function on female elderly. Table S4. Associations of single blood heavy metals (quartile conversion) and cognitive function on elderly. Table S5. Associations of single blood heavy metals (quartile conversion) and cognitive function on male elderly. Table S6. Associations of single blood heavy metals (quartile conversion) and cognitive function on female elderly. Table S7. Summary results from BKMR analysis in the whole population. Table S8. Summary results from BKMR analysis in males. Table S9. Summary results from BKMR analysis in females. Table S10. Summary results from WQS analysis in the whole population. Table S11. Summary results from WQS analysis in males. Table S12. Summary results from WQS analysis in females. Table S13. Summary results from Qgcomp analysis in the whole population. Table S14. Summary results from Qgcomp analysis in males. Table S15. Summary results from Qgcomp analysis in females. Figure S1. Spearman correlation plot of concentrations of individual metals in older people. Figure S2. Spearman correlation plot of concentrations of individual metals in males. Figure S3. Spearman correlation plot of concentrations of individual metals in females. Figure S4. Combined effects of the metal as a mixture on cognitive function in males. Figure S5. Combined effects of the metal as a mixture on cognitive function in females. Figure S6. Univariate exposure–response functions and 95% confidence interval for each heavy metal with the other metals fixed at the median in males. Models adjusted for age, race/ethnicity, education level, alcohol intake, and smoking status. Figure S7. Univariate exposure–response functions and 95% confidence interval for each heavy metal with the other metals fixed at the median in females. Models adjusted for age, race/ethnicity, education level, alcohol intake, and smoking status. Figure S8. Estimated effects (95% CIs) of single aldehydes on cognitive function of elderly people when the levels of other aldehydes were fixed at 25th, 50th, and 75th. Models adjusted for sex, age, race/ethnicity, education level, alcohol intake, and smoking status. CI, confidence interval. Figure S9. Estimated effects (95% CIs) of single aldehyde on cognitive function of males when the levels of other aldehydes were fixed at 25th, 50th, and 75th. Models adjusted for age, race/ethnicity, education level, alcohol intake, and smoking status. CI, confidence interval. Figure S10. Estimated effects (95% CIs) of single aldehydes on cognitive function of females when the levels of other aldehydes were fixed at 25th, 50th, and 75th. Models adjusted for age, race/ethnicity, education level, alcohol intake, and smoking status. CI, confidence interval. Figure S11. WQS model regression index weights for a total of four cognitive tests in males. The WQS models were adjusted by age, race/ethnicity, education level, alcohol intake, and smoking status. Figure S12. WQS model regression index weights for a total of four cognitive tests in females. The WQS models were adjusted by age, race/ethnicity, education level, alcohol intake, and smoking status. Figure S13. Qgcomp model regression joint effect (95% CI) for blood heavy metals and four cognitive tests in older people. The Qgcomp models were adjusted by sex, age, race/ethnicity, education level, alcohol intake, and smoking status. Figure S14. Qgcomp model regression joint effect (95% CI) for blood heavy metals and four cognitive tests in males. The Qgcomp models were adjusted by age, race/ethnicity, education level, alcohol intake, and smoking status. Figure S15. Qgcomp model regression joint effect (95% CI) for blood heavy metals and four cognitive tests in females. The Qgcomp models were adjusted by age, race/ethnicity, education level, alcohol intake, and smoking status. Figure S16. Qgcomp model regression index weights for blood heavy metals and four cognitive tests in males. The Qgcomp models were adjusted by age, race/ethnicity, education level, alcohol intake, and smoking status. Figure S17. Qgcomp model regression index weights for blood heavy metals and four cognitive tests in females. The Qgcomp models were adjusted by age, race/ethnicity, education level, alcohol intake, and smoking status.

Author Contributions: Conceptualization, X.L.; data curation, S.S., N.L. and Y.W.; funding acquisition, X.L.; methodology, S.S.; project administration, X.L.; software, G.W.; supervision, X.Z. (Xin Zhao), H.C. and Z.Y.; validation, G.W. and Y.W.; visualization, X.Z. (Xiaoan Zhang) and Z.Y.; writing—original draft, S.S. and N.L.; writing—review and editing, X.Z. (Xiaoan Zhang) and X.Z. (Xin Zhao). All authors have read and agreed to the published version of the manuscript.

Funding: This work was supported by the Key Project of Science and Technology of Henan province [No. SBGJ202102014], Major Projects of Collaborative Innovation of Zhengzhou [No. 18XTZX12009], Special Major Public Welfare Project of Henan Province No. 201300310800], and the Key Project of Science and Technology of Henan province [No. 222102310418].

Institutional Review Board Statement: Not applicable.

Informed Consent Statement: Informed consent was obtained from all subjects involved in the study.

Data Availability Statement: The data that support the findings of this study are openly available in NHANES at https://wwwn.cdc.gov/Nchs/Nhvanes/ (accessed on 1 February 2023). Further inquiries can be directed to the corresponding author.

Conflicts of Interest: The authors declare no conflict of interest.

References

1. GBD 2017 US Neurological Disorders Collaborators; Feigin, V.L.; Vos, T.; Alahdab, F.; Amit, A.M.L.; Barnighausen, T.W.; Beghi, E.; Beheshti, M.; Chavan, P.P.; Criqui, M.H.; et al. Burden of Neurological Disorders Across the US From 1990–2017: A Global Burden of Disease Study. *JAMA Neurol.* **2021**, *78*, 165–176. [CrossRef] [PubMed]
2. Song, S.; Stern, Y.; Gu, Y. Modifiable lifestyle factors and cognitive reserve: A systematic review of current evidence. *Ageing Res. Rev.* **2022**, *74*, 101551. [CrossRef] [PubMed]
3. Xiao, L.; Zan, G.; Qin, J.; Wei, X.; Lu, G.; Li, X.; Zhang, H.; Zou, Y.; Yang, L.; He, M.; et al. Combined exposure to multiple metals and cognitive function in older adults. *Ecotoxicol. Environ. Saf.* **2021**, *222*, 112465. [CrossRef] [PubMed]
4. Carlin, D.J.; Rider, C.V.; Woychik, R.; Birnbaum, L.S. Unraveling the health effects of environmental mixtures: An NIEHS priority. *Environ. Health Perspect.* **2013**, *121*, A6–A8. [CrossRef]
5. Clemens, S.; Ma, J.F. Toxic Heavy Metal and Metalloid Accumulation in Crop Plants and Foods. *Annu. Rev. Plant Biol.* **2016**, *67*, 489–512. [CrossRef]
6. Rehman, K.; Fatima, F.; Waheed, I.; Akash, M.S.H. Prevalence of exposure of heavy metals and their impact on health consequences. *J. Cell. Biochem.* **2018**, *119*, 157–184. [CrossRef]
7. Farina, M.; Rocha, J.B.; Aschner, M. Mechanisms of methylmercury-induced neurotoxicity: Evidence from experimental studies. *Life Sci.* **2011**, *89*, 555–563. [CrossRef]
8. Jenkitkasemwong, S.; Akinyode, A.; Paulus, E.; Weiskirchen, R.; Hojyo, S.; Fukada, T.; Giraldo, G.; Schrier, J.; Garcia, A.; Janus, C.; et al. SLC39A14 deficiency alters manganese homeostasis and excretion resulting in brain manganese accumulation and motor deficits in mice. *Proc. Natl. Acad. Sci. USA* **2018**, *115*, E1769–E1778. [CrossRef]
9. Lu, K.; Liu, T.; Wu, X.; Zhong, J.; Ou, Z.; Wu, W. Association between Serum Iron, Blood Lead, Cadmium, Mercury, Selenium, Manganese and Low Cognitive Performance in Old Adults from National Health and Nutrition Examination Survey (NHANES): A Cross-Sectional Study. *Br. J. Nutr.* **2023**, 1–11. [CrossRef]
10. Vlasak, T.; Jordakieva, G.; Gnambs, T.; Augner, C.; Crevenna, R.; Winker, R.; Barth, A. Blood lead levels and cognitive functioning: A meta-analysis. *Sci. Total Environ.* **2019**, *668*, 678–684. [CrossRef]
11. Liu, H.; Su, L.; Chen, X.; Wang, S.; Cheng, Y.; Lin, S.; Ding, L.; Liu, J.; Chen, C.; Unverzagt, F.W.; et al. Higher blood cadmium level is associated with greater cognitive decline in rural Chinese adults aged 65 or older. *Sci. Total Environ.* **2021**, *756*, 144072. [CrossRef]

12. Rita Cardoso, B.; Silva Bandeira, V.; Jacob-Filho, W.; Franciscato Cozzolino, S.M. Selenium status in elderly: Relation to cognitive decline. *J. Trace Elem. Med. Biol.* **2014**, *28*, 422–426. [CrossRef]
13. Shang, N.; Zhang, L.; Wang, S.; Huang, T.; Wang, Y.; Gao, X.; Xu, S.; Zhang, J.; Zhang, L.; Niu, Q.; et al. Increased aluminum and lithium and decreased zinc levels in plasma is related to cognitive impairment in workers at an aluminum factory in China: A cross-sectional study. *Ecotoxicol. Environ. Saf.* **2021**, *214*, 112110. [CrossRef]
14. Karalija, N.; Papenberg, G.; Wahlin, A.; Johansson, J.; Andersson, M.; Axelsson, J.; Riklund, K.; Lindenberger, U.; Nyberg, L.; Backman, L. Sex differences in dopamine integrity and brain structure among healthy older adults: Relationships to episodic memory. *Neurobiol. Aging* **2021**, *105*, 272–279. [CrossRef]
15. Langeard, A.; Fakrahnak, Z.; Vrinceanu, T.; Noriega de la Colina, A.; Pothier, K.; Berryman, N.; Vu, T.T.M.; Girouard, H.; Karelis, A.D.; Bherer, L. Sex-moderated association between body composition and cognition in older adults. *Exp. Gerontol.* **2020**, *138*, 111002. [CrossRef]
16. Ferrer-Quintero, M.; Green, M.F.; Horan, W.P.; Penn, D.L.; Kern, R.S.; Lee, J. The effect of sex on social cognition and functioning in schizophrenia. *NPJ Schizophr.* **2021**, *7*, 57. [CrossRef]
17. Subramaniapillai, S.; Almey, A.; Natasha Rajah, M.; Einstein, G. Sex and gender differences in cognitive and brain reserve: Implications for Alzheimer's disease in women. *Front. Neuroendocrinol.* **2021**, *60*, 100879. [CrossRef]
18. Sasaki, N.; Carpenter, D.O. Associations between Metal Exposures and Cognitive Function in American Older Adults. *Int. J. Environ. Res. Public Health* **2022**, *19*, 2327. [CrossRef]
19. Saxena, R.; Gamble, M.; Wasserman, G.A.; Liu, X.; Parvez, F.; Navas-Acien, A.; Islam, T.; Factor-Litvak, P.; Uddin, M.N.; Kioumourtzoglou, M.A.; et al. Mixed metals exposure and cognitive function in Bangladeshi adolescents. *Ecotoxicol. Environ. Saf.* **2022**, *232*, 113229. [CrossRef]
20. Li, C.; Xia, W.; Jiang, Y.; Liu, W.; Zhang, B.; Xu, S.; Li, Y. Low level prenatal exposure to a mixture of Sr, Se and Mn and neurocognitive development of 2-year-old children. *Sci. Total Environ.* **2020**, *735*, 139403. [CrossRef]
21. Valeri, L.; Mazumdar, M.M.; Bobb, J.F.; Claus Henn, B.; Rodrigues, E.; Sharif, O.I.A.; Kile, M.L.; Quamruzzaman, Q.; Afroz, S.; Golam, M.; et al. The Joint Effect of Prenatal Exposure to Metal Mixtures on Neurodevelopmental Outcomes at 20-40 Months of Age: Evidence from Rural Bangladesh. *Environ. Health Perspect.* **2017**, *125*, 067015. [CrossRef] [PubMed]
22. Bobb, J.F.; Valeri, L.; Claus Henn, B.; Christiani, D.C.; Wright, R.O.; Mazumdar, M.; Godleski, J.J.; Coull, B.A. Bayesian kernel machine regression for estimating the health effects of multi-pollutant mixtures. *Biostatistics* **2015**, *16*, 493–508. [CrossRef] [PubMed]
23. CDC. National Health and Nutrition Examination Survey. Available online: https://www.cdc.gov/nchs/nhanes/ (accessed on 1 February 2023).
24. Shi, Y.; Wang, H.; Zhu, Z.; Ye, Q.; Lin, F.; Cai, G. Association between exposure to phenols and parabens and cognitive function in older adults in the United States: A cross-sectional study. *Sci. Total Environ.* **2023**, *858*, 160129. [CrossRef]
25. Bobb, J.F.; Claus Henn, B.; Valeri, L.; Coull, B.A. Statistical software for analyzing the health effects of multiple concurrent exposures via Bayesian kernel machine regression. *Environ. Health* **2018**, *17*, 67. [CrossRef] [PubMed]
26. Carrico, C.; Gennings, C.; Wheeler, D.C.; Factor-Litvak, P. Characterization of Weighted Quantile Sum Regression for Highly Correlated Data in a Risk Analysis Setting. *J. Agric. Biol. Environ. Stat.* **2015**, *20*, 100–120. [CrossRef]
27. Snowden, J.M.; Rose, S.; Mortimer, K.M. Implementation of G-computation on a simulated data set: Demonstration of a causal inference technique. *Am. J. Epidemiol.* **2011**, *173*, 731–738. [CrossRef]
28. Rempelos, L.; Wang, J.; Baranski, M.; Watson, A.; Volakakis, N.; Hadall, C.; Hasanaliyeva, G.; Chatzidimitriou, E.; Magistrali, A.; Davis, H.; et al. Diet, but not food type, significantly affects micronutrient and toxic metal profiles in urine and/or plasma; a randomized, controlled intervention trial. *Am. J. Clin. Nutr.* **2022**, *116*, 1278–1290. [CrossRef]
29. Jan, A.T.; Azam, M.; Siddiqui, K.; Ali, A.; Choi, I.; Haq, Q.M. Heavy Metals and Human Health: Mechanistic Insight into Toxicity and Counter Defense System of Antioxidants. *Int. J. Mol. Sci.* **2015**, *16*, 29592–29630. [CrossRef]
30. Seale, L.A.; Ogawa-Wong, A.N.; Berry, M.J. Sexual Dimorphism in Selenium Metabolism and Selenoproteins. *Free Radic. Biol. Med.* **2018**, *127*, 198–205. [CrossRef]
31. Birla, H.; Minocha, T.; Kumar, G.; Misra, A.; Singh, S.K. Role of Oxidative Stress and Metal Toxicity in the Progression of Alzheimer's Disease. *Curr. Neuropharmacol.* **2020**, *18*, 552–562. [CrossRef]
32. Tamtaji, O.R.; Heidari-Soureshjani, R.; Mirhosseini, N.; Kouchaki, E.; Bahmani, F.; Aghadavod, E.; Tajabadi-Ebrahimi, M.; Asemi, Z. Probiotic and selenium co-supplementation, and the effects on clinical, metabolic and genetic status in Alzheimer's disease: A randomized, double-blind, controlled trial. *Clin. Nutr.* **2019**, *38*, 2569–2575. [CrossRef]
33. Akahoshi, N.; Anan, Y.; Hashimoto, Y.; Tokoro, N.; Mizuno, R.; Hayashi, S.; Yamamoto, S.; Shimada, K.I.; Kamata, S.; Ishii, I. Dietary selenium deficiency or selenomethionine excess drastically alters organ selenium contents without altering the expression of most selenoproteins in mice. *J. Nutr. Biochem.* **2019**, *69*, 120–129. [CrossRef]
34. Yu, L.; Liu, W.; Wang, X.; Ye, Z.; Tan, Q.; Qiu, W.; Nie, X.; Li, M.; Wang, B.; Chen, W. A review of practical statistical methods used in epidemiological studies to estimate the health effects of multi-pollutant mixture. *Environ. Pollut.* **2022**, *306*, 119356. [CrossRef]
35. Schmidt, C.W. Environmental Factors in Successful Aging: The Potential Impact of Air Pollution. *Environ. Health Perspect.* **2019**, *127*, 102001. [CrossRef]
36. Tan, Y.; Fu, Y.; Huang, F.; Wen, L.; Weng, X.; Yao, H.; Liang, H.; Kuang, M.; Jing, C. Association between blood metal exposures and hyperuricemia in the U.S. general adult: A subgroup analysis from NHANES. *Chemosphere* **2023**, *318*, 137873. [CrossRef]

37. Pitts, M.W.; Kremer, P.M.; Hashimoto, A.C.; Torres, D.J.; Byrns, C.N.; Williams, C.S.; Berry, M.J. Competition between the Brain and Testes under Selenium-Compromised Conditions: Insight into Sex Differences in Selenium Metabolism and Risk of Neurodevelopmental Disease. *J. Neurosci.* **2015**, *35*, 15326–15338. [CrossRef]
38. Wang, L.; Yin, Y.L.; Liu, X.Z.; Shen, P.; Zheng, Y.G.; Lan, X.R.; Lu, C.B.; Wang, J.Z. Current understanding of metal ions in the pathogenesis of Alzheimer's disease. *Transl. Neurodegener.* **2020**, *9*, 10. [CrossRef]
39. Wan, H.; Jiang, Y.; Yang, J.; Ma, Q.; Liu, L.; Peng, L.; Liu, H.; Xiong, N.; Guan, Z.; Yang, A.; et al. Sex-specific associations of the urinary fourteen-metal mixture with NAFLD and liver fibrosis among US adults: A nationally representative study. *Ecotoxicol. Environ. Saf.* **2022**, *248*, 114306. [CrossRef]
40. Gade, M.; Comfort, N.; Re, D.B. Sex-specific neurotoxic effects of heavy metal pollutants: Epidemiological, experimental evidence and candidate mechanisms. *Environ. Res.* **2021**, *201*, 111558. [CrossRef]
41. Jarup, L. Hazards of heavy metal contamination. *Br. Med. Bull.* **2003**, *68*, 167–182. [CrossRef]
42. Nguyen, T.V. Developmental effects of androgens in the human brain. *J. Neuroendocrinol.* **2018**, *30*, e12486. [CrossRef] [PubMed]
43. Zhao, D.; Huang, Y.; Wang, B.; Chen, H.; Pan, W.; Yang, M.; Xia, Z.; Zhang, R.; Yuan, C. Dietary Intake Levels of Iron, Copper, Zinc, and Manganese in Relation to Cognitive Function: A Cross-Sectional Study. *Nutrients* **2023**, *15*, 704. [CrossRef]
44. Keil, A.P.; Buckley, J.P.; O'Brien, K.M.; Ferguson, K.K.; Zhao, S.; White, A.J. A Quantile-Based g-Computation Approach to Addressing the Effects of Exposure Mixtures. *Environ. Health Perspect.* **2020**, *128*, 47004. [CrossRef] [PubMed]
45. Guo, X.; Wu, B.; Hu, W.; Wang, X.; Su, W.; Meng, J.; Lowe, S.; Zhao, D.; Huang, C.; Liang, M.; et al. Associations of perchlorate, nitrate, and thiocyanate with metabolic syndrome and its components among US adults: A cross-sectional study from NHANES. *Sci. Total Environ.* **2023**, *879*, 163083. [CrossRef] [PubMed]

MDPI AG
Grosspeteranlage 5
4052 Basel
Switzerland
Tel.: +41 61 683 77 34

Nutrients Editorial Office
E-mail: nutrients@mdpi.com
www.mdpi.com/journal/nutrients

www.ingramcontent.com/pod-product-compliance
Lightning Source LLC
Chambersburg PA
CBHW061625210926
51850CB00009B/2434
* 9 7 8 3 7 2 5 8 6 4 8 0 5 *